MARK KURZ

O9-AIF-463

Digital Electronics for Scientists

Digital
Electronics
for Scientists

H. V. MALMSTADT *University of Illinois*

and

C. G. ENKE *Michigan State University*

W. A. Benjamin, Inc. / **New York** / **Amsterdam** / **1969**

DIGITAL ELECTRONICS FOR SCIENTISTS

Copyright © 1969 by W. A. Benjamin, Inc.
All rights reserved

Standard Book Number 8053-6899-X
Library of Congress Catalog Card Number 75–80101
Manufactured in the United States of America
1234K2109

*The manuscript was put into production on December 13, 1968;
this volume was published on June 30, 1969*

W. A. BENJAMIN, INC. *New York, New York 10016*

Preface

The whole world of electronics and scientific instrumentation is developing at a feverish pace. The recent wide-scale commercial introduction of high-speed integrated circuits has been accompanied by entirely new design concepts. Together they are making possible digital and analog instrumentation undreamed of a few years ago. The computer-managed laboratory, laboratory robots, scientific observations in the nanosecond range, and the control of events within microseconds are now a reality and are being used or tested on a small scale.

The trend is obvious. The work of every scientist and engineer will be profoundly affected by the powerful new instrumentation, automation, and computation now being released or under development. These new systems can be significant in the lives of everyone. The stage is set for truly spectacular innovations; completely automated medical and chemical laboratories, remote labs on other planets, and automated scientific research centers are already in advanced planning stages.

This book provides a systematic introduction to the digital circuits, the concepts, and the systems that are basic to the new instrumentation-computation revolution. The goal is to lead the reader from the simplest discrete switching devices, such as semiconductor diodes and transistors, to the latest high-speed integrated circuits and digital instrumentation systems. Based on recent experiences in our own laboratories, we are convinced that acquaintance with these systems often will enable the scientist and engineer to set up experiments and collect and analyze scientific data in hours or days that would have required months or years in the recent past or, more likely, that would not have been possible.

The book is written for science and engineering students and for practicing scientists and engineers, and also for the chemists, physicists, pathologists, medical researchers, and many engineers who want to gain background in the new digital electronics so as to find the most efficient and effective ways to implement their investigations. It is assumed that the reader has little or no background in digital electronics or even with any electronics. Therefore, the basic concepts and circuits are introduced at the start. These are then rapidly exploited to form subsystems and systems of direct application in scientific instruments for measurement, control, or computation. We have tried to sort out the most significant of the new digital systems and arrange them in a logical way so that this text could be used entirely by itself for a course in digital electronics, or incorporated in existing courses that also present linear circuits. Suggestions for integrating digital electronics in existing electronics for scientists courses are given in the Special Notes (p. xvii).

It is possible to read and study this digital electronics book without doing any experiments. However, throughout the book experimental summaries are presented. If the reader does not do the experiments the concise summaries will at least indicate what could be done. But the real payoff to the student comes, we believe, when he does the experiments. Therefore, specific instructions are given for each experiment in a separate section at the end of the book.

The real success of any electronics course for scientists and non-electronics engineers is usually based on what can be done with the material when the course is over. Therefore, an efficient sequence of experiments is especially important. The experiments must rapidly expand the students' skills in circuits and systems. They must also provide a confidence and "feel" for what can be done, and hopefully provide insight and inspiration on how to create new systems that are unique to his own specific discipline. This means that any set of experiments for digital electronics must lead to an instrumental sophistication much greater than in the past. To accomplish this a unique system of integrated circuit cards, modules, parts, and instruments were also devised. Equipment details are presented only as they are needed for specific experiments.

A new experimental system and new methods of interconnecting parts and circuit cards were considered essential so that a student could progress rapidly from connecting the most basic circuits to studying logic gate and flip-flop applications and then to building and experimenting with digital and analog-digital instruments such as frequency meters, timers, counters, DVM's, and many others. The requirements, ideas, and concepts of the authors for specific digital circuit cards, parts, modules, and instruments have been implemented by a development program of the Heath Co. (Benton Harbor, Michigan).

The circuits and instruments that the student builds are representative of those now used for research-quality scientific instrumentation. Therefore, when the set of experiments is completed the experimenter will have gained a working ability with the electronic digital circuits and instruments that are directly applicable for scientific measurements, computation, and control functions.

The significance of a background in digital and analog electronics is so great that most science and engineering students will probably receive such training in the future. Ideally this training would come in the junior or senior years in various physics, chemistry, medical, or engineering curricula. However, many students do not discover their need for such training until they are in graduate school, and our own courses enroll undergraduates, graduates, and postdoctorals in the same course.

The authors are grateful for many comments of students and staff who used some of the materials in the preliminary form. Dr. Stan Crouch was especially helpful with his suggestions and class testing of certain experiments. Marcia Wachter, Elaine Evans, Pat Ryan, Celia Miller, Bonnie Robbins, and Deanna White were all very helpful in typing, correcting, and assembling various parts of the manuscript. The art work was under the direction of Mr. Stan Parnell of Smith Associates, who not only did an excellent job but cooperated in checking diagrams at the odd times convenient to the authors. The scientific instrument engineering staff at Heath Co., in particular, Charles Gilmore and LeRoy Hiltgen and their engineering groups, developed the digital modules and instruments that enable the experiments to be performed so efficiently. Dr. Julian Kateley made helpful comments on the computation section of Chapter 7. Finally, we'd like to thank our families for the extra burdens they carried while we worked on the manuscript.

H. V. MALMSTADT

C. G. ENKE

Urbana, Illinois
East Lansing, Michigan
April 1969

Contents

Chapter **2** **Switching Concepts and Diode Circuits**

Chapter **3** **Transistors, Relays, and Other Switches**

Chapter 5   **Flip-Flops and Multivibrators**

Chapter 7   Digital and Analog-Digital Instruments and Systems

Special Notes

Digital Electronics for Scientists should be applicable in a wide variety of courses. Some of the ways that we have used or recommend the use of this material according to certain course objectives are described here.

Courses in digital electronics. The sequence of material in the table of contents has been class tested for efficient presentation of digital electronics. It is recommended that students beginning this sequence have some knowledge of ac and dc circuits and electrical measurements. The type of background material presented in many elementary modern college physics courses should be adequate preparation. For students who have had courses in linear electronics and have also studied semiconductor switching devices most sections of Chapters 2 and 3 can be readily skipped.

Courses in digital logic. For those courses in which the emphasis is on digital logic the student could start with Chapter 4, Sections 4–1 to 4–3 and 4–8, and proceed to Chapters 5 through 7.

Courses in analog and digital electronics. For scientists and engineers who have had little or no experience in electronics it is frequently important to provide a broad background in both analog and digital electronics in a one- or two-semester course. For these purposes we have tested various sequences that utilize materials from *Digital Electronics for Scientists* (DEFS) and *Electronics for Scientists* (EFS). One of the most successful sequences starts with a study of electrical measurements and ac and dc circuits (Chapter 1 and Supplements 2 and 3 of EFS) and proceeds to the first four chapters of DEFS, as follows:

Electrical Measurements (Chapter 1 and Supplements 2 and 3, EFS)
Digital Measurements (Chapter 1, DEFS)

Switching Concepts and Diode Circuits (Chapter 2, DEFS)
Transistors, Relays, and Other Switches (Chapter 3, DEFS)
Switching Logic and Logic Gates (Chapter 4, DEFS)
Amplifiers and Oscillators (Chapters 3, 4, and 5, EFS)
Flip-Flops and Multivibrators (Chapter 5, DEFS)
Counters, Registers, and Readouts (Chapter 6, DEFS)
Servo Systems and Operational Amplifiers (Sections 7–1 and 7–2 of
    DEFS, supplemented with selected sections of Chapters 6, 7,
    and 8 of EFS)
Digital and Analog-Digital Instrumentation (Chapter 7, DEFS)

One of the big advantages of this sequence of topics is that many of the basic principles and characteristics of semiconductor devices and circuits can be introduced with the relatively simple and intuitive switching concepts. Then when the more complex analysis of linear circuits is introduced the student has already gained a background in semiconductor devices and switching circuits. In courses where we have tested this approach we found that the students learn linear circuits quickly and easily.

H.V.M.
C.G.E.

chapter one

Digital Measurements

Whenever scientific measurements are made, a *digital* number is the desired result. The position of the recorder pen or meter needle is a function of the information signal from the *transducer*[1] and is read off the meter scale or recorder chart in terms of a digital number. Although this type of measurement and readout has many laboratory applications, the many recent developments in electronics have made automated measurements that provide *direct digital readout* of either analog signals or discrete events increasingly common. The advantages of direct digital measurements are numerous: The signal readout is not subject to a "reading error" of the scale. The digital information can be handled directly and processed by computers. Discrete information in the form of pulses does not have to be averaged but can be measured directly. There is often less sensitivity to noise. Relative precision and accuracy can be greatly improved.

Many discrete phenomena are frequently averaged by slow-responding transducers and readouts to provide voltages which correspond to the average of many discrete events. In those cases where the discrete events can be observed by existing transducers it is possible, by using high-speed digital instrumentation and computation, to use the time or amplitude relations among the discrete events to advantage: to determine the mechanism by which events occur or how an event occurring in 1

---

[1] A *transducer* is a device for converting one form of energy into another form. Transducers can convert information about physical quantities (temperature, pressure, light, pH, etc.) into related electrical signals (voltage, current, resistance, etc.) that can be readily measured; transducers also convert electrical quantities to physical quantities, such as electrical current to indicator position.

1

μsec interval influences the event during a subsequent microsecond, or in some cases to manipulate these events within microsecond or shorter-time intervals.

For those scientific measurements where the discrete information would be irrelevant or economically infeasible, the digital voltmeter and other analog-digital devices are very useful instruments. However, it is to be expected that as new transducers and techniques are developed, many present analog measurements will be superseded by digital measurements of the discrete events. Some of the most dramatic scientific advances of the future will probably be a consequence of new and better high-speed digital instrumentation and computation enabling rapid discrete events to be observed and manipulated and the data rapidly analyzed.

In this chapter some of the basic concepts of digital measurements are presented, and the elementary ideas of binary and decimal counting with electronic devices are introduced. This material is intended to provide a perspective and background for making frequency, period, count, time interval, ratio, and voltage measurements with electronic digital instruments.

1-1 Concepts of Digital Measurements

Perhaps the most widely applied digital measurement technique is basically high-speed electronic counting with digital readout of data. In general the question of how many events or items N that occur within specified boundary conditions D (e.g., revolutions per mile, cycles per second, particles per gram, or in general N/D) can be answered by utilizing digital electronic counters in conjunction with suitable transducer-sorting systems as illustrated by the block diagram in Fig. 1-1.

Electrical signals are produced by the *transducer-sorting system*.[2] These signals now contain the information about the phenomenon to be measured and are shaped into a form and magnitude necessary for operating the specific type of digital circuitry. The electrical signals are controlled by an *electronic gate*[3] which directs the signals at the proper

[2] A *transducer-sorting system* converts or encodes the desired information about a complex physical system into various physical quantities (such as emitted light of specific wavelengths), selectively sorts or isolates the information and converts it to electrical signals. The system often will consist of several types of transducers.

[3] An *electronic gate* is a circuit that functions like a gate or door. It allows electrical signals to pass through when OPEN and stops signals from passing through when it is CLOSED.

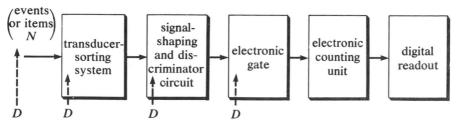

Fig. 1-1 Electronic counting measurement.

times into the *electronic counting unit.* The counting unit advances one step for each input pulse. The *digital readout* provides a numerical visual display which is related to the specific measurement of interest. The dotted arrows leading from the letter D indicate that there may be several boundary conditions D_1, D_2, D_3, etc., over which the counting is performed; these conditions can be imposed on or by the event, the transducer, the *signal shaper*[4] and *discriminator,*[5] and/or the electronic gate. The ingenuity used in determining and setting reliable boundary conditions will frequently determine the significance of the data, and these concepts of selectivity for discrete signals will be elaborated in subsequent sections and chapters.

A specific example illustrating several boundary conditions for a single measurement is shown in Fig. 1-2. Radiation of various wavelengths is emitted throughout a flame. To determine the number of photons of a specific wavelength emitted from a segment of the flame between the times t_1 and t_2, a space selector (lens-aperture system) is used to focus on the desired flame segment, a wavelength selector (monochromator) is used to isolate the desired wavelengths, and an electronic gate control is used to open the gate to allow signals to the counting unit only between times t_1 and t_2.

It is important to note that the accuracy of the measurement is determined equally by numerator and denominator in the ratio of N/D. Even if 10^6 counts are accumulated in the readout for the selected boundary conditions, the accuracy could be considerably less than 1 part per million if the overall accuracy of D is not good. In the above example if the overall accuracy of selecting the flame segment, isolating

---

[4] A *signal shaper* is a circuit that will shape input voltage signals to the amplitude, duration, rise time, or other characteristics required for specific applications.

[5] A *discriminator circuit* selects voltage signals of certain preset amplitudes, duration, or other conditions and allows only those that have the desired characteristics to pass through to its output.

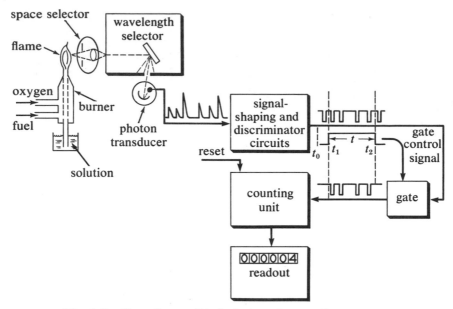

Fig. 1-2   Counting emitted photons from a flame source.

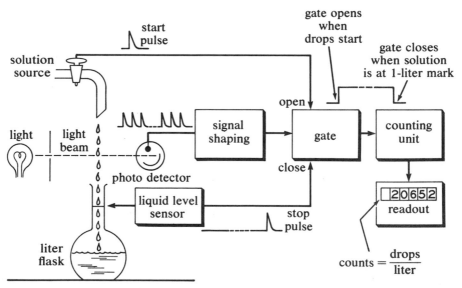

Fig. 1-3   Measurement of drops per liter by electronic gating of events counter.

the desired wavelength and setting the gate time is only 10%, then the maximum measurement accuracy is only 10%. Also, if the measured numerator N is based on random statistical processes, as in photon counting, the reliability of N depends on the specific statistics. Typically the precision of the numerator for random events would be $1/N^{1/2}$. Therefore, if the number of photons counted is 10^6 the precision of N would be about 1 part per thousand or 0.1%.

Events Counter

The example in Fig. 1-2 illustrates that multiple boundary conditions might be set for counting the discrete events, or the limits D might be only one specific condition such as mass, volume, area, distance, event, or time. For example, units such as the following might be encountered: N drops of solution per liter; N bacteria per square centimeter; N ions per spark; N photons per second.

The use of an events counter is illustrated in Fig. 1-3 for the measurement of the number of drops per liter. The counter is started by an electronic pulse at the instant that drops start falling into the flask and is stopped by a pulse when the flask is filled to the 1-liter mark. The readout of counts at the end of the measurement has the units, drops per liter.

Frequency Meter

When a boundary condition D is unit time t, such as seconds or microseconds, the mode of measurement is *events per unit time* or frequency, N/t. Frequently the unit time intervals are available within the electronic counting instrument. In such instruments the frequency mode of measurement and the specific unit time base for the measurement are switch selected. For example, to measure the frequency output of a signal generator, the instrument is switched to the frequency mode so that the gate is opened and closed by an accurate time base from the internal time standard or clock, as illustrated in Fig. 1-4. If the frequency of the signal generator is 372.435 kHz and the gate is opened by the clock for 1 sec, then 372,435 oscillations will be counted while the gate is open. Longer or shorter gate times could be selected which can increase or decrease the number of significant figures displayed by the readout.

The unit time base is often provided from an extremely stable crystal oscillator clock. The long-term (years) accuracy of the clock is typically 1 part in 10^5 to 1 part in 10^7, and short-term (hours) stability is sometimes about 1 part in 10^9.

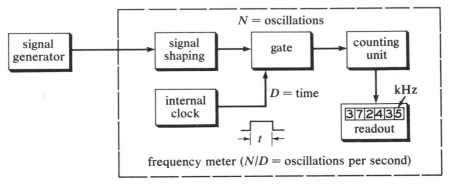

Fig. 1-4   Frequency meter (N/D = oscillations per second).

In making frequency measurements there is an uncertainty of ±1 count in the readout data in addition to any instability of N/D. This is because the clock gate time t is not synchronized with the incoming signal, as illustrated in Fig. 1-5. For identical gate times and input frequency, it can be seen that the number of pulses counted per unit time varies by 1 depending on the relative phase of N and D.

Since the denominator D (time) is so accurate in good frequency meters the measurement accuracy is usually limited by the number of events N in the selected measurement time. Therefore, the accuracy decreases for lower frequencies when measuring in the frequency mode. Because of the loss in resolution for low-frequency measurements the accuracy of measuring the period in microseconds can be much greater than the measurement of frequency. Measurements of period, multiple period, and frequency ratio will be discussed in Section 1-5.

Digital Voltmeter

Many scientific measurements involve quantities which are the net result of numerous discrete phenomena, and various transducers have been developed to provide voltage outputs which are the analog of the quantity to be measured; a few examples include temperature, pressure, absorbed and emitted light. Frequently these analog voltage signals are plotted on chart recorders versus some function, e.g., temperature versus time or percent light absorbed versus wavelength, and the magnitudes of the signals are subsequently read from the chart at certain selected points and the data are then analyzed. For convenience and practicality of data handling and computation, it is often desirable to have direct digital readout of the analog voltage at selected points. Electronic digital voltmeters are ideal for this purpose.

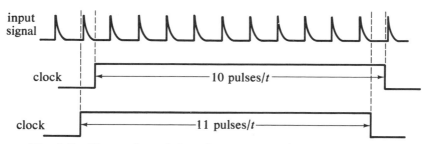

Fig. 1-5 Illustration of the ±1 count error in nonsynchronized measurement modes.

If the same basic units illustrated in Fig. 1-4 for the frequency meter are used in conjunction with a voltage-to-frequency (V-F) converter an unknown voltage will be displayed as a numerical readout. If the V-F converter provides 100,000 Hz/V, then for an input voltage of 0.35423 V and a gate period of 1 sec, there would be 35,423 counts in the counting unit, and the range selectors would set the decimal point so that the digital readout would be 0.35423 V. The accuracy is usually limited by the linearity and stability of the V-F converter which is typically 0.1–0.005%. Specific types of digital voltmeters and analog digital instruments will be described and investigated in later sections and chapters.

Experiment 1-1   *Introduction to the Universal Counter*

A familiarity with the physical arrangement of the basic digital measurement instrument is obtained by locating and identifying the controls and connections. Manual time measurements are then made to check the operation of the instrument and to learn the function of several basic controls.

1-2 Counting

It was shown in the previous section that the counting process is the basis of a wide variety of digital measurements. In general terms the counting of physical phenomena involves the following steps: (a) detecting the discrete events, (b) discriminating against counting any but the events of interest, (c) deciding and controlling when the counting is to start and stop, (d) counting the events, and (e) presenting the resulting count in a useful way. For the measurement to be accurate, the counting interval must be exact and the transducer and shaping circuits must produce a pulse for every event of interest while perfectly discriminating against other events and noise. The signal-shaping circuit is often a critical element in discrimination because it must ignore the electrical noise signals coming from the transducer circuit.

Input Comparator

One way for the signal-shaping circuit to discriminate against noise would be by amplitude, i.e., only signal peaks exceeding a certain "threshold" level would result in output pulses. A comparator commonly used for this application is shown in Fig. 1-6. The characteristics of the comparator are such that the output is always one of two levels: LOW or HIGH, sometimes called **0** and **1,** respectively. When the comparator input voltage e_C is less than the reference voltage e_R, the output is HIGH, but when e_C is greater than e_R, the output is LOW. Thus whenever a pulse at e_C exceeds the threshold level e_R, the output e_o makes a transition from HIGH to LOW and back again as in Fig. 1-7a. The comparator output voltage levels are designed to operate the counting unit reliably. The attenuator and ac connection for the input signal are often included to bring the signal amplitude within the adjustment range of e_R and to avoid overdriving the comparator.

For signals such as those in Fig. 1-7a the setting of the threshold level can be very critical. If it is set too low, noise pulses will be counted; if it is too high, some events will be missed. The higher the signal-to-noise ratio of the input signal, the better the discrimination can be. Figure 1-7b shows the comparator output for a sinusoidal input signal. In this case the capacitor coupling was used and the threshold set at 0 V. The zero-crossing setting, sometimes called "auto" used with the ac

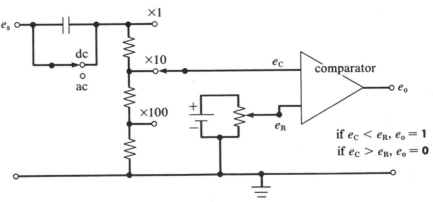

Fig. 1-6   Input comparator.

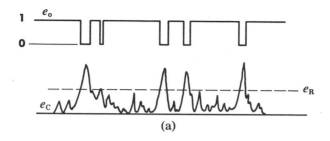

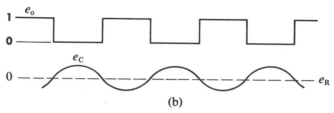

Fig. 1-7   Input and output waveshapes for comparator.

input, automatically insures that this type of signal will cross the threshold (trigger the transition of e_o from **1** to **0**) at the maximum rate of change of e_o.   This results in the greatest reproducibility of trigger time with respect to a given point on the signal waveshape.

Note that the transition of e_o from **1** to **0** in Fig. 1-7b occurs during the positive-going slope of the input signal.   An additional switch is often used with the comparator circuit which inverts the input or output signal so that the **1–0** transition[6] of e_o can occur on the negative-going slope of the input signal.   This would be useful if, for instance, the pulses of Fig. 1-7a were negative instead of positive.

Once the comparator is adjusted, it reacts to the input signal decisively according to its instructions.   Whether or not the HIGH–LOW output transitions correspond exactly to the desired events depends on the quality of the input signal and the care in setting the comparator controls. In any case, as far as the counting circuit connected to the comparator output is concerned, when a HIGH–LOW transition occurs and the gate is open, it is an event and will be recorded.

In some types of measurement, the comparator output opens and closes the count gate in response to the signal at the comparator input.

---

[6]The abbreviations **1–0,** HIGH–LOW, ON–OFF, and the like will be used to indicate transitions *from* the first-mentioned level *to* the second, that is, *from* the **1** level *to* the **0** level, *from* HIGH *to* LOW, and *from* ON *to* OFF for the examples given.

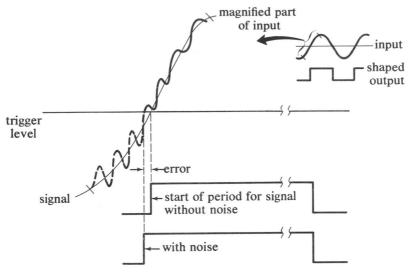

Fig. 1-8   Trigger time error due to noise.

The number of events counted may depend directly on the gating interval. In this case, the time accuracy of the comparator response is critical. One common source of time error is noise on the input signal as shown in Fig. 1-8. If the noise component is not synchronized to the signal waveform, the comparator will not trigger at the same pure signal input voltage each time and, thus, not at the same time in the waveform cycle. This type of error can be minimized by triggering on the part of the signal with the maximum slope, by noise filtering, or by making the total gate period long compared to the time error.

Experiment 1-2   *Input Comparator*

The characteristics and response of the input comparator and its setting for reliable measurements, including slope, level, and attenuator controls, are studied by using an oscilloscope to display the input signal and comparator output.

Binary Number System

The number 348 in the common decimal (base 10) system means 300 plus 40 plus 8 or, written in exponents $3 \times 10^2 + 4 \times 10^1 + 8 \times 10^0$. Each digit in the number represents the coefficient of some power of

10. This is a natural consequence of having 10 numerals (0 through 9) for each digit. For the decimal system, one could say that each digit has 10 distinguishable states. A counting device to be used as one digit in the decimal system would then have to have 10 distinct states (like one wheel of the odometer in an automobile). In electronic circuits, the fewer the required number of unambiguously distinguishable states, the greater the reliability and simplicity of the circuit.

The minimum number of distinguishable states a circuit can have and still be useful is 2. It may seem unnecessary to go to the extreme, but other advantages of two-state circuits over circuits with more than 2 distinguishable states are that their output levels are easy to distinguish, arithmetic operations are simple, and many components such as switches, relays, and diodes which only have 2 states (ON or OFF) can be used as active devices. A 2-state circuit is called a "binary circuit" or often simply a "binary."

If each binary circuit is to represent a digit, then by analogy with the base-10 number system, each digit will represent the coefficient of some power of 2.

Counting in the base 2 (binary) system goes as follows: 0, 1, 10, 11, 100, 101, etc. The number 1001 is $1 \times 2^3 + 0 \times 2^2 + 0 \times 2^1 + 1 \times 2^0$ which is equivalent to 9 in the decimal system. Four binary circuits (A, B, C, and D) of which the outputs are either 0 or 1 can be used to represent all the numbers between 0000 and 1111 (0–15) as shown in Table 1-1.

Thus two binary circuits have 4 different output combinations. Three binary circuits have 8 different combinations, four binaries have 16 combinations, n binary circuits have 2^n easily distinguishable combinations of output levels. By using enough binary circuits, the accuracy or the number of significant figures can be made as large as desired. For 1%, seven binary circuits have 128 states; for 0.1%, ten binary circuits have 1024 states; for 0.01%, fourteen binary circuits 16,384 states, and so on.

Binary Counting

Four binaries arranged for counting will respond to successive events according to Table 1-1. After each event the states of binaries A, B, C, and D will represent the next higher binary number. A group of four binaries can thus count to 15, five binaries to 31, and n binaries to $2^n - 1$.

The binary most often used in counting applications is one which changes its output to the opposite state in response to each input signal. This circuit is like the pull-chain light switch which turns a light ON if it was OFF, or OFF if it was ON, each time the chain is pulled. An electronic binary circuit can be made to change its output level whenever the level at its input changes from 1 to 0. The change in output is then triggered by a negative-going input signal. A binary that changes state with every input pulse is often called a "toggle" or T binary. Figure 1-9 shows a series of T binaries being used to count the rotations of a shaft. Once each rotation, a magnet attached to the shaft induces a current in the sensor coil (S) which is converted by the comparator to a pulse (I) suitable for triggering binary A. Prior to the first rotation to be counted, all binaries are "reset" to the 0 state. On the first rotation, when I goes from 1 to 0, binary A goes from 0 to 1 giving a count of 0001. On the second rotation, binary A changes from 1 to 0 triggering binary B to the 1 state giving a count of 0010. On the fourth rotation, A goes from 1 to 0 triggering B from 1 to 0, which, in turn, triggers C from 0 to 1, giving a count of 0100, and so on.

Lights connected to the outputs of the binaries go ON when the binary is in the 1 state. To determine what decimal number the outputs

Table 1-1   Binary Numbers to 1111

Binary Numbers				
$\times 2^3$ D	$\times 2^2$ C	$\times 2^1$ B	$\times 2^0$ A	Equivalent Decimal Numbers
0	0	0	0	0
0	0	0	1	1
0	0	1	0	2
0	0	1	1	3
0	1	0	0	4
0	1	0	1	5
0	1	1	0	6
0	1	1	1	7
1	0	0	0	8
1	0	0	1	9
1	0	1	0	10
1	0	1	1	11
1	1	0	0	12
1	1	0	1	13
1	1	1	0	14
1	1	1	1	15

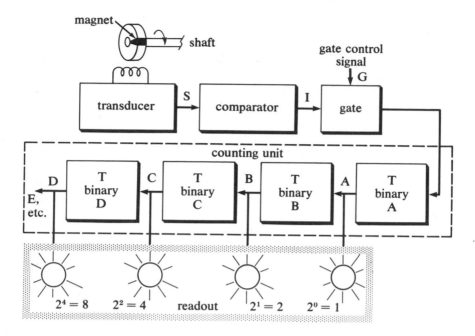

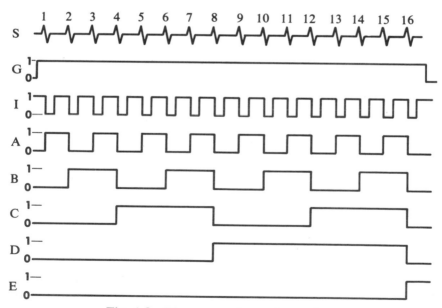

Fig. 1-9 Binary counter and waveforms.

of A, B, C, and D represent, it is necessary to observe their output states and look it up in Table 1-1 or sum the powers of two shown at the top of the table for those binaries with a 1 output.

Decimal Counting

The conversion of a number from the binary system to the more familiar decimal system is a fairly simple process for small numbers, but for larger numbers the process becomes unwieldy and time-consuming. "Binary-coded decimal" numbers are often used when dealing with large numbers which will eventually be converted to decimal presentation. Four binary circuits are used in combination to represent the numbers 0000 to 1001 (0 through 9 in Table 1-1). This set of four binaries then represents one digit of a decimal number. Another set of four binaries is used to represent the next digit and so on. Using Table 1-1, the binary-coded decimal number 0011 1000 0110 is the number 386 in decimal notation. This technique is somewhat wasteful since only 10 of the possible 16 states of each group of four binaries is being used, but the ease of conversion can more than make up for the waste in many cases.

In the binary-coded decimal form of counting, a four-binary counter is connected to reset to 0000 on the tenth count rather than the sixteenth. The output of the D binary has a 1–0 transition on the tenth count and is connected to the next set of four binaries which will advance on each tenth count.

In the example given above, a literal translation between the number represented by the four binaries and the decimal number was used (0011 = 3). This is the simplest of the binary-coded decimal techniques to understand and is often used. There is no reason, however, why each decimal numeral could not be assigned to any one of the 16 possible states of the four-binary group. In chapter six we shall see how various other codes are particularly suited to certain special operations. To distinguish it from the others, the code described above and shown in the top ten lines of Table 1-1 is called the 1-2-4-8 or *natural* code. This code identification is used because a 1 output level from each of the four binary circuits in the group A, B, C, and D is translated to the decimal equivalents 1, 2, 4, and 8, respectively. Since each set of binaries is used for one decade of the decimal count it is called a "decade counting unit" (DCU). Figure 1-10 shows a series of DCU's being used to count the radioactive decays in a nuclear experiment. The outputs of the DCU's

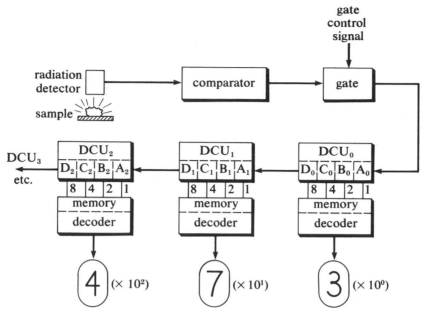

Fig. 1-10 Counter using decade counting units and decimal display.

provide a binary-coded decimal number which is then converted or "decoded" and connected to the decimal display.

A *memory* is often included in the decoder display circuitry. The memory holds the result of the completed count so that this result can be displayed while the counting circuits are reset to zero (cleared) and the next count is begun. At the end of the count, the new result is transferred into the memory and decoder circuits for display, and so on.

1-3 Timing

Digital counting measurements have been described as determining the number of events occurring within certain prescribed boundaries. One of the most common boundaries is time. For the measurement of the frequency of a repetitive signal, the number of cycles during a precise time interval is counted. Measurements of this type depend on the ability to generate a gate control signal of accurately known duration, such as 1.000000 or 0.001000000 sec. In standard instrumentation this is accomplished by using a high-frequency crystal-controlled oscillator and decade scaling circuits.

The circuit of Fig. 1-11 uses a 1-MHz crystal-controlled oscillator as the time standard generator (sometimes referred to as the clock). A comparator is used to convert the oscillations into discrete events. These events which occur at a rate of exactly 1 million/sec are then connected to a series of decade counting units to obtain lower frequencies which have an accuracy equal to that of the clock source. Since a pulse appears at the output of a decade counting unit for every 10 pulses at the input, the output repetition rate of the DCU is exactly one-tenth the repetition rate of the input signal. Used in this way, without decoding or readout, the DCU is called a "decade scaler" since the input frequency is scaled down by a factor of 10.

The time base selector switch is used to choose the desired gating interval from among signals with repetition periods ranging in decade

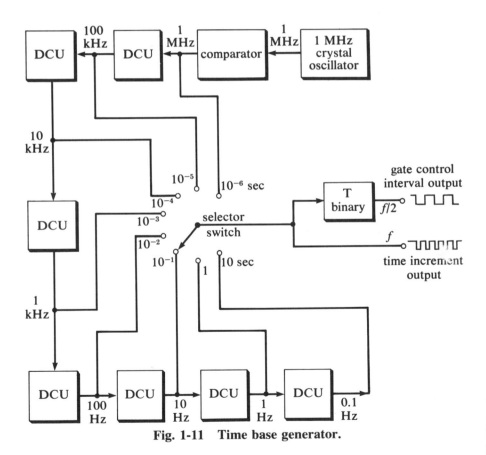

Fig. 1-11   Time base generator.

steps from 1 μsec to 10 sec. To obtain a gate control signal, a toggle binary is used. The T binary changes state for each input pulse so its output is HIGH for exactly one timing period, then LOW for one period, and so on. When connected to the gate, the gate will open for one period and then close. At the end of the closed period, the counter will be reset to zero to prepare for the measurement during the next gating period. If more than one time base period is desired for the observation of the measurement, the gate may be held closed by a signal from a display time circuit to delay resetting the counter.

The time base shown in Fig. 1-11 is also useful in the measurement of the time period between recurring events or the time interval between one event and another. In this case the pulses from the time-increment output are counted for the time interval prescribed by the boundary conditions. Putting the time base selector in the 10^{-6} sec position provides 1-μsec time increments and, therefore, allows the measurement of time interval to the nearest microsecond. Longer time increments might be selected for the measurement of periods longer than 1 sec.

1-4   General Purpose Counting-Timing Instruments

Modern digital counting instruments often contain a time base unit such as described in the previous section. The principal functional circuits of such an instrument are shown in block form in Fig. 1-12. The comparators, clock, scaler, counter, and readout functions have been described earlier in this chapter, and the gate functions for a versatile, general purpose instrument are described here.

The gate shown in Fig. 1-12 has a direct gate control input D_1 which is a logic-actuated input with, e.g., logic 1 to OPEN and logic 0 to CLOSE. In addition there is a gate control binary supplying the second gate control signal D_2. The binary has three inputs. A pulse at the start input opens the gate and a pulse at the stop input closes the gate. Pulses at the T (toggle) input alternately open and close the gate. If a repetitive signal is applied to the T input, the gate will open for one cycle, close for the next cycle, and so on.

The particular kind of measurement performed by the instrument depends on how the blocks are interconnected. A general purpose instrument uses a function switch to make the interconnections necessary to perform the desired function. Six common measurement modes and their interconnections are described below.

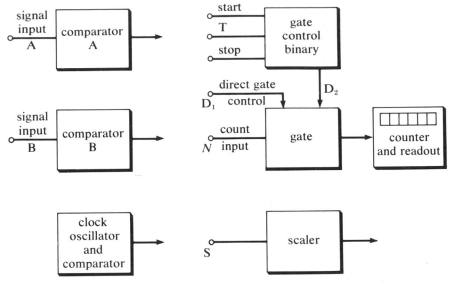

Fig. 1-12   Basic blocks for a general purpose counter-timer.

Events Counter Mode

The measurement is counts within a given boundary.   Let A be the signal to be counted and B be the signal to define the count boundary. Provision is generally made to scale down by n decades (10^n) the A input events if the counter cannot hold the total number of events which might occur within the measurement boundary.   The measurement ratio displayed is thus (A/F) events per B boundary, where A is the number of events at the A input and $F = 10^n =$ the scaling factor.   The block diagram connections are shown in Fig. 1-13.   The A signal goes to the scaler, then to the count input.   Comparator B controls the gate directly at D_1.

Experiment 1-3   *Events Counting*

Counting measurements are made in order to become familiar with the counting measurement and to experiment with manual and remote start and stop and with the scaling of input events.

Frequency Mode

This measurement is A events per time.   Thus A is connected in Fig. 1-14 to the numerator input N and the clock is connected to S.   The

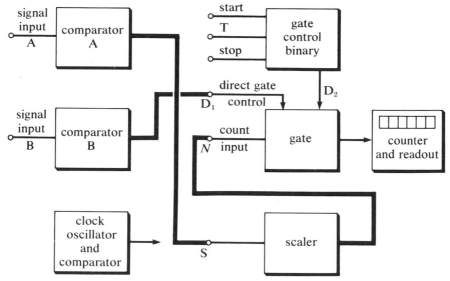

Fig. 1-13   Block diagram of events counter.

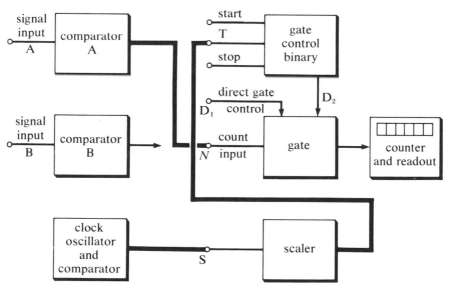

Fig. 1-14   Block diagram of frequency meter.

clock output is scaled to provide a range of accurate measurement times. The scaler is connected to T so that the gate is open for one complete clock cycle.

Experiment 1-4   *Frequency Measurement*

The frequency of a signal generator output is measured. The effects of input frequency and time base on precision are studied. The effect of the comparator setting on measurement accuracy is observed. The frequency of the crystal oscillator and scaler outputs is measured.

Period Mode

In the period mode the number of unit time increments are counted during one complete cycle (period) of the input signal. If the crystal clock is a 1-MHz oscillator, the unit time increment is $1/10^6 = 1$ μsec. For a 100-Hz input signal the period is $1/100$ sec $= 10^4$ μsec. Therefore, the counter would accumulate 10,000-μsec time increments during one period of the input signal. The resolution of this measurement made in 0.01 sec is 100 times greater than a frequency measurement of the same signal taking 1.00 sec. The period measurement is thus preferred for low-frequency signals whereas the frequency measurement is best for high frequencies.

The period mode is the inverse of the frequency mode, i.e., time per cycle. Therefore, the inputs to the gate are simply reversed, as illustrated in Fig. 1-15.

Experiment 1-5   *Period Measurements*

Period measurements of a sine-wave source are made for a variety of input frequencies and time base settings. The effect of these variables and the comparator setting on the measurement precision and accuracy are studied.

Period Average Mode

The accuracy of the period measurement may be limited by the time error in the comparator triggering which directly affects the gate time.

To improve the accuracy of period measurements the average time of 10, 100, 1000, or more periods can be measured by inserting a scaler

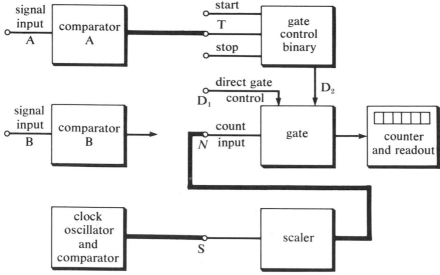

Fig. 1-15   Block diagram of period meter.

to divide the input signal so that the gate is open for 10, 10^2, ... , 10^n periods. Since for a given noise the absolute trigger error is constant the relative error of a multiple-period measurement will be reduced 10^n times, where $10^n = F$ the number of periods averaged.

As shown in the block diagram, Fig. 1-16, the signal input (B) is scaled down to provide, 10, 100, 1000 or more periods for the gate

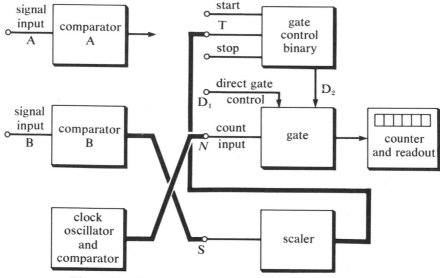

Fig. 1-16   Block diagram of period average meter.

interval. The T input is used so the gate will be open over the entire scaled period.

Experiment 1-6 *Multiple Period Average Measurement*

This experiment demonstrates how the comparator B input signal can be used in place of the internal crystal oscillator and how this is used to make multiple period measurements with 1 μsec resolution. The advantages of the multiple period measurement for certain signals are determined.

Frequency Ratio Mode

The measurement of the ratio of two frequencies is similar in principle to the period-average mode except that another unknown input frequency is substituted for the internal clock. Therefore, the ratio,

$$\frac{N}{D} = \frac{A \text{ cycles/sec}}{B \text{ cycles/sec}} = \frac{f_A}{f_B}$$

is measured. To increase the accuracy of the ratio measurement it is

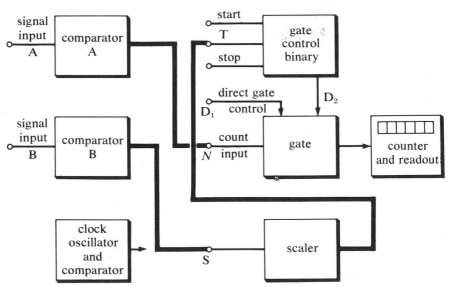

Fig. 1-17   Block diagram of frequency ratio meter.

usually desirable to introduce a scaler so that

$$\frac{N}{D} = \frac{A \text{ (hertz)}}{\dfrac{B \text{ (hertz)}}{F}} = F\left(\frac{f_A}{f_B}\right)$$

The block diagram Fig. 1-17 shows this to be identical to that of the period average mode except that comparator A instead of the clock is connected to N.

Experiment 1-7 *Frequency Ratio*

The frequency ratio measurement is seen to be the same as the multiple period except that both signals to the comparator inputs are from external sources. The ac power line frequency is compared with the frequency scale markings on a signal generator.

Time A-B Mode

Time is measured in the interval bounded by the A signal and the B signal (time/A-B boundary). Precise time pulses from the clock and scalar are connected to N in Fig. 1-18. The A signal starts the gate in-

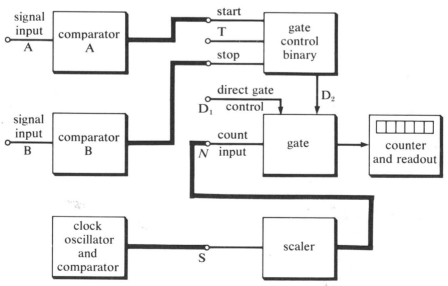

Fig. 1-18 Block diagram of time interval meter.

terval and the B signal stops it. The time increment measured (micro-seconds, milliseconds, or seconds) is determined by the scaling factor/ time base selector switch.

Experiment 1-8 *Time A-B Measurements*

The time A-B mode which was introduced in Experiment 1-1 is also used with electrical start and stop signals to measure positive and negative half-cycles of a square wave to check for symmetry.

Summary

The interconnections of the functional circuits required to provide the above six measurements are summarized in Table 1-2.

Table 1-2 Modes, Count Ratios, and Interconnections for a General-Purpose Counter-Timer

| Mode | Count Ratio (N/D) | Interconnections | | | |
		Comparator A Output to:	Comparator B Output to:	Clock Output to:	Scaler Output to:
Count	$\dfrac{A/F \text{ events}}{B \text{ boundary}}$	S	D_1	—	N
Freq.	$\dfrac{A \text{ events}}{\text{time}}$	N	—	S	T
Time A-B	$\dfrac{\text{time}}{A\text{-}B \text{ boundary}}$	Start	Stop	S	N
Period	$\dfrac{\text{time}}{A \text{ cycle}}$	T	—	S	N
Period Avg.	$\dfrac{\text{time}}{F \times B \text{ cycles}}$	—	S	N	T
Freq. Ratio	$F\left(\dfrac{\text{Freq. A}}{\text{Freq. B}}\right)$	N	S	—	T

1-5 Digital Voltage Measurement

The measurements described in the previous sections demonstrate the special ability of digital instruments to count accurately events and to measure the time relationship among events. In every case the event to be measured is converted to an electrical signal and then to a pulse compatible with the requirements of the gate and counting circuits. The amplitude of the input signal is relatively unimportant since the measured parameters may vary considerably and still give a clear indication as to whether or not the event occurred within the prescribed boundary. However, for many signal sources, it is the amplitude of the signal which is related to the desired information. When the advantages of digital measurement, data transmission, or computation are required, it is necessary to convert such "amplitude analog" signals into a two-level form where the information is presented by a group of pulses or logic level outputs (digital form) or by the time relationships among pulses from a single source (time-analog form). Several techniques have been devised to perform this voltage-to-time conversion for various applications, many of which will be described in Chapter 7. Two of the more common voltage-to-time converters are described briefly below.

A ramp-type converter is shown in Fig. 1-19. The output of a linear ramp generator is compared with the input signal amplitude (attenuated, if necessary). The output of a stable oscillator is counted for the period required for the ramp voltage to rise to the amplitude of the input signal. The larger the signal, the longer the time needed for the ramp voltage to match it and the more oscillations that will get through the gate to the counter. If the ramp is linear the count will be proportional to the input voltage. The digital output is in the form of a "burst" of pulses, and the information which was previously an amplitude is carried by the number of pulses in the burst. When the counter and display are included and the oscillator frequency and ramp slope are adjusted properly, the pulse count will be equal to the input voltage directly and the instrument is one type of digital voltmeter (DVM).

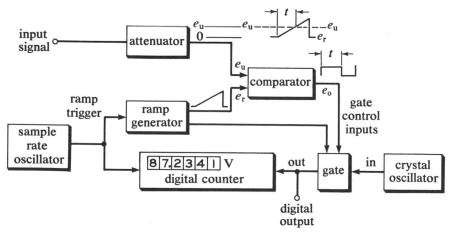

Fig. 1-19   Ramp-type digital voltmeter.

Let us look a little more closely at the sequence of events in the conversion shown in Fig. 1-19. The *ramp generator* has two output signals, a ramp and a pulse of duration equal to the ramp. The pulse is used to open the *gate* only when the ramp is being generated. A signal pulse from the *sample-rate oscillator* resets the counter and triggers the start of the ramp. Both gate-control inputs are now **1** so the count of the crystal oscillator output begins. As long as the ramp output e_r remains less than the signal e_u, the comparator output e_o will be **1**. At the instant e_r exceeds e_u by a very small increment, the comparator output e_o goes to **0**, closing the gate, and stopping the count, which is displayed on the digital counter. When the ramp resets, the comparator output is again **1**, but the pulse from the ramp generator is now **0** which keeps the gate closed. This circuit is a voltage-to-time converter which converts the input voltage amplitude to a gate time and then simply uses the circuits of Fig. 1-12 connected to measure time.

Another popular type of voltage-to-time converter is the voltage-to-frequency converter shown in Fig. 1-20. The input signal, attenuated if necessary, is connected to the input of a voltage-controlled oscillator (VCO). A VCO circuit is described in detail in Chapter 7, but to understand the conversion and measurement concepts it is only necessary to know that the output frequency of this VCO is proportional to the input voltage. Thus the frequency which is directly proportional to the signal amplitude can be measured accurately using the frequency meter of Fig. 1-4. If the ratio of frequency to voltage (f/V) of the VCO is exactly

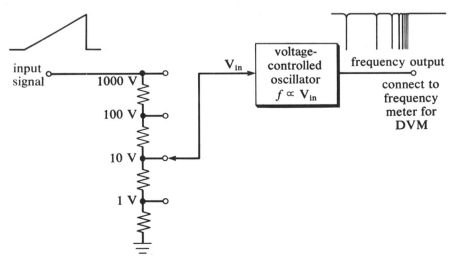

Fig. 1-20 Voltage-to-frequency converter.

some power of 10 the digital display will read directly in volts if the decimal point is appropriately placed.

In both cases illustrated above and, indeed, in all cases involving analog-to-digital (A-D) conversion, the accuracy of the final result is usually limited by the accuracy of the A-D converter. By the nature of the analog part of their function, A-D converters use amplifiers which are subject to all the drift, nonlinearity, and noise associated with non-digital or "linear" circuits. The frequency meter can measure the output frequency of a VCO to six or more places, but the linearity and stability characteristics of the VCO might limit its accuracy to only three or four places. Also, the errors because of input impedance, noise, thermal emf's, etc., must be considered at the input as for any analog measurement. The accuracy of some A-D converters is much less susceptible than others to errors from noise. These differences will be discussed when considering the specific circuits.

Experiment 1-9 *Voltage Measurement with the Digital Voltmeter*

The DVM is used to check the calibration of a voltage reference source. The effect of source resistance on the accuracy of a voltage measurement is studied. To measure voltage, the Universal Digital Instrument uses a very accurate voltage-to-frequency (V-F) converter. Since the output of the V-F converter is a frequency proportional to the voltage, a digital frequency measurement of this output is made for known input voltages, and the frequency-to-voltage ratio of the V-F converter is determined.

---

chapter two

Switching Concepts and Diode Circuits

It was shown in the previous chapter how the repeated use of simple 2-state circuits forms the basis for most of the functional units in a complex digital instrument. Groups of any 2-state circuit or device could, with sufficient ingenuity, be used to perform the basic digital functions of counting, gating, storing, and decoding. In practice, a great variety of electronic circuits as well as electromechanical, optoelectronic, mechanical, hydraulic, and pneumatic devices have been used as binary elements, and there is much in the digital system design philosophy which is common to all classes of binary elements. In this chapter the general concepts of *electronic switching* are considered. Then one of the most basic and important electronic switching elements, the *diode*, is described, and the most important diode switching circuits are presented.

2-1 Introduction to Switching Circuits

An electronic binary must have 2 stable and easily distinguishable electrical states. An ordinary mechanical switch has 2 such states; OPEN and CLOSED; i.e., OFF and ON, respectively. The ideal switch contacts have zero resistance when closed and zero conductance (infinite impedance) when OPEN. In practice, neither ideal is achieved.

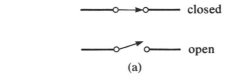

(a)

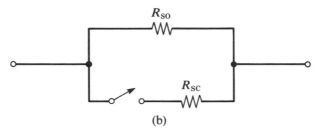

(b)

Fig. 2-1 (a) Switch symbol; (b) schematic representation of nonideal switch.

Nonideal Switches

The schematic symbols that are often used for OPEN and CLOSED switches are shown in Fig. 2-1a. However, in considering the action of a real switch in an electronic circuit it is frequently important to represent the nonideal OPEN and CLOSED switch resistances, as illustrated by the schematic for a nonideal switch in Fig. 2-1b.

In the generalized switching circuit shown in Fig. 2-2, the current in the circuit (through the load) is controlled by the switch. Another way to view the circuit, in consideration of the nonideal switch, is that the source voltage e_s is divided between the resistance of the load and the resistance of the switch. To transfer the maximum signal to the load the closed switch resistance, R_{sc}, should be very small compared to R_L. Similarly, if the open switch resistance is not very much greater than R_L the current will not be effectively turned off by the switch. For

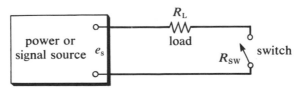

Fig. 2-2 Generalized switching circuit.

purely resistive switches, when R_{SO}, R_{SC}, and R_L are known, the effectiveness of a given switch in a desired application can be evaluated using Kirchhoff's laws and the circuit of Fig. 2-2.

Switch contacts can be actuated in several ways: mechanically (toggle and rotary switches), electromechanically (relays and solenoids), and magnetically (magnetic reed relays) and these are discussed in Chapter 3. Also, purely electronic devices including many types of transistors and semiconductor diodes are currently being used as switching elements. Their speed, small size, and ease of actuation by electrical signals are making them the designer's choice for an increasing range of switching applications. The requirements are the same: the device must have two terminals between which an electrical current can be conducted, the device or the circuit must be designed so that there are only two stable states of conductivity, and the device must be capable of being "switched" from one state to the other.

Most electronic switches have nonlinear characteristics, i.e., the current through the switch is not a linear function of the voltage across the switch terminals. Therefore, the dc analysis of how the switch performs in a circuit is made most readily by a graphical solution which utilizes the *current-voltage curve* of the device and the *load line* of the circuit. The graphical solution is first illustrated for a simple resistive linear switch and later in the chapter for the nonlinear diode.

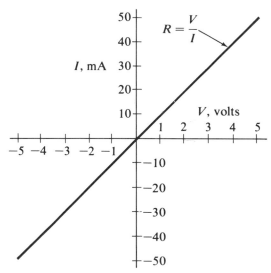

Fig. 2-3   Current-voltage (*I-V*) curve for a resistor.

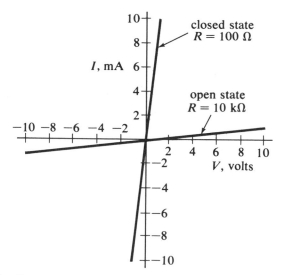

Fig. 2-4   Current-voltage curves for a less-than-ideal switch.

Current-Voltage Curves and the Load Line

The electrical characteristics of a resistive device for which the resistance is not constant over its operating range are often best shown by a current-voltage curve. A current-voltage (I-V) curve is a plot of the voltage V across a device vs the current I through it. The I-V curve of a resistor is shown in Fig. 2-3. This line is straight as expected from Ohm's law. A large resistance has a more horizontal I-V curve, whereas a lower resistance has a more vertical curve. The current-voltage curve of a perfect *conductor* would coincide with the vertical coordinate, and the line for an infinite resistance would coincide with the horizontal coordinate.

A switch has two I-V curves — one for the resistance of each state. The I-V curves for a perfect switch, which has zero ON resistance and infinite OFF resistance, will fall on the vertical and horizontal coordinates, respectively. Exaggerated deviations from the ideal switch characteristics are shown in Fig. 2-4.

To determine the *operating points* for a switch, i.e., the voltage across and current through a switch in its ON and OFF states, it is only necessary to superimpose the I-V curves for the switch, such as in Fig. 2-4, on the *load line* for the switch circuit. This process is illustrated for the switch circuit of Fig. 2-5 by utilizing the I-V curves illustrated in Fig. 2-4.

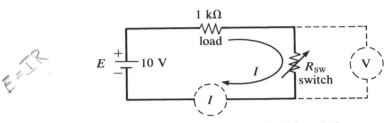

Fig. 2-5 Simple circuit with switch.

Note that the switch is shown as a device with a variable resistance R_{SW}. From Ohm's law

$$I = E/(R_L + R_{SW}) \qquad (2\text{-}1)$$

Rearranging yields $IR_{SW} = E - IR_L$, where IR_{SW} is equal to V the voltage across the device.   Therefore,

$$V = E - IR_L \qquad (2\text{-}2)$$

This is a linear equation relating I and V, and a current-voltage line can be drawn as shown in Fig. 2-6.   It is called "load line" because it is determined only by the source voltage E and the load resistance R_L.   An easy way to obtain the load line is to locate the intercepts and join them by a straight line.   From Eq. (2-2) it is seen that if $V = 0$, then $I = E/R_L$ and if $I = 0$, then $V = E$.   For any value of R_{SW}, the operating values of I and V must be located on the load line.

Figure 2-7 shows a superposition of the current-voltage curves of Fig. 2-4 with the load line of Fig. 2-6.   Since the switch I-V curve and

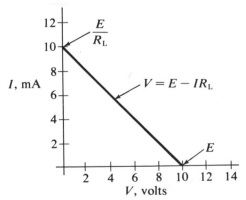

Fig. 2-6 Load line for the circuit of Fig. 1-3.

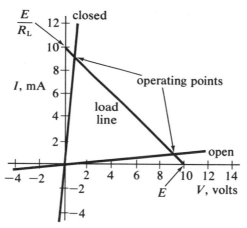

Fig. 2-7 **Load line and characteristic curve combined to give operating points.**

the circuit load line both define the possible current-voltage relationship for the system, the actual circuit current and switch voltage must be given by the intersection of the *I-V* curve and the load line. Each intersection is called an *operating point* of the switch in that circuit. Since the switch has an *I-V* curve for each state, there are two operating points as shown in Fig. 2-7. *The action of each new switching device can be studied using this graphical method.*

Experiment 2-1 *Characteristic Curves and Load Line for an "Ideal" Switch*

A curve tracer is constructed so as to display the characteristic curves of an ideal switch on the screen of an oscilloscope. The same curve tracer will be especially useful in studying diode and transistor characteristics in subsequent experiments.

Actuating Switch Contacts Electrically

When an electrical signal is used to actuate a switch, there are generally two electrical circuits involved—the actuating circuit and the switched circuit as shown in Fig. 2-8. The degree of interaction between the actuating and switched circuits depends upon the type of switch used and the circuit. Some switching devices have inherently one or two connections in common between the switch and the actuating element. The three possible degrees of interconnection for simple switches are

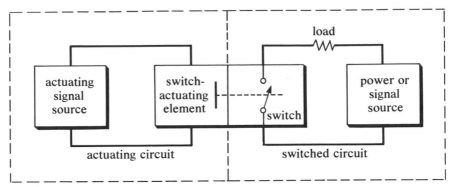

Fig. 2-8   Electrically actuated switching circuit.

shown in Fig. 2-9. Examples of 4-, 3-, and 2-terminal devices are the relay, transistor, and diode, respectively. In the relay the actuating signal is applied to a coil which can be electrically isolated from the switching contacts. The transistor is an example of a 3-terminal switch where the emitter connection is common to both the actuating and the switched circuits. The diode is a 2-terminal switch. If the polarity of the signal applied to a diode is reversed, a conducting diode becomes a poor conductor.

The diode and its common circuits are discussed in subsequent sections, and the 3- and 4-terminal switches are described in the next chapter.

Switch Capacitance

In the previous discussion of the nonideal switch we only considered the reduction in amplitude of the voltage across the switch caused by the

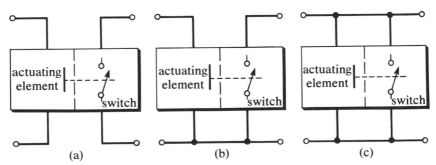

Fig. 2-9   Switching devices: (a) 4-terminal, (b) 3-terminal, (c) 2-terminal.

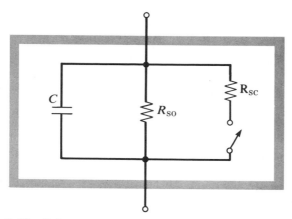

Fig. 2-10   Schematic representation of a nonideal switch.

fact that when it was ON it was not a perfect conductor and when OFF it was not an infinite resistance. However, there are also factors inherent in the switch device or circuit which affect the maximum speed of switching across all devices, and in all circuits there are capacitances which are unavoidable in the construction of the device or circuit. These capacitances result in a certain inertia in the switch and affect the time required to go from the ON to OFF states or vice versa.

To account for the switch capacitance the schematic of the nonideal switch given in Fig. 2-1 can be modified to have an equivalent capacitance C as shown in Fig. 2-10.

The determination of the maximum switching speed of the nonideal switch of Fig. 2-10 in the switch circuit of Fig. 2-5 requires an analysis of the charge and discharge of the capacitance C through the effective circuit resistances. In the next section there is a general treatment of the response characteristics of RC circuits to switch or pulse signals, and the section is concluded with a specific analysis of the nonideal switch of Fig. 2-10 in a typical switch circuit.

2-2   Resistor-Capacitor (RC) Circuits

The RC circuit is frequently introduced into switch circuits for signal shaping or coupling, but resistance and capacitance can also be an inherent and unwanted part of the switch components or circuit construction. In either case, an understanding of the response characteristics of the RC circuit to typical switch signals is important.

Series RC Circuit

There are several basic factors that govern the response of the simple RC circuit to a sudden change of voltage. In Fig. 2-11 a source of voltage E (of negligible internal resistance) can be suddenly impressed across the series RC circuit by turning switch S ON (assumed to be an ideal switch). These factors hold true for the circuit of Fig. 2-11:

a. At every instant, the voltage across the capacitor is directly proportional to the stored charge Q; that is, $e_C = Q/C$, where C is the capacitance (from the definition of capacitance).

b. The voltage across the resistor is $e_R = iR$ (Ohm's law)

c. The sum of the voltage drops around the circuit must equal the impressed voltage at every instant (Kirchhoff's law); that is,

$$E = e_R + e_C = iR + \frac{Q}{C} \tag{2-3}$$

d. The current is the same in all parts of a series circuit at a given instant, and it is equal to the net voltage divided by the resistance; that is,

$$i = (E - e_C)/R$$

Suppose that the switch has been OFF for sufficient time so that $Q = 0$. Therefore $i = 0$, $e_R = iR = 0$, $e_C = Q/C = 0$. Now turn the switch ON. At the instant of closing, the capacitor has not had a chance to charge, and $e_C = 0$. Therefore $i = (E - e_C)/R = E/R = 10/100 = 0.1$ A, and $e_R = 10$ V. Immediately the flow of electrons starts to charge the capacitor; so Q and e_C increase and i and e_R decrease, each exponentially, as shown in Fig. 2-12. Note that the time scale is calibrated in units of RC. The product RC has the units of seconds

$$\frac{\text{volts}}{\text{coulombs/seconds}} \times \frac{\text{coulombs}}{\text{volts}} = \text{seconds}$$

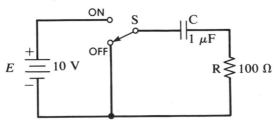

Fig. 2-11   Change of voltage across the RC circuit by a switch.

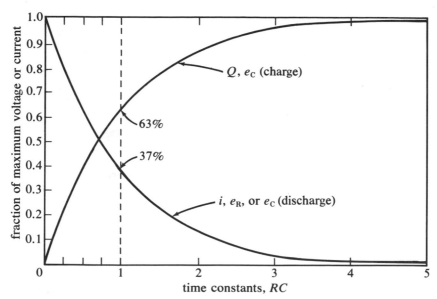

Fig. 2-12   Charge and discharge curves in the RC circuit.

and is called the *time constant*.   The current i at any time t after turning the switch ON is given by[1]

$$i = \frac{E}{R} e^{-t/RC} \tag{2-4}$$

After a time $t_1 = RC$, the current $i_t = (E/R)e^{-RC/RC} = (E/R)e^{-1} = (E/R) \times$ 0.368.   In other words, at time $t_1 = RC = 100 \times 1 \times 10^{-6} = 10^{-4}$ sec, the

[1]Note: The exact expression for the voltage e_C across the capacitor as a function of time can be obtained by substituting dQ/dt for i in Eq. (2-3) and integrating.   The result is

$$e_C = E(1 - e^{-t/RC}) \tag{2-5}$$

and substituting Eq. (2-5) into Eq. (2-3),

$$e_R = E - e_C = Ee^{-t/RC} \tag{2-6}$$

and, since $e_R = iR$, it follows that

$$i = \frac{E}{R} e^{-t/RC} \tag{2-4}$$

current is 36.8% of the current at the instant of impressing voltage E. This means that at time $t_1 = RC$ the voltage across the resistor is only 36.8% of its initial value and the capacitor is charged to 63.2% of the impressed voltage. The time $t_1 = RC$ which is referred to as the time constant of the RC circuit is given the symbol τ. Table 2-1 tabulates values of e_C and e_R for different multiples of τ, both for the charging of the capacitor by impressing voltage E and for discharging a capacitor that is charged to a voltage E. Note that when $t = 4.6RC$ the capacitor is charged to 99.0% of the impressed voltage and e_R is only 1% of its initial value. For practical purposes the capacitor is often considered to be fully charged when $t \approx 5\tau$.

Table 2-1   Output Voltages Across Capacitor and Resistor in Series RC Circuit

Time	Capacitor Charging		Capacitor Discharging	
	e_C % E applied	e_R % E applied	e_C % E initial	e_R % E initial
RC	63.2	36.8	36.8	36.8
2RC	86.5	13.5	13.5	13.5
2.3RC	90.0	10.0	10.0	10.0
3RC	95.0	5.0	5.0	5.0
4RC	98.2	1.8	1.8	1.8
4.6RC	99.0	1.0	1.0	1.0

It is important to keep in mind that the voltage across a capacitor cannot change instantly—capacitor voltage changes exponentially with time. From one point of view, the capacitor may be considered a short circuit at the first instant of impressing a sudden voltage change; from another viewpoint, an instantaneous change of voltage represents an infinitely high frequency so that the reactance of the capacitor is zero.

By observing the output of the series RC circuit, first across the resistor and then across the capacitor, various waveshapes similar to those in Fig. 2-13 can be observed if the switch of Fig. 2-11 is turned ON and OFF, or a rectangular pulse is fed to the input. It is immediately obvious that the output waveform is greatly dependent on the relationship of RC time constant τ to pulse width T_p. It is interesting to observe that the leading edge of the output across the resistor is always steep, assuming a steep leading edge for the input voltage. In contrast, the leading edge of the capacitor output is always changing exponentially. Note that the sum of voltages across the capacitor and resistor equals the input voltage at each instant for a given RC time constant. This can be observed by comparing the pairs of curves in (c), (d), and (e) of Fig. 2-13.

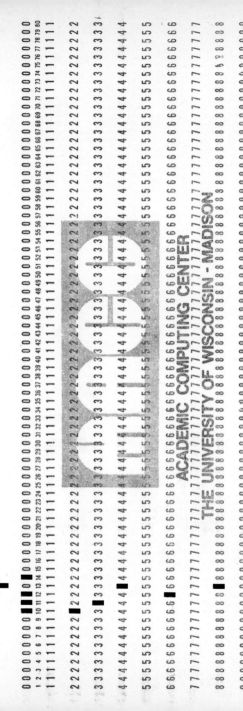

ACADEMIC COMPUTING CENTER
THE UNIVERSITY OF WISCONSIN - MADISON

WIS. 769   ISC/PRYOR 50954

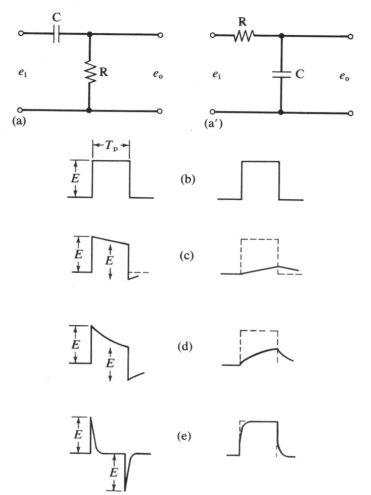

Fig. 2-13 The RC series circuit: (a) output taken across resistor; (a') output taken across capacitor; (b) input voltage; (c) output voltages across R and C, respectively, when $\tau \gg T_p$; (d) $\tau \approx T_p$; (e) $\tau \ll T_p$.

It is worth while to observe that sharp positive and negative pulses can be obtained across the resistor when the RC time constant is much shorter than the pulse width. This finds application in many circuits. In Fig. 2-13c the voltage across the capacitor is a rather linear sawtooth voltage. This suggests that the circuit can act as an integrator when the time constant is long compared with pulse width. In fact, when the output is taken across the capacitor, the RC circuit is often referred to as an "integrator," because $e_o = e_C \approx (1/RC) \int e_i \, dt$.

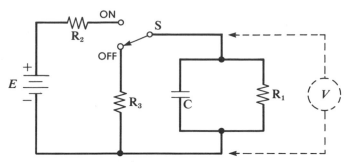

Fig. 2-14 Parallel RC switching circuit.

Parallel RC Circuit

To determine the shape and time constant of voltage changes in a switching circuit, it is only necessary to determine whether the output signal is across the capacitance or the resistance and what components determine the RC time constant in the ON and OFF cases. To take another example, consider the parallel RC circuit of Fig. 2-14. When switch S is thrown to the ON position, C will begin to charge toward the final voltage established by the voltage divider R_1 and R_2, i.e., $ER_1/(R_1 + R_2)$. The rate of charge is determined by C and by the output resistance of the charging circuit which is the source E and the voltage divider. The output resistance of a voltage divider is the parallel combination of the divider resistors[2] or $R_1R_2/(R_1 + R_2)$; thus the charging time constant is $CR_1R_2/(R_1 + R_2)$. When switch S is turned OFF, capacitor C discharges toward zero through the parallel discharge paths R_1 and R_3. Thus the discharge time constant is $CR_1R_3/(R_1 + R_3)$. Since the voltage across the capacitor is being observed, the waveforms will resemble those in the right-hand column of Fig. 2-13.

Experiment 2-2   *Charge and Discharge Curves*
for an RC Circuit

A square wave is applied across an RC circuit and the output signals across the resistor and capacitor are observed with an oscilloscope. The frequency, resistance, and capacitance values are varied to show the effects of period and time constant on the output signals.

---

[2] H. V. Malmstadt, C. G. Enke, and E. C. Toren, Jr., *Electronics for Scientists*, p. 543, Benjamin, New York, 1963.

Switch Inertia

Returning now to the nonideal switch represented by the schematic of Fig. 2-10 and placing this switch in the circuit shown in Fig. 2-15, the effect of the equivalent switch capacitance will be noted. The resistances R_{SO} and R_{SC} are the open and closed values for switch S, and R_L is the load resistance of the circuit.

Assume that the switch S is initially closed so that the voltage V_{SC} across the switch is

$$V_{SC} = E \frac{R_{SC}}{R_L + R_{SC}}$$

as determined by the voltage divider R_L and R_{SC}. Now if the switch is opened, the voltage across the open switch will rise exponentially as the equivalent capacitance C is charged to the maximum open switch voltage determined by the voltage divider R_L and R_{SO},

$$V_{SO} = E \frac{R_{SO}}{R_L + R_{SO}}$$

as illustrated in Fig. 2-16.

The charging time constant τ_0, will be determined by C and the output resistance of the voltage divider R_L and R_{SO}. Thus,

$$\tau_0 = C R_L R_{SO}/(R_L + R_{SO}) \qquad (2\text{-}7)$$

If $R_{SO} \gg R_L$ as is usually the case, then $\tau_0 \approx C R_L$.

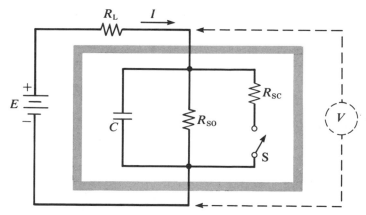

Fig. 2-15   Switch circuit with nonideal switch.

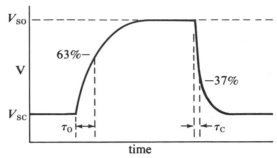

Fig. 2-16   Effect of capacitance on transition time.

The discharge time constant τ_C is determined by C and the parallel discharge paths R_{SO}, R_{SC}, and R_L. Since it is assumed that for an effective switch, $R_{SO} \gg R_{SC}$,

$$\tau_C = C R_L R_{SC}/(R_L + R_{SC}) \tag{2-8}$$

In most cases $R_{SC} \ll R_L$ so that $\tau_C \approx C R_{SC}$.

In the above circuit, if $C = 100$ pF, $R_L = 10$ kΩ, $R_{SO} = 1$ MΩ, and $R_{SC} = 100$ Ω, then $\tau_0 = C R_L = 100 \times 10^{-12} \times 10^4 = 10^{-6}$ sec $= 1\mu$sec, and $\tau_C = C R_{SC} = 100 \times 10^{-12} \times 10^2 = 0.01$ μsec. Note that $\tau_C \ll \tau_0$, which is true for the typical switch. Therefore, a circuit with significant switch capacitance reaches a steady state more rapidly when the switch is closed than when it is opened. There are other factors inherent in certain devices such as diodes and transistors which determine switching speed, and these will be discussed in the last section of this chapter and in the next chapter.

2-3   Semiconductor Diode Characteristics

Semiconductor diodes are some of the most useful switching devices in a wide variety of circuits. Before considering the switching characteristics of diodes, however, it is important to summarize some of the properties of semiconductor materials.

Semiconductors

As the name implies, a pure semiconductor material is neither a good conductor nor a good insulator, but is somewhere in between, as illustrated by Fig. 2-17.

Whereas the conductor silver might have a resistivity of only 10^{-6} Ω-cm, and the insulator polyethylene, 10^{15} Ω-cm, the semiconductor germanium would be about 50 Ω-cm and the semiconductor silicon would be about 50 kΩ-cm. There are charge carriers in a semiconductor that conduct a current when voltage is applied across it, but, since the number is small, the resistance is relatively high. Most of the electrons are strongly attached to the parent atoms and are not free to move about in response to the small electric fields.

Pure semiconductors. There are many types of semiconductors, including elements, intermetallic compounds, and even organic compounds, but at present the elements silicon and germanium are the most important commercially. These well-known Group IV elements have four electrons in their outer orbitals that are available to form valence bonds with other atoms. Assume now that an absolutely pure germanium metal bar is melted and a small portion is withdrawn and allowed to cool in such a way that each atom in the solidified material is equidistant from four adjoining atoms. The four covalent bonds that join each atom to four others are difficult to disrupt, and, thus, crystalline germanium is a semiconductor that exhibits a high resistance. In fact, it takes 0.75 eV to disrupt the bonds for pure germanium, and it takes even more energy, 1.12 eV, to break the valence bond for pure silicon. Since this amount of energy must be applied over lengths of about an atomic distance, the applied field strength necessary to break a bond for a germanium crystal is very high. At absolute zero the pure germanium crystal is an insulator, but at room temperature there are some free charge carriers that can conduct current because electrons will occasionally gain sufficient energy to break their bonds and become free. The free electrons can move randomly through the crystal lattice and can act as a negative charge carrier. When an electron breaks away, it leaves a vacancy or "hole." A hole, in turn, can be filled by an electron. As an electron moves in to fill the hole, another hole is created so there is an apparent random movement of the hole.

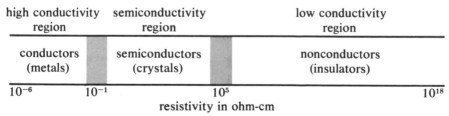

Fig. 2-17   Materials resistivity spectrum. (From *GE Transistor Manual*, 1964.)

Being "the absence of an electron," the hole behaves as a positive charge carrier. Both electrons and holes contribute to the conductivity of a semiconductor.

As the temperature of a crystal increases, the rate of producing free electrons and holes increases. Note that the crystal is electrically neutral and that, at a given temperature, an equilibrium is reached wherein the number of holes and electrons being formed is equal to their recombination rate. The "lifetime" of an individual electron or hole before recombination can vary over a range of less than a microsecond to more than a millisecond. Now, if a voltage is applied across the germanium crystal there will be a net drift of electrons toward the plus terminal and holes toward the negative terminal. The average velocity toward the terminal of each type of carrier is known as its "drift velocity," v_n for electrons, and v_p for holes. The current density i_n that is carried by the electrons is

$$i_n = nqv_n$$

and the current density carried by the holes is

$$i_p = pqv_p$$

where n and p are the number of electrons and holes, respectively, per cubic centimeter, and q is the charge of the electron. The total current is the sum of the individual current densities, so that

$$I = i_n + i_p = q(nv_n + pv_p) \tag{2-9}$$

and since the velocities of electrons and holes are proportional to the applied field $\mathscr{E}$ (that is, $v_n = \mu_n\mathscr{E}$, and $v_p = \mu_p\mathscr{E}$), then

$$I = q(n\mu_n + p\mu_p)\mathscr{E} \tag{2-10}$$

From Ohm's law it is apparent that $q(n\mu_n + p\mu_p)$ is the "conductivity" σ of the semiconductor (that is, σ is the reciprocal of the resistivity, ρ).

In terms of energy bands, the valence band is nearly full of electrons, but a few electrons have sufficient energy to attain the conduction band. The inherent ability of a pure semiconductor to conduct current is known as its "intrinsic conductivity." It is dependent on the number of free charge carriers, and this number is a Boltzmann function of temperature, as given by the expression

$$np = n_i^2 = AT^3e^{-E_g/kT} \tag{2-11}$$

where n_i is the "intrinsic carrier concentration," A and k are constants, E_g is the energy of the forbidden gap between valence and conduction bands, and T is the absolute temperature. For germanium $n_i \approx 2.5 \times 10^{13}$ carriers per cubic centimeter at room temperature, and for silicon $n_i \approx 1.5 \times 10^{10}$ carriers per cubic centimeter.

Doped semiconductors. Although the intrinsic properties of a pure semiconductor are important considerations, the desirable characteristics of semiconductors for use in diode rectifiers and transistors are obtained by purposely adding small amounts of selected impurities. Assume that a small amount of a Group V element, such as antimony, is added to germanium, so that the ratio of antimony to germanium is about 1 ppm.[3] Each antimony atom is, therefore, completely surrounded by germanium atoms, and four of its five valence electrons form covalent bonds, as the antimony atom becomes part of the crystal lattice. The extra electron is now only loosely bound to the atom, and an energy of only about 0.01 eV is required to free it.

The antimony atoms are called "donors" because they increase the free electron concentration n which greatly increases the conductivity of the germanium semiconductor. A semiconductor doped with donor atoms is referred to as an "n-type" semiconductor. It is likewise possible to dope germanium with a Group III element, such as indium, so as to increase the free hole concentration p. The resulting impure crystal is known as a "p-type" semiconductor, and the impurity is called an "acceptor."

Note that when donor atoms contribute excess electrons to the semi-condutor there is a greater probability of the intrinsic holes recombining with electrons. The concentration of holes, therefore, is reduced considerably below the intrinsic number present at room temperature for a pure semiconductor. Since the electrons are now, by far, the "majority carrier," the application of a voltage to the n-type semiconductor results in a current carried primarily by electrons and Eq. (2-10) reduces to

$$I_n = qn\mu_n\mathscr{E} \tag{2-12}$$

In an analogous way in a p-type semiconductor,

$$I_p = qp\mu_p\mathscr{E} \tag{2-13}$$

Note that in a p-type semiconductor the holes move in the semiconductor but electrons flow in the connecting wires.

[3] Parts per million, by weight, is abbreviated ppm.

The pn-Junction Diode

Now consider what happens when p-type and n-type semiconductors are joined together at a junction. As might be expected, the electrons and holes combine, but only for a very narrow region right at the junction. The reason for the narrow region of recombination is that the n type becomes positively charged as electrons move into the p type, and the p type becomes negatively charged as the holes move into the n type. The result is a "barrier field," or "built-in potential," ϕ_0, of about 0.3 V which repels most of the majority carriers of each type from approaching the junction. The movement of the intrinsic minority carriers, however, is aided by the barrier field. The resulting current from the minority carriers is exactly balanced by the few majority carriers that do gain sufficient energy to cross the barrier, so that at equilibrium a net current of zero is maintained for the unbiased diode, which is represented in Fig. 2-18a. The narrow region at the junction is called the "depletion

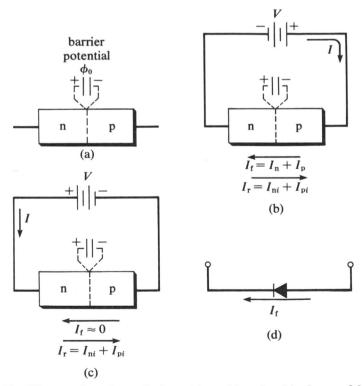

Fig. 2-18 The pn-junction diode: (a) unbiased; (b) forward-biased; (c) reverse-biased; (d) schematic symbol.

layer" because it contains very few mobile charge carriers. The electric field within the depletion layer is a result of the ionized acceptors and donors that exist there. Outside this narrow region the electric field is essentially zero, so that the net charge within the depletion region must be zero.

Biasing the pn junction. If a voltage V is applied across the pn diode, so that the p-type semiconductor is made positive with respect to the n-type, it opposes the built-in potential ϕ_0, as shown in Fig. 2-18b. In effect, then, the electrons of the n type and the holes of the p type are forced closer together. Therefore, the probability of majority carriers passing across the barrier is increased, and a "forward current" I_f is obtained. If the forward current is small, so that the voltage drops in the semiconductor material are small, the potential drop across the depletion layer is $\phi_0 - V$. There is also a reverse current I_r due to minority carriers, but it is small compared with I_f, and the net current is in the forward direction.

If the diode is now reverse-biased, as shown in Fig. 2-18c the potential barrier at the junction is increased to $\phi_0 + V$, and the width of the depletion layer is increased. The effect is to reduce the probability of majority carriers crossing the barrier. The reverse current, due to minority carriers, remains the same as before, assuming the same temperature. Therefore, as the reverse voltage is applied, the forward current becomes insignificant and the reverse current remains constant at a small value I_r, even though the reverse voltage V is changed over a wide range. It is thus apparent that the pn junction diode is a good switch since the applied signal can change it from a good conductor to a poor conductor or from an effectively low resistance to a very high resistance.

The diode I-V curve. The current-voltage curve for a representative diode is shown in Fig. 2-19. A slight forward voltage bias causes the current to increase exponentially. With reverse voltage bias the reversed current I_r remains constant until the breakdown region is reached wherein the reverse current increases rapidly with increased voltage.

The pn-junction characteristics are determined by the carrier concentrations which are determined by charge diffusion across the barrier, rate of charge generation, charge lifetime, and the temperature. Based on the Boltzmann distribution the well-known pn-junction diode equation can be obtained which is

$$I = I_r \left(e^{qV/kT} - 1 \right) \qquad (2\text{-}14)$$

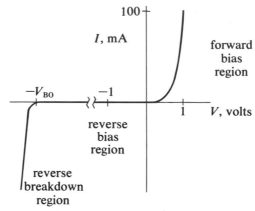

Fig. 2-19   Current-voltage curve of a silicon diode.

where I is the diode current, I_r the reverse current, q the charge on the electron (1.6×10^{-19} coulomb), V the applied bias voltage, k the Boltzmann constant (1.38×10^{-23} W-sec/°K), T the absolute temperature (300°K at room temperature).

Forward bias corresponds to positive values for V in Eq. (2-14), and at room temperature (300°K) the value of $q/kT = 39$.   For small values of V (about 0.1 V or greater), the exponential term is more than 100 so the quantity (-1) is negligible and Eq. (2-14) is approximately

$$I_f = I_r e^{qV/kT} \tag{2-15}$$

and

$$\log I_f = \log I_r + KV \tag{2-16}$$

Thus the current rises as an exponential function of the voltage.   For real diodes, a plot of log I_f vs V often gives a straight line over several orders of magnitude.   Equation (2-15) is based on a value for V which is the inner barrier potential from the zero-bias condition.   If V is taken as the voltage across the diode terminals the equation is only approximate at high currents because of the ohmic voltage drop in the semiconductor material.

When a reverse bias voltage is applied, V is negative in Eq. (2-14), and the exponential term diminishes and becomes negligible at small reverse bias so that

$$I = -I_r \tag{2-17}$$

With reverse bias the barrier is widened and charge movement is at a minimum and will be small and quite constant at a fixed temperature, as indicated by Eq. (2-17). For this reason the reverse current I_r is often called the *saturation* current and given the symbol I_s or I_o.

Many factors characteristic of the pn junction determine the reverse current I_r, and, although very small, it increases exponentially with increasing temperature. The reverse current is doubled for a 10°C temperature rise in a germanium diode and for a 6°C rise in a silicon diode.

Silicon diodes have a forbidden gap energy that is greater than that for germanium diodes. This causes the reverse current for silicon to be about one-thousandth of that for germanium diodes. Equation 2-14 shows that because of the lower I_r, the forward voltage needed for significant conduction in a silicon diode is considerably higher than for the same conduction in a germanium diode. For example, about 0.6 to 0.7 V is required for a silicon diode to conduct as compared to about 0.2 V for germanium as indicated in the *I-V* forward characteristic curves in Fig. 2-20.

Experiment 2-3   *Diode Characteristic Curves*

A silicon diode characteristic curve is displayed utilizing the oscilloscopic curve tracer of Experiment 2-1, and significant diode parameters are measured from the curve. The experiment may be repeated for a germanium diode.

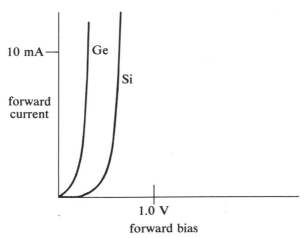

Fig. 2-20   Forward characteristic curves for germanium and silicon.

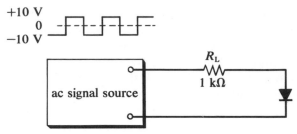

Fig. 2-21   Alternating current switching circuit.

Load lines superimposed on diode I-V curves.   The circuit of Fig. 2-21 shows the junction diode acting as a switching device in a circuit with an alternating potential signal source.   Since the source voltage has two values (+10 V and −10 V) there will be two load lines as shown in Fig. 2-22.   The two operating points are at the intersections of the load lines and the current-voltage curve.   From the operating points it is clear that a diode is a switch which is OFF or ON depending upon the polarity of the applied signal.   The voltages across the load and switch for the circuit of Fig. 2-21 for each half-cycle can be obtained from Fig. 2-22.

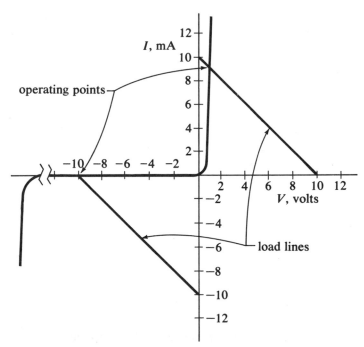

Fig. 2-22   Operating points for circuit of Fig. 2-12.

Diode resistance. Since the *I-V* curve of the diode is not linear, several parameters are required to characterize a particular type of diode. Of particular interest in switching applications are the ON and OFF resistances. Consider first the ON resistance. If the circuit is known, the ON operating point can be determined as in Fig. 2-22. The *static forward resistance* R_{fs} is, then, simply

$$R_{fs} = V/I \qquad (2\text{-}18)$$

When it is known that the diode is forward biased, R_{fs} can be substituted for the diode in the circuit as an *equivalent* component. The use of equivalent linear components (resistances, capacitances, and signal sources) which approximate a given device's characteristics is often a valuable aid in circuit analysis. Since the operating point must be known fairly accurately in order to calculate the static resistance, the usefulness of the static resistance in the early stages of circuit analysis is limited.

A characteristic which is useful over a wider range is the *dynamic resistance* R_{fd}, which is the reciprocal of the slope of the current-voltage curve at the operating point,

$$R_{fd} = dV/dI_f \qquad (2\text{-}19)$$

The static and dynamic resistance approximations are shown in Fig. 2-23 along with the circuit of equivalent linear components for each approximation method. Note that the dynamic resistance line intercepts the voltage axis at V_γ and that a voltage source of this value must be included in the complete equivalent circuit.

A useful relationship for estimating the dynamic resistance of a for-

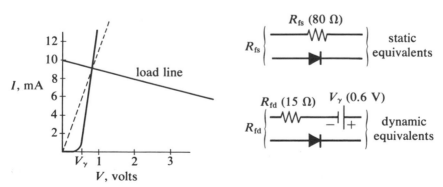

Fig. 2-23 **Static and dynamic equivalent resistance approximations for forward-biased diode.**

ward-biased diode can be obtained by differentiating Eq. (2-14) with respect to V, which gives

$$\frac{dI_f}{dV} = \frac{q}{kT} I_r \, e^{qV/kT} = \frac{q}{kT} I_f$$

or inverting,

$$\left(\frac{kT}{q}\right)\frac{1}{I_f} = \frac{dV}{dI_f} = R_{fd} = \frac{0.026}{I_f} = \frac{26}{I_{mA}} \tag{2-20}$$

For example, according to Eq. (2-20), if $I = 0.1$ mA, then the dynamic resistance is 260 Ω, but at 1 mA it is about 26 Ω, etc.

In practice, the dynamic resistance is more nearly expressed by the empirical equation, $R_{fd} = 45/I_{mA}$. Again, this difference comes about because of the ohmic resistance of the semiconductor. The static and dynamic resistance approximations for the reverse-biased diode are shown in Fig. 2-24. In this case the dynamic resistance curve intersects the current axis and a current generator of output I_o is used in the equivalent circuit.

Experiment 2-4 *Resistance of a Diode*

Values obtained from the forward-biased characteristic diode curve are used to plot log I vs E so as to demonstrate the logarithmic relationship over a few decades. The dynamic resistance is measured and compared with the static resistance at a current of 1 mA. The problems of using an ohmmeter to measure diode resistance are determined.

Reverse Breakdown and Zener Diodes

In Fig. 2-19 it was noted that when the reverse voltage exceeded a certain value the reverse current suddenly increases, and this is called the *breakdown* or *Zener voltage*. The large reverse bias produces an electric field (greater than 10^6 V/cm) that is sufficient to break the covalent bonds and produce free mobile electron-hole pairs, and this is known as *Zener breakdown*. It occurs in abrupt junctions between highly doped regions. Another type of breakdown can occur which is called *avalanche breakdown* and is caused by carriers that gain sufficient kinetic energy to ionize other atoms with which they collide, thus producing additional mobile electron-hole pairs. The mechanism of breakdown depends on the volt-

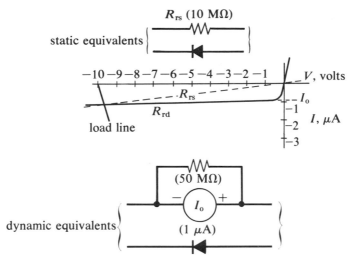

Fig. 2-24   Static and dynamic equivalent resistance approximations for reverse-biased diode.

age region of breakdown, and both mechanisms can be present in the same diode. However, the term *Zener diode* is usually used without regard for the mechanism. Silicon diodes that break down above 8 V probably depend on the avalanche mechanism, and those in the region below 5 V on the Zener mechanism. For those diodes that break down between 5 and 8 V the dominant mechanism depends on the impurity distribution at the junction.   Two common symbols for a junction diode specifically designed for superior characteristics in the Zener region are shown in Fig. 2-25.   The symbol is meant to suggest the shape of the *I-V* curve.

In a similar manner, the reverse bias region before breakdown is sometimes used as a source of very small constant currents.   A very high impedance is a desirable characteristic in constant current sources.   This application of the diode is not too common, however, because of the large effect of temperature on the reverse current I_r and the limited range of possible currents.

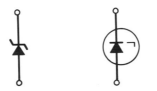

Fig. 2-25   Zener diode symbols.

The use of a Zener diode for regulating the output voltage of a power supply is described in Section 2-4.

Experiment 2-5 *Reverse Breakdown and Zener Diodes*

The reverse characteristic curve for a Zener diode is observed and the significant parameters of the Zener diode are measured from the curve.

Classes of Diodes

The two general classes of diodes are *signal* diodes and *rectifier* diodes. The signal diodes are used for many functions in digital instrumentation and detector circuits where the forward current and reverse voltage requirements are not high (e.g., 50 mA and 50 V) but where fast response times, low capacitance, and low leakage currents are desirable characteristics.

The rectifier diodes are capable of high current, starting at about 100 mA and ranging to several hundred amperes. The reverse voltage ratings typically vary from about 50 to 600 V, but some newer designs exceed 1000 V. All pn junctions can be damaged by excessive reverse voltages unless the current is limited to a safe value, and, consequently, they have a *peak reverse voltage* (PRV) rating.

2-4   Power Supply Circuits

The switching characteristics of the semiconductor diode, i.e., its capability of changing from a conductor to a nonconductor, depending on the polarity of the applied voltage, makes it a basic component of electronic power supplies that utilize ac line voltage to provide the dc voltages necessary to operate specific electronic circuits.

Power supplies used for the operation of transistor and integrated circuits usually provide only low dc voltages of about 3 to 30 V. The operation of vacuum-tube circuits generally require dc voltages of a few hundred volts. Many measurement devices such as photomultiplier tubes, ionization detectors, oscilloscope tubes, and digital readouts require dc supply voltages of hundreds or even thousands of volts. The stability of the power supply will often determine the stability of the entire instrument.

An ideal dc power supply would have (1) a constant output voltage regardless of variations in the current required by the load (good regula-

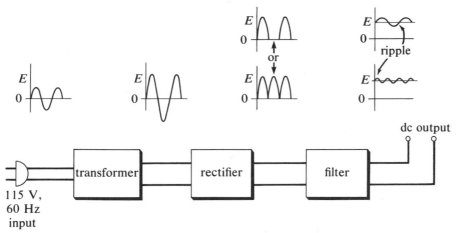

Fig. 2-26 Block diagram showing conversion of ac input to dc output voltage.

tion or low output resistance); (2) a constant output voltage regardless of variations in temperature, ac line voltage, age of power supply, etc. (good stability); and (3) no noise voltage of line or other frequency superimposed on the dc output (low ripple). In addition to these characteristics, the dc output voltage and the current capability of the power supply must meet the operational requirements of the electronic devices.

Scheme for Converting ac Line Voltage to dc Voltage

The block diagram in Fig. 2-26 illustrates the basic components and functions usually required to convert an ac to a useful dc voltage. First, a transformer is used to convert the ac line voltage to more nearly the desired voltage. Next, a rectifier circuit converts the ac to a pulsating dc voltage.[4] To smooth out the pulsating dc voltage, a filter network is employed. Sometimes a regulator circuit is added to improve the stability and regulation and further to reduce the ripple of the output voltage.

Transformers

A transformer is used to provide a step-down or step-up ac voltage from the line voltage. The line voltages in the United States are usually 115 V or 220 V, 60 Hz and in Europe 220 V, 50 Hz. One transformer

[4] Direct current is a net flow of charge in one direction. If a direct current or voltage is discontinuous or varies in magnitude, it is called "pulsating dc."

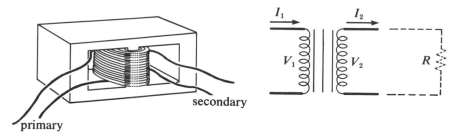

secondary

primary

Fig. 2-27   Pictorial and schematic representation of transformer.

can have several windings for different voltage requirements. A schematic diagram and pictorial of a transformer are given in Fig. 2-27.

The voltage is proportional to the magnetic flux $d\phi/dt$. Since the primary and secondary windings are in close proximity the flux is the same in each winding, and the voltage V_2 induced in the secondary is proportional to the number of turns on the secondary winding. Therefore,

$$\frac{V_1}{V_2} = \frac{n_1}{n_2} \tag{2-21}$$

It is thus possible to have the correct step-down or a step-up voltage on the secondary (compared to the primary line voltage) by utilizing a transformer with the correct turns ratio.

Experiment 2-6   *Measurement of Transformer Voltage*

The rms and peak voltages of a low-voltage transformer are measured and the turns ratio is determined.

Rectifier Circuits

Since a diode conducts current effectively in only one direction, the current in a circuit which is in series with a diode must necessarily be dc. The simplest rectifier circuit is shown in Fig. 2-28. On the positive half-cycle of the ac voltage, the diode can conduct, allowing current to pass through R_L. The R_L is the "load" or the circuit which is to be supplied with direct current. On the negative half-cycle, the diode is reverse-biased and, therefore, nonconducting. This rectifier circuit is called a "half-wave" rectifier because only half of the ac current wave is present in the load circuit.

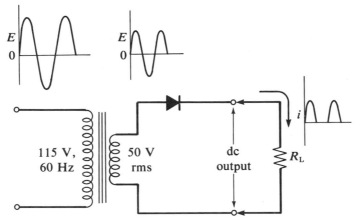

Fig. 2-28   Half-wave rectifier circuit.

In considering rectifier circuits, it is important to consider the ratings of the rectifiers, which include: the *maximum average forward-current rating*, which is $\frac{1}{2} \times E_{av}/R_L$, because the diode conducts only half the time; and the *peak inverse voltage*, which is the maximum voltage of the reverse or nonconduction polarity that should be applied to the rectifier. The diode of Fig. 2-28 has to withstand a peak inverse voltage of $50 \times 1.4 = 70$ V. The effective resistance of a conducting diode is not constant but depends on the current, but an approximate knowledge of the forward resistance allows one to calculate the power loss in the rectifier ($I^2 R_{fs}$).

For many applications it is desirable to have a rectifier circuit which supplies current during both half-cycles of the ac power and, thus, provides a more continuous current to the load. A "full-wave" rectifier circuit is shown in Fig. 2-29. This circuit is essentially two half-wave

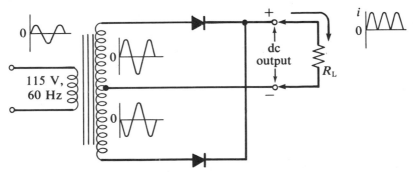

Fig. 2-29   Full-wave rectifier circuit.

rectifiers in parallel with inputs that have a phase difference of 180°. The voltage output of the full-wave rectifier is equal to the voltage developed by each half of the transformer secondary. For a 30-V peak output from a full-wave rectifier, one would use a $(60/1.4) + 5 = 48$-V center-tapped transformer. The extra 5 V is to compensate for the drop in the rectifiers and filters. Note that the rectifiers must withstand an inverse voltage of twice the peak value of the source. For the case above, the peak inverse voltage would be $1.4 \times 48 = 67$ V.

A way to obtain full-wave rectification which does not require a center-tapped transformer is shown in Fig. 2-30. This circuit is called the "bridge rectifier." Trace the current path through the rectifiers. On the positive half-cycle, D_2 and D_4 conduct. On the negative half-cycle, D_1 and D_3 conduct. In each case, the direction of electron flow through the load R_L is the same.

Having two rectifiers in series with the load drops the peak inverse voltage that each rectifier must withstand to the peak value of the supply voltage.

Two rectifiers can be connected to a single ac source and wired so that their outputs are in series as in Fig. 2-31. The output voltage available from such a circuit is twice that which is available from the ac source with a half-wave or bridge rectifier. For this reason, this kind of circuit is called a "voltage-doubler" rectifier. On the positive half-cycle, capacitor C_1 is charged to the peak value of the supply voltage (in this case $115 \times 1.4 = 160$ V). On the negative half-cycle, C_2 is charged to the same potential. Since C_1 and C_2 are in series across the load, the output voltage is twice the peak voltage of the ac source. The capacitors are essential to the operation of the circuit, because they maintain the potential

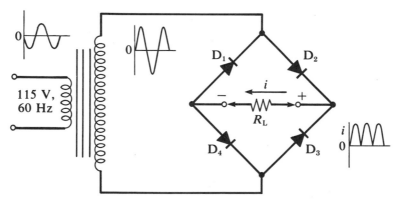

Fig. 2-30   Bridge-rectifier circuit.

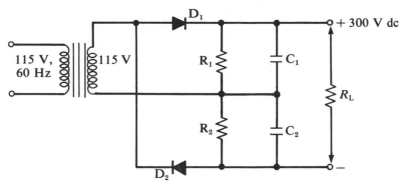

Fig. 2-31   Voltage-doubler rectifier circuit.

developed during one half-cycle so that the potential developed during the next half-cycle can be added to it.   The capacitors C_1 and C_2 have a filtering action which is described in the next section.   Since current is supplied to the load during both half-cycles, this voltage-doubler circuit is considered to be full-wave.   The peak inverse voltage applied to each rectifier is twice the peak value of the supply voltage, in this case 320 V. Where the peak inverse voltage rating of the diode is insufficient, two diodes may be put in series so that their peak inverse voltage ratings are additive.

The power dissipated by an alternating current in a load R_L is equal to $(I_{rms})^2 R_L$.   When the alternating current is passed through a rectifier, the waveshape is changed.   Half- and full-wave rectifiers supply current to the load during half the cycle and over the entire cycle, respectively. In both cases the current through the load varies between zero and some peak value.   Often the rectified current is passed through a filter to reduce the current variation through the load.   With an effective filter, the current through the load is the current supplied by the rectifier averaged over at least one cycle.   The power dissipated in the load is thus $(I_{av})^2 R_L$.   The average value of alternating current is commonly less than the root-mean-square (rms) value.   This results in a loss in the effectiveness of the current through the process of filtering.   If a half-wave rectifier supplies current to a filter and load over the entire half-cycle, the rms current is $I_{peak}/2$, but the average current is only I_{peak}/π.   Thus, the power dissipated in the load by the filtered current is only $(2/\pi)^2 \times 100 = 40.6\%$ of the power dissipated by the same current unfiltered.   The evaluation of this loss in effectiveness is complicated by the fact that the filter also affects the fraction of the cycle during which the rectifier conducts.   However, the loss is generally less for full-wave than for half-wave rectifiers.

Experiment 2-7 *Diode Rectifier Circuits*

All the rectifier circuits commonly found in power supplies are connected and their characteristics observed including half-wave, full-wave, bridge, and voltage-doubler rectifier circuits. The ac and dc components are measured and compared.

Filtering the Rectified Voltage

The output voltages of the rectifier circuits discussed in the previous section vary with time as shown in Fig. 2-32. The output can be considered as a voltage that varies about the average dc potential. The average dc potential is called the *dc component* of the output. For the half-wave rectifier output, the lowest and most predominant frequency of the ac component is the frequency of the ac line as shown by the dotted line in Fig. 2-32a. From Fig. 2-32b, it can be seen that the lowest and most predominant frequency of the output from a full-wave rectifier is twice the ac supply frequency. For most applications, it is necessary to reduce the magnitude of the ac component of the rectifier output to a value which

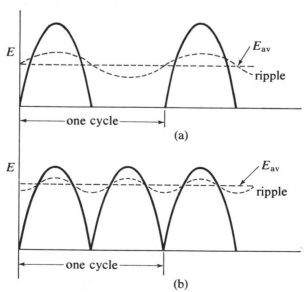

Fig. 2-32   Rectified voltage and fundamental ripple frequency: (a) half-wave and (b) full-wave rectifier outputs.

is very small compared with the average dc potential. The electrical device that accomplishes this task is called a "filter." The effectiveness of the filter is called the "ripple factor," which is defined as the rms value of the ac component or ripple divided by the average dc potential.

A rather effective filter is simply a large capacitance in parallel with the load R_L. The capacitor may be thought of as storing charge when the ac component is positive and as discharging through the load when the ac component is negative. Another way to consider the capacitor filter is that its impedance for the ripple frequency is very low compared with R_L, and it thus diverts most of the alternating current away from the load circuit.

A more detailed picture of the action of a capacitor filter is presented in Fig. 2-33. The capacitor charges to the peak value of the input voltage. If R_L were infinite, the voltage across the capacitor would quickly reach a constant value equal to the peak value of the alternating current supplying the rectifier. In the practical case where R_L is not infinite, the capacitor begins to discharge through R_L as soon as the input voltage decreases to that voltage to which the capacitor has charged. This results in the output waveshapes shown in Fig. 2-33a.

It can be seen from Fig. 2-33 that the magnitude of the ripple voltage will be decreased if R_L, C, or the frequency is increased. The expression for the ripple factor, $r = 1/(2\sqrt{3}fCR_L)$, bears this out. Here f is the frequency of the main ac component, equal to the line frequency for half-wave rectifiers and twice the line frequency for full-wave rectifiers. It can also be seen that, as the ripple increases in magnitude, the average dc output will decrease, $E_{dc} = 1.4E_{rms} - I_{dc}/2fC$, where E_{rms} is the rms value of the rectifier supply voltage and I_{dc} is the average dc current through R_L. The regulation of the capacitor filter improves also with larger values of C and f.

The above equations have been derived on the assumption that the components used are more or less ideal. Also, some simplifying assumptions have been made with regard to waveshape. Although the equations may be accurate only to about 10%, they are very useful in determining the effectiveness and general characteristics of the several filters in common use.

The output voltage and ripple voltage are determined here for the low-voltage power source in Fig. 2-33b. Note that, since this is a full-wave rectifier, $f = 2 \times$ line frequency $= 120$ Hz. The output voltage is $E_{dc} = (1.4 \times 18 \text{ V}) - I_{dc}/2fC$. Since

$$I_{dc} = \frac{E_{dc}}{R_L}$$

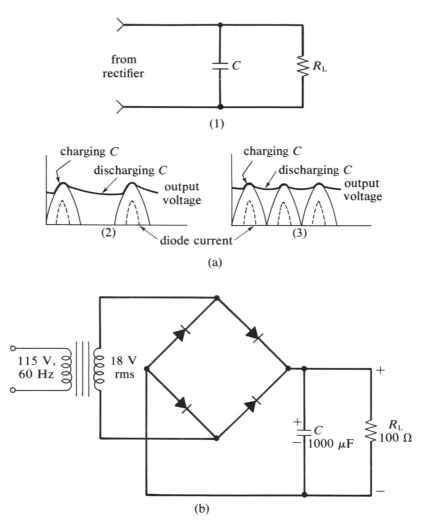

Fig. 2-33   **(a) Capacitor filter: (1) circuit; (2) half-wave input; (3) full-wave input.   (b) Low-voltage, capacitor-filtered power supply.**

and, substituting for I_{dc},

$$E_{dc} = \frac{(1.4 \times 18)}{(1 + \frac{1}{2}fCR_L)}$$

then, substituting for f, C, and R_L,

$$E_{dc} = \frac{(1.4 \times 18)}{1 + 1(2 \times 120 \times 1000 \times 10^{-6} \times 100)} = 24 \text{ V}$$

The ripple factor r is

$$r = \tfrac{1}{2}\sqrt{3}fCR_L = \frac{1}{(3.5 \times 120 \times 1000 \times 10^{-6} \times 100)} = 0.024$$

The rms ripple voltage is

$$E_{ripple} - rE_{dc} = 0.6 \text{ V rms} = 1.7 \text{ V peak-to-peak}$$

Another filter device is simply an inductance in series with the load. The inductance, by opposing the changes in current through it, tends to maintain a more constant current through the load. This results in a more constant output voltage. The current tends toward the average value under the influence of the inductance. The voltage across the load then becomes the average voltage output of the rectifier. For a full-wave rectifier and a choke filter, $E_{dc} = 2/\pi E_{peak} = 2\sqrt{2}/\pi E_{rms} \approx 0.9 E_{rms}$. This expression assumes that the dc resistances of the transformer, rectifier, and choke are negligible compared with R_L. Notice that the output voltage is quite independent of the load resistance, making the choke-filtered supply well regulated. The output voltage, however, is considerably lower than that with the capacitor filter.

To be effective with a half-wave rectifier, the inductance value must be rather large. The sharp increase in current which occurs at the beginning of the conduction cycle in a half-wave rectifier causes such a large voltage to be induced in the coil that a severe strain is put on the insulation of the windings. For this reason, it is not practical to use chokes to filter the output of a half-wave rectifier.

Combinations of chokes and capacitors make the best passive filters, but the current trend is toward active filtering and regulating circuits for power supplies in precision instruments.

Experiment 2-8   *Power Supply Filters*

The ability of capacitance and RC filters to reduce ac ripple on the dc output voltage of a power supply is investigated, and the effectiveness of different filter values under various conditions is determined.

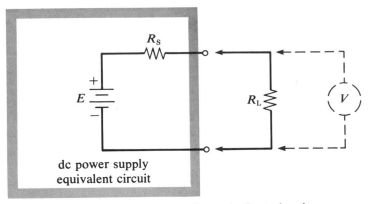

Fig. 2-34   Power supply equivalent circuit.

2-5   Zener Diode Regulator Circuit

Any power supply could be represented by the Thévenin equivalent circuit[5] shown in Fig. 2-34, with the load R_L connected at its output. The voltage V across the load is

$$V = E - IR_S \qquad (2\text{-}22)$$

Therefore, anything that changes the current I changes the voltage V. Obviously, if the load changes the current changes, and the effect on voltage V depends on the relative values of R_L and R_S. Also, anything that changes E will change V. Therefore, if the line voltage changes, the voltage across the load changes. In other words, to regulate fully (control) the voltage V it is necessary to compensate for changes of E and/or changes of load current I.

One solution is to introduce a variable resistor R_R in series with the load that would automatically vary to maintain V constant. This type of *series regulator* is quite common. Another type of regulator is the *shunt regulator*. The shunt regulator equivalent circuit is shown in Fig. 2-35.

Part of the total supply current I goes through the load $R_L(I_L)$ and part through a variable resistance $R_R(I_R)$. Since $V = E - IR_S$ and since $I = I_L(R_R + R_L)/R_R$, it follows that

$$V = E - \left(I_L \frac{R_R + R_L}{R_R} \right) R_S \qquad (2\text{-}23)$$

[5] Thévenin's theorem, see H. V. Malmstadt, C. G. Enke, and E. C. Toren, Jr., *Electronics for Scientists*, p. 542, Benjamin, New York, 1963.

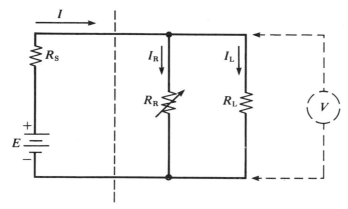

Fig. 2-35   Shunt regulator.

It can be seen that a device that would effectively vary R_R so as to main-tain V constant at all times would be a valuable regulator. This is the role of a Zener diode in certain regulated supplies. A Zener-regulated supply is shown in Fig. 2-36.

Since the Zener voltage remains essentially constant for a wide range of currents its effective resistance $R_R = V_Z/I$ changes to compensate for changes of supply voltage E or load current. The diode continues to regulate until the circuit conditions require the Zener current to fall to a value in the region of the knee on the diode I-V curve. The power dis-sipation rating determines the upper limit for the Zener diode current $(W_{max} = I_{Z(max)}V_Z)$.

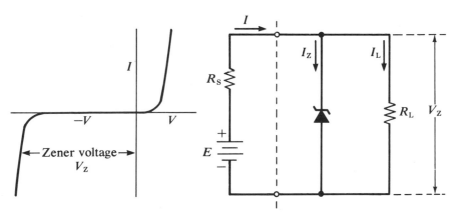

Fig. 2-36   Zener-regulated supply and characteristic curve.

Experiment 2-9 *Zener Diode Regulator*

The effectiveness of a Zener diode to regulate a power supply against changes of input voltage and load is measured. Conditions for optimum regulation are determined.

2-6 Diode Clipping Circuits

The so-called *clipper circuit,* or *limiter,* is useful in a variety of ways in wave-shaping circuits. Limiters are used to prevent voltages from swinging too far in either the positive or the negative direction, to cut off either a positive or a negative pulse, to convert a sine wave into a rectangular wave, and to isolate or eliminate various sections of wave-forms.

When a diode is in series with a resistor, as shown in Fig. 2-37, it is apparent that the circuit may be analyzed like a voltage divider. When the diode is forward-biased (Fig. 2-37a), most of the input voltage appears across the resistor. When it is reverse-biased (Fig. 2-37b), only a small fraction of the voltage appears across the resistor. From the assumed forward and back resistances for a semiconductor diode, 100 Ω and 100 MΩ, respectively, it can be seen that this diode is not an ideal switch but is quite adequate in most cases.

The diode-resistor circuit of Fig. 2-37 can be used as a limiter by connecting the load either in *series* with the diode or in *shunt* with the diode. The diode-resistor circuit of Fig. 2-37 is redrawn in Fig. 2-38a to illustrate the *series limiter* and in Fig. 2-38b to illustrate the *shunt limiter.* By reversing the diodes the positive peaks can be clipped off. Note that a dc return through the input voltage source e_i is necessary so that the clipper does not become a clamp (see Section 2-7).

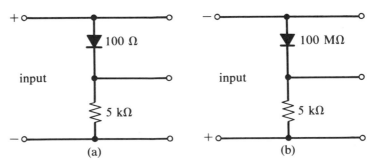

(a)   (b)

Fig. 2-37   Diode-resistor series circuit: (a) forward-biased; (b) reverse-biased.

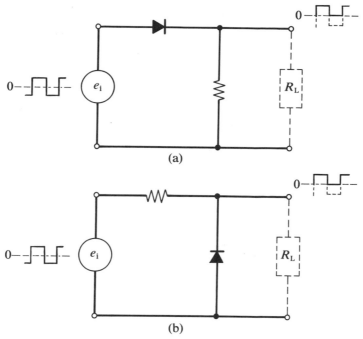

Fig. 2-38   Limiters: (a) series; (b) shunt.

In some cases it is desirable to clip off the part of the signal that is either below or above a fixed bias voltage E. This can be accomplished by the circuit illustrated in Fig. 2-39. The voltage e_i and E are mixed at the output e_o. When the diode is forward-biased ($R_{diode} \ll R$), e_i will predominate at the output. Conversely, when the diode is reverse-biased ($R_{diode} \gg R$) the output voltage will be approximately equal to E. When the diode is connected as in Fig. 2-39a, it will not conduct until the input voltage swings more positive than E. Therefore all the input signal more negative than E appears clipped off in the output waveform. Likewise, by reversing the diode (Fig. 2-39b) all the signal more positive than E will be clipped off, because the diode will not conduct when the input voltage swings more positive than E.

A limiter that clips signals above and below fixed voltage limits is shown in Fig. 2-40a. It is sometimes called a *slicer,* because it slices out a section of the input voltage. The same type of clipping can be obtained by using two Zener diodes in a back-to-back configuration, to replace both the regular diodes and bias batteries, as shown in Fig. 2-40b.

During the positive half-cycle, Zener diode Z_1 appears as a short circuit and Z_2 conducts only when the breakdown potential V_Z is exceeded.

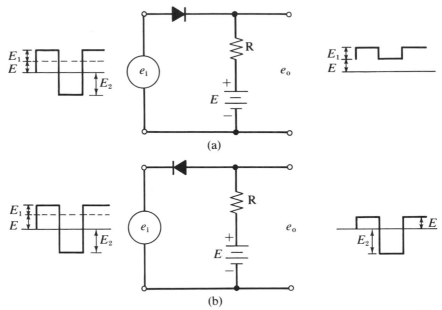

Fig. 2-39 Clipping above and below a fixed voltage level: (a) clipping below E; (b) clipping above E.

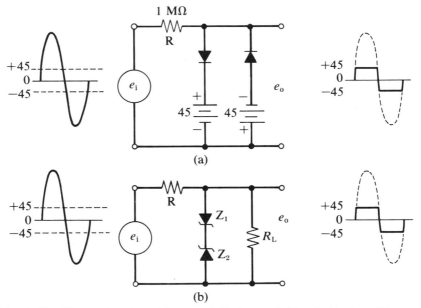

Fig. 2-40 Slicer circuits: (a) use of diodes and bias batteries; (b) use of Zener diodes.

During the next half-cycle this condition reverses. Zener diodes can also be used as replacements for the bias batteries and regular diodes in other clipper circuits.

Experiment 2-10 *Diode Clipping*
The use of diodes to clip or slice input signals is investigated.

2-7 The Diode Clamping Circuit

The *clamping circuit* is usually designed to "clamp" the top or bottom of a waveform to a fixed dc level, which may be zero, while preserving its shape and amplitude. It is sometimes called a "dc restorer," because it can restore the dc component that is lost in capacitor coupling circuits. A simple dc restorer is illustrated in Fig. 2-41 together with the equivalent circuits for positive and negative applied voltages and the initial waveforms during dc restoration.

The input voltage is a 500 Hz square wave with amplitude varying from +10 to −10 V, or 20 V peak to peak. Assume that the capacitor is uncharged and that the square wave is suddenly applied as shown by the input waveform in Fig. 2-41. The +10 V is applied across the diode so that it conducts. The effective resistance is now 1 kΩ in parallel with 1 MΩ if it is assumed that the ON diode resistance is 1 kΩ. Therefore the time constant for charging the capacitor is short ($1 \times 10^{-6} \times 10^3 = 10^{-3}$ sec). Since the period between alternations of the square wave is 0.001 sec, the capacitor will charge to 63% of the applied voltage, or about 6.3 V, before the input voltage drops to −10 V. The instant the voltage drops, the diode stops conducting and the effective resistance is the 1-MΩ resistor of the circuit. Now the time constant for the discharge of the capacitor is very long ($1 \times 10^{-6} \times 10^6 = 1$ sec), and it does not discharge significantly during the 0.001 sec of this input negative half-cycle.

On the next positive half-cycle the diode does not conduct until the voltage has exceeded 6.3 V, because this is the voltage retained on the capacitor from the previous cycle, and it opposes the applied input. The net voltage applied across the diode at the start of the second cycle is $10 - 6.3 = 3.7$ V. During the second positive half-cycle the capacitor will charge some more and will add 2.3 V (63% of 3.7 V) to the previous 6.3 V, to give a total of 8.6 V. Again during the negative cycle the discharge of the capacitor is insignificant so that the 8.6 V on the capacitor opposes the positive swing of the third half-cycle. The capacitor continues to charge to 63% of the remaining voltage until the charge is essentially 10 V.

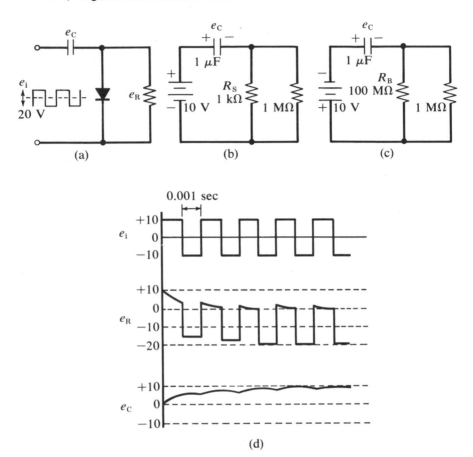

Fig. 2-41 **Direct current restorer or clamping circuit: (a) clamping circuit; (b) equivalent circuit with diode conducting; (c) equivalent circuit with diode not conducting; (d) waveforms.**

After the capacitor is charged to 10 V, it is apparent that, during each positive half-cycle of the input, the output will be 0 V. During each negative half-cycle of the input, the output will be $-10 - 10 = -20$ V. In effect the entire input waveform has shifted downward at the output so that the top is on the zero axis. The bottom of the waveform could be clamped to the zero axis by simply reversing the diode. In the case of narrow pulses it is necessary to clamp to the base and not the peak.

A dc restorer can be biased to clamp the bottom or top of the waveform at some preselected voltage by placing a battery in series with the cathode of Fig. 2-41a. The circuit will then clamp at the battery voltage.

Experiment 2-11 *Diode Clamps*

A diode clamp is used to restore the dc component of an input square wave after it has been coupled through an RC circuit.

2-8 Diode Switching Circuits

An ideal diode switching circuit transmits an input signal as an exact reproduction to the output during a selected time interval only, and with a zero output at other times. The selected time interval for transmission is controlled by an actuating signal source.

The basic idea of the diode switch is illustrated in Fig. 2-42. The switch S is closed by the gating signal in the series circuit and opened in the shunt circuit when it is desired to transmit the signal.

Even though the diode is a 2-terminal device, it is possible to turn signals ON and OFF in response to an actuating source. The circuit of Fig. 2-43 shows one way this could be done. When the actuating source output is 0 V, the diode acts as a shunt rectifier (Fig. 2-42b) shorting the output terminals for any signal source voltage greater than 0.6 V.

When the actuating source output is +10 V, the diode becomes re-verse-biased and will not affect the transmission of signals of 10 V or less to the output terminals. As far as the output is concerned, the signal source has been switched ON and OFF. This circuit could be used as the *gate* in a digital instrument as described in Chapter 1. The circuit is somewhat similar to a clipping circuit where the actuating source is *E*. In the OFF position, all signals are clipped off at 0.6 V and in the ON position the clipping level is set at 10 V.

The switching circuit of Fig. 2-43 has a weakness. In order to maintain no variation in output in the OFF position, the actuating source

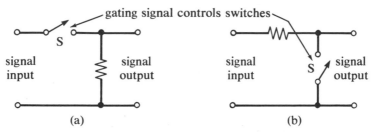

Fig. 2-42 Equivalent circuits for a sampling gate: (a) series switch; (b) shunt switch.

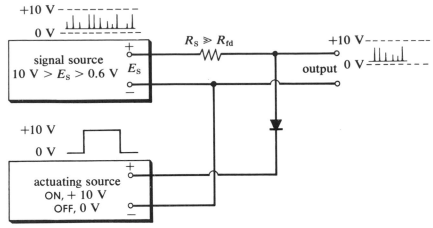

Fig. 2-43   Switching circuit using one diode.

must be capable of accepting the full signal current at its output without that output changing potential i.e., the actuating source output impedance must be nearly zero in the 0-V state.   Another circuit which provides better isolation between the signal and actuating sources uses two diodes as shown in Fig. 2-44.   When the actuating source output is 0 V, diode D_2 conducts the current from the 10-V source through R_S to ground. The output potential is held at 0.6 V, the forward drop across D_2.   Positive signals from the signal source give D_1 a reverse bias resulting in no

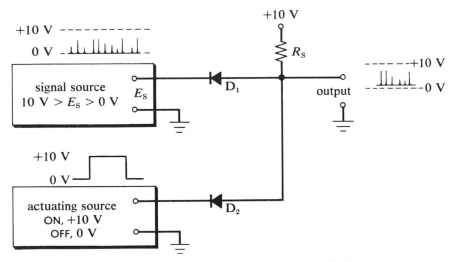

Fig. 2-44   Switching circuit using two diodes.

change in current through D_2 or output potential. However, when the actuating source potential is more positive than the signal source, D_1 is conducting and D_2 is reverse-biased. The output will now follow E_s and be equal to E_s plus the forward drop across D_1. This circuit is a basic diode switching circuit which, as subsequent chapters will show, is used in a wide variety of digital circuits.

Experiment 2-12 *Diode Switching Circuits*

The diode switching circuit of Fig. 2-44 is constructed and the appearance of the input signal at the output is controlled with an actuating signal from a suitable generator.

2-9 High-Speed Diode Switching

Most electronic switching devices are inherently several orders of magnitude faster than mechanical switches. The development of signal-switching techniques operable in the microsecond time region was a major breakthrough for digital instruments. As we continue in our efforts to measure time more precisely and to resolve events separated by shorter and shorter times, the problem of switching time continues to limit the accuracy and speed of our measurements.

A reverse-biased junction diode has very few charge carriers in the junction region. If the diode is suddenly forward-biased, the resistance will remain high until majority carriers, accelerated by the new potential gradient, cross the junction boundary. This requires only a few nanoseconds. Having crossed the junction, they are now minority carriers and are neutralized eventually by majority carriers. This reduces the majority carrier concentration near the junction which causes a net diffusion of majority carriers toward the junction from each side. Once established, this diffusion maintains the supply of charge carriers at the junction necessary for the flow of charge across the junction. But until the concentration gradient is established, an electric potential gradient is required to maintain the necessary rate of transport of majority carriers to the junction. Thus, even though forward current is flowing, the forward potential drop across the diode will be higher than its steady-state value. The attainment of steady state is usually accomplished within a tenth of a microsecond. In most applications the turn-on time would be just the time required for the forward current to reach a particular fraction of the steady state level after the forward bias was applied.

The forward-biased junction diode has relatively large concentrations of minority carriers near the junction. These carriers have just crossed the junction but have not yet been neutralized by the majority carriers. When a reverse bias is suddenly applied to a forward-biased diode, the minority carriers in the junction region will be attracted back across the junction by the new gradient. That is, electrons near the junction in the p region will be accelerated back toward the nonpositive n region and similarly for the holes near the junction in the n region. In other words, the junction is still conducting. This gives rise to a substantial reverse current until the minority carriers in the junction region are depleted to their usual low value for the reverse-biased diode. This is called the *stored-charge effect* because of the excess of minority charge carriers which are stored in the junction region during conduction and which need to be swept out of the junction region before the diode is OFF. This effect will be discussed again in more detail in the next chapter.

The diode current behavior following the sudden application of a reverse bias is shown in Fig. 2-45. The storage time, t_s, is the time required to sweep the minority carriers from the junction. During the transition interval, the diode resistance is increasing. This occurs gradually because excess minority carriers farther from the junction are still crossing the junction and because the capacitance of the reverse-biased pn junction must charge through R_L to the reverse bias potential $-V_r$. The storage time t_s is typically one- or two-tenths of a microsecond, having longer times for higher values of forward current. The transition time t_t may require several more tenths of a microsecond.

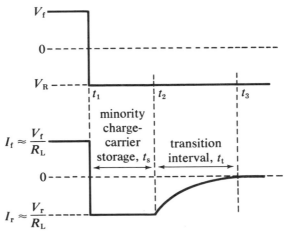

Fig. 2-45   Diode current behavior after application of step reverse bias.

Diode switching times can, if necessary, be much faster than the typical examples above. Low currents and low circuit impedances help to reduce the turn-off time. Special high-speed switching diodes are made which greatly reduce the amount of the stored charge and the width of the junction region. Such devices in appropriate circuits have achieved nanosecond switching times.

chapter three

Transistors, Relays, and Other Switches

Seldom has a component in any apparatus received the general publicity as has the *transistor,* a word known to every school child. This remarkable device has brought about major advances in electronics and has pointed the way to the development of other devices and the new integrated circuits.

One of the most important applications of the transistor is as a *switch.* Before introducing the basic transistor switching circuits, the general characteristics of the junction transistor are discussed. In this way, the circuit considerations can be related to the basic phenomena inherent in the transistor.

Several other 3- and 4-terminal switches are also presented in this chapter, including the field-effect transistor, unijunction transistor, silicon-controlled rectifier, light-sensitive (optoelectronic) devices, and electromechanical relays. In all cases the basic characteristics of the devices are presented and typical basic switching circuits are described.

3-1   Junction Transistor Characteristics

The operation of the pn-junction transistor is based on many of the principles described in Chapter 2 for the pn-junction diode. In fact, the transistor is essentially made up of two pn-junction diodes coupled together by a very thin common *base,* either of p-type or n-type semiconductor material. Therefore, there are two types of junction transistors depending on the material of the base, either the npn or the pnp transistor, as shown by the pictorial representation in Fig. 3-1.

In Fig. 3-1 a lead is shown attached to the *base*. Leads are also shown connected to the two regions of semiconductor material that are adjacent to and of opposite type to the base material. These regions are called the *emitter* and the *collector*. Although it would appear from the figure that the emitter and collector are identical, the fabrication techniques provide each with certain practical characteristics described later in this section. However, in principle, the emitter-base and collector-base junctions are of similar type, and the operation of the transistor depends on the applied voltages between the terminals.

Principles of Transistor Operation

The internal operation for the npn transistor is represented schematically in Fig. 3-2. It should be noted that the base is very thin, only about 10^{-3} to 10^{-4} cm for a typical transistor. If the emitter-base junction is forward biased the energy barrier is reduced so that the majority carriers in the emitter (electrons for the npn transistor) cross the barrier and are *emitted* or *ejected* into the base, where they become minority carriers. The emitted electrons that cross the junction rapidly diffuse through the very thin base region. When they reach the region of the base-collector junction they are rapidly accelerated toward the collector. Note that because the collector-base junction is reverse biased, the n-type collector is positive with respect to the p-type base. The electrons that reach the collector junction are, therefore, attracted to the collector and into the external circuit. In effect, then, the movement of minority carriers through the base (i.e., electrons in the p-type base) primarily determines the collector current I_C.

Some of the electrons combine with holes as they diffuse through the base, and, thus, a small base current is required to maintain the hole concentration. Also contributing to the base current is the need to com-

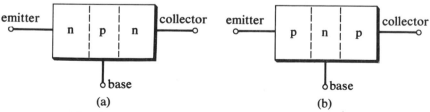

Fig. 3-1   Basic transistor types: (a) npn; (b) pnp.

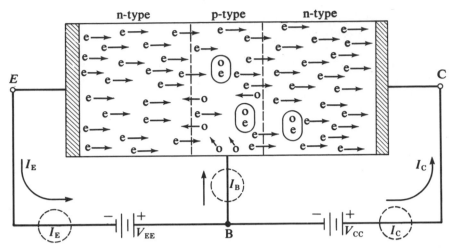

Fig. 3-2 **Representation of npn transistor in operation with forward-biased emitter-base junction and reverse-biased collector-base junction.** e = electrons, o = holes, and (oe) = recombination of holes and electrons.

pensate for the difference between hole currents across the two junctions. The holes in the base cross the forward-biased emitter-base junction, but the number of these holes is much smaller than the number of electrons crossing from emitter to base, because the emitter has a much higher impurity content (charge carrier density) than the base. The number of holes that cross the collector-base junction is usually very small because the collector is heavily doped and the junction is reverse biased. Therefore, the base current I_B is primarily a result of the emission of holes from base to emitter and the electron-hole recombination that occurs in the base.

Because $I_E = I_B + I_C$, from applying Kirchhoff's law to junction B of Fig. 3-2, and because I_B is usually small compared to I_E, the collector current I_C will be just a little less than I_E. Typical relative values might be $I_C = 99$, $I_E = 100$, and $I_B = 1$. For the configuration shown in Fig. 3-2 it is useful to define a forward current gain α_N as

$$\alpha_N = \frac{I_C}{I_E} \tag{3-1}$$

the fraction of the emitter current which appears in the collector circuit.

The collector current I_C is assumed here to be a fraction θ of the injected electron current I_n,

$$I_C = \theta I_n \tag{3-2}$$

and because the emitter current I_E is the sum of the electron current I_n and the hole current I_p,

$$I_E = I_n + I_p \tag{3-3}$$

it follows from Eqs. (3-1), (3-2), and (3-3) that

$$\alpha_N = \frac{I_C}{I_E} = \frac{\theta}{1 + I_p/I_n} \tag{3-4}$$

Therefore, even if there is no recombination in the base so that $\theta = 1$, it can be seen from Eq. (3-4) that α_N would be less than unity. As noted previously, the fabrication procedures for the emitter and base regions of the npn transistor result in very few holes crossing the emitter-base junction, as compared to the number of electrons that cross, so that the ratio I_p/I_n is small, and from Eq. (3-4) it follows that α_N is approximately equal to θ.

In the discussion above it was assumed that the collector current I_C was a fraction θ of the emitter electron current I_n. However, it is possible that the kinetic energy of the electrons as they are accelerated through the base-collector junction will be sufficient to cause additional electrons to be released from the atoms in the collector. The collector current I_C in the external circuit will, thus, exceed the rate of electron flow into the collector through the base-collector junction. Therefore, θ can be expressed as the product of two terms—there is a term M to account for any collector-current multiplication, and a term β_t (called the "base transport factor") that is the ratio of carriers which flow from base into collector to the number of carriers entering the base from the emitter, so that $\theta = \beta_t M$. The current gain relationship of Eq. (3-4) can thus be written

$$\alpha_N = \left(\frac{1}{1 + I_p/I_n}\right)\beta_t M = \epsilon\beta_t M \tag{3-5}$$

In practice, transistors are generally designed to keep α_N close to unity.

Common-Base Collector Characteristics

In Fig. 3-2 the connection of one lead of V_{EE}, one lead of V_{CC}, and the base lead to a *common* point provides the so-called "common-base" circuit configuration. If the collector current I_C is plotted against the collector-base voltage V_{CB} for fixed values of emitter current I_E, the characteristic curves of Fig. 3-3 are obtained for the npn transistor described above.

It can be seen from Fig. 3-3 that when $I_E = 0$ there is a collector current, which is a result of collector-base reverse leakage current I_{CO}. The collector current can be written

$$I_C = \alpha_N I_E + I_{CO} \tag{3-6}$$

If the base-emitter and collector-base junctions are both reverse biased the collector current can be decreased below I_{CO}. In this case the sign of I_E in Eq. (3-6) is negative. If the collector-base and emitter-base junctions are both forward biased, then the operating characteristics are in the region left of the vertical axis, and this is the saturation region for the common-base connection.

In Fig. 3-3, the region that is bounded by the two vertical lines and the line for $I_E = 0$, the base-emitter junction is forward biased and the collector-base reverse biased. In this region the value of the collector-base voltage V_{CB} has very little effect on the collector current I_C. This is

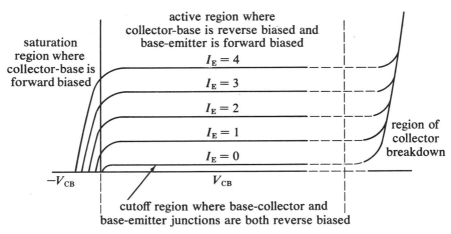

Fig. 3-3 Collector characteristics for common-base transistor configuration.

often called the "active" region, which is utilized for small-signal linear circuits.

As V_{CB} is increased the depletion layer of the collector-base junction increases so that the effective width of the base decreases. This decrease of effective base width means that for the npn transistor the electrons from the emitter need to move an even shorter path before they move into the collector. Therefore, there is less chance for recombination and less loss of electrons in the base. When the value of V_{CB} gets too high an effect called *punchthrough* occurs. The large collector-base voltage causes the effective base width to approach zero so that the emitter electrons (for the npn transistor) move directly to the collector and breakdown occurs. In this case the value of M in Eq. (3-5) increases rapidly and/or Zener breakdown occurs. The breakdown region is shown in Fig. 3-3 to be to the right of the dotted vertical line.

Saturation. The saturation effect is very important for switching circuits and it is considered here. If a load resistor R_L is connected in the collector circuit as shown in Fig. 3-4, then the collector-base voltage $V_{CB} = V_{CC} - I_C R_L$. If the voltage drop across the load, $I_C R_L$, is less than V_{CC}, then the collector-base junction remains reverse biased, and the emitter electrons which cross the collector junction flow into the external circuit. However, when the emitter current I_E is increased sufficiently the collector current I_C increases proportionately, and the voltage drop $I_C R_L$ across the load can exceed V_{CC}. In this case the collector-base junction becomes forward biased. Now the collector can inject electrons into the base, although at the same time it is receiving emitter-injected electrons to provide the collector current I_C. The net result is

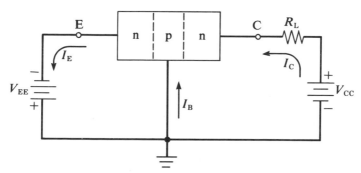

Fig. 3-4   Effect of voltage drop across load resistor on the transistor operation.

that there is a buildup of electron concentration at the collector-base junction, and the transistor is said to be in *saturation*.

If the emitter current is suddenly decreased the collector current continues to flow for a short time because of the electrons stored at the collector junction. The collector current remains relatively constant for a brief period after cutoff of I_E until the electron concentration at the collector junction returns to zero. This time delay between emitter and collector current changes can be very important in switching circuits, and is often called the *storage-time delay*. It is discussed in Section 3-2. The storage time can be reduced if the collector can be made to act as a poor emitter when forward biased. By fabricating the collector from a high resistivity material, its electron injection efficiency is low, which decreases the storage-time delay.

Common-Emitter Collector Characteristics

When the transistor emitter lead is in *common* with the base and collector voltage sources, a so-called *common-emitter* circuit exists as illustrated in Fig. 3-5. The base-emitter junction is shown forward biased so that electrons are injected into the base, and the collector-base junction is reverse biased so that the electrons that diffuse to the collector junction are accelerated into the collector and move through the external circuit. Again, from Kirchhoff's law and the direction of the electrode currents shown in Fig. 3-5,

$$I_E = I_C + I_B \qquad (3\text{-}7)$$

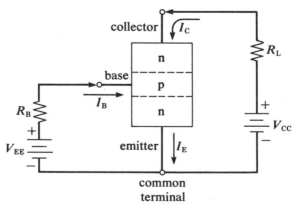

Fig. 3-5 **Common-emitter connections for npn transistor.**

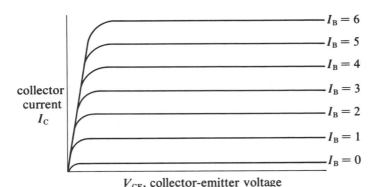

V_{CE}, collector-emitter voltage

Fig. 3-6 Collector characteristic curves for the common-emitter connection of npn transistor.

Typical collector characteristic curves are shown in Fig. 3-6 for the common-emitter connection of the transistor. These curves illustrate the relationship of I_C, V_{CE}, and I_B, and are useful in evaluating several dc parameters and for circuit analysis as shown in Section 3-2.

Experiment 3-1 *Collector Characteristic Curves for Transistors*

The collector characteristic curves for one of the npn-type transistors connected in a common-emitter circuit are obtained using the same oscilloscopic curve tracer as in Experiment 2-1. These curves are recorded so that they can be used to evaluate several dc parameters as described in Section 3-2.

The voltage from collector to emitter, V_{CE}, is equal to the sum of V_{CB} and V_{BE}, as illustrated in Fig. 3-7. If the transistor goes into saturation the polarity of V_{CB} reverses but always remains smaller than V_{BE}. Since the voltage across a forward-biased silicon junction is about 0.6 V, it would be expected that a saturated silicon transistor would have a collector-emitter voltage $V_{CE(sat)}$ less than 0.6 V, and it is typically 0.1–0.2 V. It should be noted that the relatively high ohmic resistance R_C in the collector material can cause a voltage drop, $I_C R_C$, so that a more accurate expression for the collector-emitter voltage in saturation, $V_{CE(sat)}$, is given as

$$V_{CE(sat)} = V_{CB} + V_{BE} + I_C R_C \qquad (3\text{-}9)$$

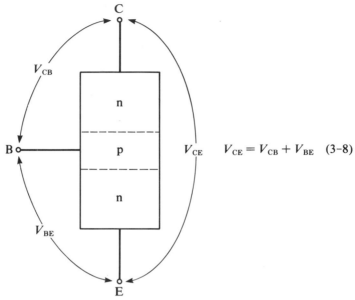

$$V_{CE} = V_{CB} + V_{BE} \quad (3\text{-}8)$$

Fig. 3-7 **Voltage relationship for common-emitter connection of npn transistor.**

Experiment 3-2 *The Collector-Emitter Voltage for a Transistor Switch* ON *(in saturation) and* OFF

The values of $V_{CE(sat)}$ and $V_{CE(OFF)}$ are measured and compared with the expected values.

Fabrication of Transistors

It was indicated in the previous discussions that the method of fabrication for the transistor influences its properties. To obtain the desired properties the location and concentration of the n- and p-type impurities must be carefully controlled. The most popular ways of forming the desired transistor pn junctions are by the grown-junction, alloy-junction, and diffused-base (mesa and planar) techniques.

Grown-junction transistors. The first commercial transistors were produced by the grown-junction technique. With this method a rather large semiconductor crystal is grown as it is continuously pulled from the

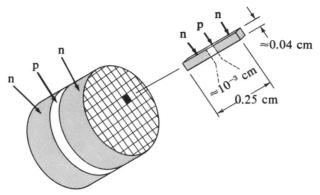

Fig. 3-8   Sectioning of grown-junction doped semiconductor to obtain small bars of npn material for transistors.

molten semiconductor material, e.g., silicon.   As the crystal grows the composition of the silicon melt is changed from n-type to p-type and back to n-type by doping the melt while continuously turning and pulling the crystal from the melt.   By cutting the doped crystal into short cylinders and then into small bars, many segments of npn material are obtained as illustrated in Fig. 3-8.   Leads are connected to the two n and the p sections of the semiconductor bar and it is then mounted on a header as shown in Fig. 3-9.

When using the grown-junction technique it is impossible to control the base-layer thickness very well, and the base thickness is relatively large, about 10^{-3} cm.   Because of the base thickness, the transistors made by this process do not have good high-frequency response, and the

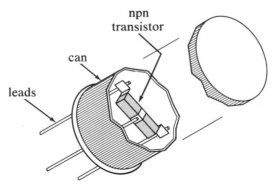

Fig. 3-9   Mounting for a grown-junction transistor.

process is quite complicated. Therefore, the grown-junction process has generally been superseded by other fabrication techniques.

Alloy-junction transistors. A thin slice (about 20 mils) of semiconductor material, e.g., germanium, can have emitter and collector dopants (indium) alloyed to either side of the wafer, as shown by Fig. 3-10. When the indium melts it dissolves some of the germanium and during the subsequent cooling the dissolved germanium recrystallizes on the base with many indium atoms incorporated in the crystal structure. Therefore, the recrystallized germanium is now doped as p type, and a pnp transistor results.

The larger size of the collector pellet as compared to the emitter enables the carriers injected at the emitter to be collected more efficiently. The dimensions of the junction regions are also hard to control by the alloy process so that the high-frequency response is generally not good.

Diffused-base transistors. It is possible to control the junction fabrication process precisely by diffusion. By a suitable sequence of masking and exposing silicon semiconductor material to a hot gas of antimony or boron dopants and etching, it is possible to produce the diffused silicon mesa or planar transistors, and cross sections of these two types as shown in Fig. 3-11.

Circuit Symbols

The junction transistor symbols used in schematic diagrams are shown in Fig. 3-12. The arrowhead indicates the direction of conventional collector current flow through the transistor and is always located

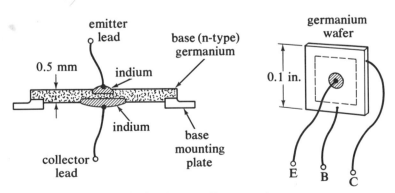

Fig. 3-10   A pnp alloy transistor.

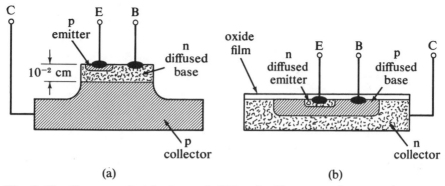

(a)   (b)

Fig. 3-11   Cross-sectional view of diffused silicon transistors: (a) mesa; (b) planar.

on the emitter.   The symbols may or may not be enclosed in circles, but the circle has no electronic significance.

Transistor Circuit Types

The common-base and common-emitter connections for the transistor were illustrated in Fig. 3-2 and 3-5, respectively.   Another method of connecting the transistor is with the collector lead common to the input and output.   The common-collector configuration is usually called an

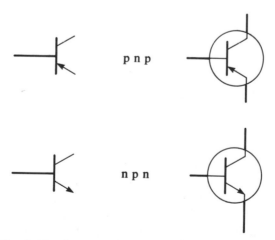

Fig. 3-12   Symbols for pnp and npn transistors.

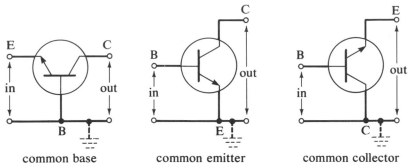

common base   common emitter   common collector

Fig. 3-13   Types of circuit configurations for the transistors.

emitter-follower circuit.   The three basic configurations are illustrated in Fig. 3-13.

If the common connection in each of the circuits is *grounded* (as indicated by the dotted ground symbol) the circuits in Fig. 3-13 are referred to as grounded-base, grounded-emitter, and grounded-collector circuits, respectively.

3-2   Junction Transistor Switching Circuits

In Chapter 2 it was shown that a diode could be operated as a switch that is either ON or OFF depending on the polarity of the applied signal. A junction transistor can act as a switch that is ON or OFF depending on the input current.   Since there are 3 terminals, only 1 terminal needs to be common with the *actuating* and *switched* circuits, as illustrated in Fig. 3-14.

Circuit Analysis

The dc analysis of the common-emitter switch circuit of Fig. 3-14 is done as outlined in Chapter 2 by superimposing the load line and I-V curves for the collector-emitter circuit (I_C vs V_{CE}).   This is shown in Fig. 3-15 for a typical transistor and a 1-kΩ collector load resistor R_L.   In the transistor case, there is a separate I-V curve for each value of the base current I_B.   It is customary to show a few representative curves and interpolate if any intermediate curves are required.   By varying I_B, the operating point can be placed almost anywhere along the load line.   In the "active" region, a small change in I_B results in a much larger change in I_C.   This is the mode of operation for amplifier circuits.   The ratio

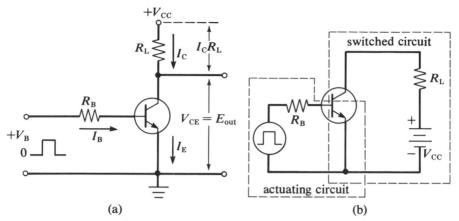

Fig. 3-14 A common-emitter npn transistor switch circuit: (a) typical schematic; (b) outline of actuating and switched circuits.

I_C/I_B is the *dc current gain*, sometimes called the "dc forward current transfer ratio" or the "dc beta" for short. The symbol for this characteristic is h_{FE} or β_N, where

$$\beta_N = h_{FE} = \frac{I_C}{I_B} \tag{3-10}$$

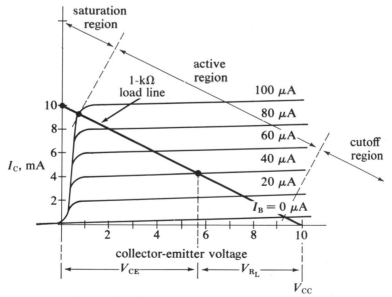

Fig. 3-15 Transistor characteristic curves.

Because $\alpha_N = I_C/I_E$ and $I_B = I_E - I_C$, it follows that, by substitution into Eq. (3-10), $\beta_N = I_C/I_B = \alpha_N/(1 - \alpha_N)$, which can be a useful relationship if only β_N or α_N is given. For example, if $\alpha_N = 0.99$, then $\beta_N = 0.99/0.01 = 99$.

Experiment 3-3 *The dc Current Gain*

The dc beta β_N or h_{FE} is determined from the characteristic curves and compared with typical values for the specific transistor type.

In switching circuits there is more interest in the regions of maximum and minimum conductance than in the active region. If the ON transistor were a perfect conductor, i.e., $V_{CE} = 0$, the collector current I_C would have the maximum possible value.

$$I_{C(max)} = \frac{V_{CC}}{R_L} \tag{3-11}$$

If the OFF transistor were a perfect insulator the collector current I_C would be 0, and

$$V_{CE(max)} = V_{CC} \tag{3-12}$$

Two points $[V_{CE} = 0, I_C = V_{CC}/R_L]$ and $[I_C = 0, V_{CE} = V_{CC}]$ are used to draw the load line. Since $I_C R_L + V_{CE} = V_{CC}$, the collector current can be expressed as

$$I_C = \left(-\frac{1}{R_L}\right) V_{CE} + \frac{V_{CC}}{R_L} \tag{3-13}$$

That is, the collector current I_C can be determined from the slope $(-1/R_L)$ and intercept (V_{CC}/R_L) of the load line. As shown in Figs. 3-14 and 3-15, the sum of V_{CE} and V_{R_L} equals V_{CC} at any point along the load line.

If a voltage of V_B is applied to the input of the circuit in Fig. 3-14a the base current I_B will flow, and the switch is turned ON if I_B is sufficiently large. Whether the switch can be considered ON can be determined by where a specific value of I_B crosses the load line in Fig. 3-15. Usually it is desired that the switch go into *saturation* so that V_{CE} is very small (about 0.1 to 0.2 V).

To ensure that the transistor goes into saturation and the switch is truly ON, the base network should provide a turn-on base current I_{B1} such that

$$h_{FE}I_{B1} > I_{C(max)} \qquad (3\text{-}14)$$

where

$$I_{C(max)} = \frac{V_{CC} - V_{CE(sat)}}{R_L} \qquad (3\text{-}15)$$

and

$$I_{B1} = \frac{V_B - V_{BE(ON)}}{R_B} \qquad (3\text{-}16)$$

The voltage $V_{BE(ON)}$ is the voltage drop across the forward-biased base-emitter junction.

Inverter

In Fig. 3-14a it can be seen that $E_{out} = V_{CC} - I_C R_L$. When the input voltage signal to the base is 0 V, there is only a small collector leakage current I_{CO} and the output voltage E_{out} is at its maximum of nearly $+V_{CC}$ volts. When the base input signal goes to its maximum of $+V_B$ volts so that the transistor saturates, then E_{out} drops to a minimum value of $V_{CE(sat)}$ or about 0.1 to 0.2 V. In other words, the switch of Fig. 3-14 inverts the input signal. It is, therefore, often called an *inverter*. The grounded-emitter switch always provides an output voltage change of opposite direction with respect to the input voltage change.

Experiment 3-4 *The Transistor Inverter Circuit*

The ability of a typical transistor switch circuit to provide an output voltage change of opposite direction to the input voltage change is observed.

Voltage and Current Limitations in Transistors

Transistor switching circuits are routinely being used in switched circuits where the supply voltage is hundreds of volts or where the current switched can be several amperes. Despite this great versatility, there are a number of ways of destroying a transistor through careless circuit design. Three limitations which can be encountered under the normal bias conditions are shown in Fig. 3-16, the maximum collector dissipation in watts ($I_C \times V_{CE}$). The safe (if adequately ventilated) region is the shaded area. The load line should normally stay within the safe boundaries. It is possible, in switching applications, for the load line to go through the area outside the maximum dissipation line if both operating points are well within the maximum limitations and the transition time is very short.

In addition to the above limitations, there are two breakdown voltages that must not be exceeded unless there is some protection against the damaging currents which could result from the breakdown. The breakdown voltages are BV_{CBO}, the maximum inverse collector-base voltage, and BV_{CEO}, the maximum collector-emitter voltage.

Transistor Switching Time

The transistor switch cannot respond instantaneously to a turn-on or a turn-off actuating signal. In many applications it is important to be aware of the magnitude of possible errors caused by a time delay in the switching circuit and how to minimize the time lags. In Fig. 3-17 the response of a transistor switching circuit is shown when a turn-on and turn-off actuating signal V_B is applied to the input circuit of Fig. 3-14.

Each segment of the turn-on and turn-off times illustrated in Fig. 3-17 will be considered briefly. First, the delay time, t_d: the base-emitter input capacitance must charge through R_B before the BE junction is actually forward biased; there is a finite transit time for the first minority

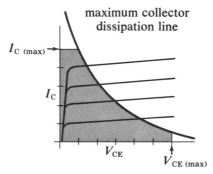

Fig. 3-16   Limiting regions for a transistor.

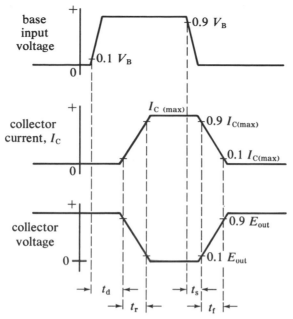

Fig. 3-17   Transistor switching time.

carriers to cross the base region; and some time is required for the collector current to reach 10% of its final value.

Next, there is the rise time, t_r. This is the time needed to establish the concentration of minority carriers in the base region (i.e., the electron density in the p-type base of the npn transistor) which is required to carry 90% of the final value of I_C. The combination of t_d and t_r is the turn-on time, t_{on}. The rise time can be substantially reduced if the base current is larger than the minimum needed for saturation. The collector current rises toward $h_{FE}I_B$ with a characteristic time constant, but cannot, of course, exceed $I_{C(max)}$ as given in Fig. 3-15. Therefore, the larger I_B is, the shorter the time required to reach $I_{C(max)}$. However, increasing I_B greatly beyond the current required to saturate the transistor, substantially increases the steady-state excess minority carrier concentration in the base region during saturation.

Assuming now, that the actuating signal was of sufficient duration and magnitude to saturate the transistor, it is important to consider the turn-off behavior. First, there is the *storage time* t_s. This is the time necessary to clear the collector-base junction of excess minority carriers, as discussed under Saturation in Section 3-1, and to decrease I_C to 90% of

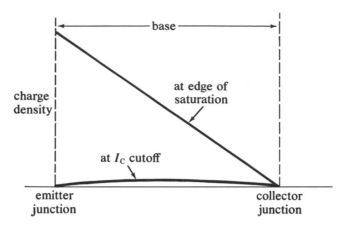

Fig. 3-18   Charge density gradients in the base.

its maximum value.   Since saturation of the transistor occurs as soon as the collector-base junction becomes forward biased, any further increase of base current causes excess charge to accumulate in the base region. Then, when the forward-bias on the base-emitter junction is suddenly removed, the excess base charge flows into the collector.   At the instant the potential gradient at the collector-base junction reaches zero, the collector no longer emits carriers into the base region, and the transistor comes out of saturation.

The fall time t_f is the time for the output current to fall from 90 to 10% of its maximum value and depends on the time to discharge the collector-emitter capacitance and the time required for the minority carriers in the base to be collected so that there is zero density gradient at the emitter junction as well as at the collector junction.   This is illustrated in Fig. 3-18 where the charge density gradients in the base between the emitter and collector junctions are drawn.

To summarize, then, the important switching times are as follows:

t_d — Delay time or "turn-on delay," the finite time that elapses between application of the base input voltage and the start of collector current flow in the transistor.   The time t_d is measured at values of input voltage and output current that are 10% of the maximum values.

t_r — Rise time, the time required for the collector current to increase from 10 to 90% of its total change.

t_s — Storage time, the time required for the collector current to decrease to 90% of its maximum value after the input has decreased to 90% of its maximum value.

t_f — Fall time, the time required for the output waveform to fall from 90 to 10% of its maximum value.

t_{ON} — Turn-on time, the sum $t_d + t_r$.

t_{OFF} — Turn-off time, the sum $t_s + t_f$.

It should be noted that the various transistor switching times are usually measured by using V_{CE} output waveform because it is easy to display on the oscilloscope. This is only valid for cases where the load is resistive.

Speed-up capacitor. The switching circuit shown in Fig. 3-19 is identical to that discussed above except for the capacitor C_B. This capacitance adds its charge to the base during turn-on which increases i_B briefly and reduces t_r. During turn-off, its action is opposite; it aids the discharge of the stored charge. When the value of C_B is correctly chosen, the turn-off time is greatly shortened as shown by the output waveforms in Fig. 3-19.

Experiment 3-5 *Transistor Switching Times*

The delay, rise, storage, and fall times for a transistor are measured and the net turn-on and turn-off times determined.

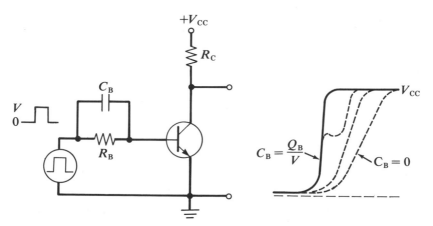

Fig. 3-19   Transistor circuit with a speed-up capacitor.

Connection of Load

A basic application of the transistor switch is to control current in an external load. There are two ways to connect the load, as illustrated in Fig. 3-20. When the load is placed in shunt with the transistor, a resistance R is connected in series between the parallel network and the supply voltage. When the switch is OFF, current flows through the load and series resistance R. When the switch is ON the load is effectively short-circuited and the current is steered from the load through the transistor.

When the external load is in series with the transistor, the current through the load is the same as the collector current I_C.

If the external load requires a higher current than can be supplied by one transistor that is controlled by a current-limited base control voltage source, it is possible to cascade transistors as shown in Fig. 3-21.

The base-emitter junction of Q_2 is the shunt load of Q_1. The base current of Q_2 is dependent on the supply voltage V_{CC}, the Q_1 load resistance R_{L1}, and $V_{BE(ON)}$. The current $I_{B(Q2)} = (V_{CC} - V_{BE(ON)})/R_{L1}$.

Whenever Q_1 is ON the base current for Q_2 is cut off, and when Q_2 is OFF there is a base current $I_{B(Q2)}$ for Q_2. The current through the load is equal to the collector current for Q_2 because it is in series with Q_2. If resistor R_{L1} is chosen such that the base current of Q_2 causes saturation [see Eq. (3-14) to (3-16)], the current in the load $I_{C(Q2)} = (V_{CC} - V_{CE(sat)})/R_{load}$.

Note that with the circuit of Fig. 3-21 the current in the load is OFF when the base voltage is positive and vice versa. If it is desirable to have the current in the load go ON when the input goes positive, it is necessary to add a transistor between the two transistors in Fig. 3-21 which acts as another inverter.

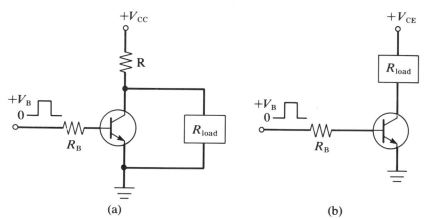

(a)   (b)

Fig. 3-20   Connection of a load to transistor switch.

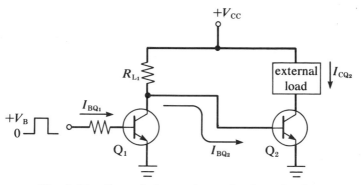

Fig. 3-21   Cascaded transistors for large loads.

Experiment 3-6   *Connection of Loads to Transistor Switch Circuits*

A load is connected to the circuits of Fig. 3-20 and 3-21 and the characteristic outputs are measured.

Emitter Follower

The common-collector or emitter-follower connection of the transistor was illustrated in Fig. 3-13. A very practical and important circuit that has this basic configuration is shown in Fig. 3-22. The input and output terminals are connected to the transistor collector through the collector supply voltage V_{CC}. However, the 3 terminals are in common for ac signals because from the viewpoint of the varying signal levels, the collector voltage is a fixed constant value just as surely as if it were grounded. It is considered to be in common with the common input and output terminals because the potential difference is a constant.

The emitter-follower circuit is seldom used as a switch because of the high input voltage required to saturate the transistor. However, the circuit has many desirable characteristics such as high input impedance, low output impedance, and excellent linearity over a wide dynamic range so it is frequently used in conjunction with switching circuits. An understanding of this linear amplifier circuit is, therefore, relevant to the study of switching circuits.

When a positive voltage V_S is applied to the base, the transistor conducts causing a voltage drop V_O across R_E which increases the transistor emitter voltage. If the input voltage V_S then decreases, I_C de-

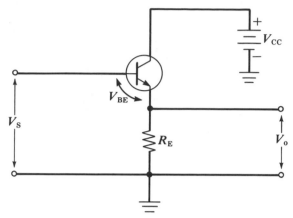

Fig. 3-22   Emitter-follower circuit.

creases, decreasing the emitter potential. Thus the emitter voltage V_O "follows" the input voltage. The characteristics of this circuit are analyzed in greater detail below.

When the input signal voltage V_S is greater than the turn-on voltage for the base-emitter junction V_{BE}, the output voltage V_O is

$$V_O = V_S - V_{BE} \tag{3-17}$$

and the voltage gain A_V of the circuit is

$$A_V = \frac{dV_O}{dV_S} = \frac{d(V_S - V_{BE})}{dV_S} = 1 - \frac{dV_{BE}}{dV_S} \tag{3-18}$$

Since dV_{BE}/dV_S is always positive, it can be seen from Eq. (3-18) that the voltage gain is always less than 1, but it approaches unity in the linear amplification region where dV_{BE}/dV_S is small. However, the input V_S cannot exceed the supply voltage V_{CC} by much before the collector-base junction becomes forward biased and the voltage between collector and base $V_{CB(ON)}$ is only a few tenth volts. The input would thus be clamped at $V_{CC} + V_{CB(ON)}$.

If the transistor of Fig. 3-22 is not saturated,

$$I_C = h_{FE} I_B \tag{3-19}$$

Since $I_E = I_C + I_B$, it follows that

$$I_E = h_{FE} I_B + I_B = I_B(h_{FE} + 1) \tag{3-20}$$

Therefore, from Eq. (3-20) the current gain A_1 is

$$A_1 = \frac{dI_E}{dI_B} = h_{FE} + 1 \tag{3-21}$$

Input impedance. One of the important characteristics of the emitter follower is its relatively high input impedance, which means that it will not "load down" the input signal source significantly. There is both an ac and dc input impedance, and in this section only the dynamic dc impedance is considered. It is defined as

$$R_{in} = \frac{dV_S}{dI_B} \tag{3-22}$$

If the input voltage signal V_S is applied directly to the base terminal, it can be seen from Eq. (3-17) that

$$V_S = V_0 + V_{BE} \tag{3-23}$$

Since the voltage drop V_0 across the emitter resistor R_E is equal to the emitter current I_E times R_E (i.e., $V_0 = I_E R_E$), it follows by substitution in Eq. (3-23) that

$$V_S = I_E R_E + V_{BE} \tag{3-24}$$

By substituting the equivalent of I_B from Eq. (3-20), $I_B = I_E/(h_{FE} + 1)$, and of V_S from Eq. (3-24), into Eq. (3-22), it follows that

$$R_{in} = \frac{d(I_E R_E + V_{BE})}{d(I_E/(h_{FE} + 1))} = \left(R_E + \frac{dV_{BE}}{dI_E} \right)(h_{FE} + 1) \tag{3-25}$$

If the emitter current I_E is larger than a few milliamperes, dV_{BE}/dI_E will typically be negligible compared to R_E, and

$$R_{in} \approx R_E(h_{FE} + 1) \tag{3-26}$$

Equation (3-26) shows that the dc input impedance R_{in} is greater than the emitter load R_E by the factor $h_{FE} + 1$.

Output impedance. The emitter-follower stage is often connected between an inverter circuit and the external load because of its ability to act as an impedance transformer. This is illustrated by the circuit of Fig. 3-23. A low output impedance for a voltage source means that

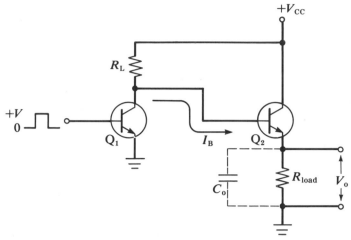

Fig. 3-23 **Emitter-follower as impedance transformer between inverter and output load.**

relatively heavy loads can be connected to the output of the source without changing the output voltage. The low output impedance of the emitter follower is one of its most important characteristics. The output impedance of the emitter follower is seen to be only a fraction $[1/(h_{FE} + 1)]$ of the input source impedance $(Z_{source} = dV_S/dI_B)$ by considering the following equations obtained by substitution from Eqs. (3-17) and (3-20):

$$Z_{output} = \frac{dV_O}{dI_E} = \frac{d(V_S - V_{BE})}{dI_B(h_{FE} + 1)} = \frac{1}{h_{FE} + 1} \times \frac{d(V_S - V_{BE})}{dI_B}$$

and if V_{BE} is negligible compared to V_S,

$$Z_{output} = \frac{Z_{source}}{h_{FE} + 1} \qquad (3\text{-}27)$$

Since the maximum output voltage $V_{O(max)}$ can be given as $V_{O(max)} = V_{CC} - I_B R_L - V_{BE}$, it can also be readily shown, by substitution for $I_B = I_E/(h_{FE} + 1)$ and for $I_E = V_{O(max)}/R_{load}$, that

$$V_{O(max)} = \frac{V_{CC} - V_{BE}}{1 + R_L/[R_{load}(h_{FE} + 1)]} \qquad (3\text{-}28)$$

If V_{BE} is assumed to be 0.6 V and the supply voltage $V_{CC} = 5$ V,

$R_L = 1$ kΩ, and $h_{FE} = 50$ for the selected transistor Q_2 in Fig. 3-23, then it can be seen in Eq. (3-28) that the second term in the denominator is negligible for a wide range of load resistance R_{load}. A plot of output voltage vs load resistance for this example is shown in Fig. 3-24.

Experiment 3-7   *Emitter-Follower Circuit*

The voltage and current gains of an emitter-follower circuit are measured and checked with expected values. The input impedance is also determined.

Nonsaturating Switching Circuits

The storage time which is a result of allowing a transistor to saturate significantly limits the switching speed. The problem of storage time does not arise if the transistor is not allowed to saturate, and an improvement in switching speed is possible.

One method of providing a nonsaturating transistor switch is to clamp the collector in such a way that the collector-base junction cannot become forward biased. One method of clamping is shown in Fig. 3-25. The diode D becomes forward biased as the transistor begins to leave the active region, at the edge of saturation. Therefore, the collector is always held sufficiently negative with respect to the base so that the collector-base junction remains reverse biased and the transistor cannot saturate. The excess base current is conducted away through the diode.

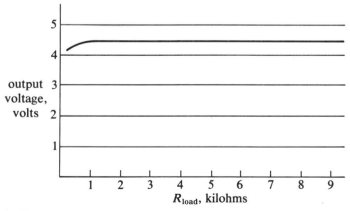

Fig. 3-24   Plot showing small change of output voltage vs load resistance for circuit of Fig. 3-23.

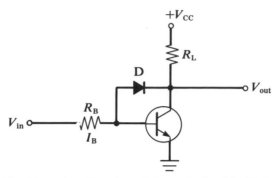

Fig. 3-25 Nonsaturating transistor switch with diode clamp.

Another method of providing nonsaturated switches is to control the collector current. A circuit utilizing transistors operated in the common-base configuration is shown in Fig. 3-26. When the input signal V_B is negative transistor Q_1 is cut off, but the base-emitter junction of Q_2 is forward biased so Q_2 is ON and the emitter current $I_E \approx V_{EE}/R_E$. Now when the input signal goes positive the transistor Q_1 goes ON and the voltage drop across R_E causes the base-emitter junction of Q_2 to become reverse biased and Q_2 goes OFF. The current is in effect "steered" to either Q_1 or Q_2 by the input signal. The steering depends on both polarity and amplitude of the input signal. The collector-base junction will always be at some value of reverse or zero bias if $I_C R_L \leqslant V_{CC}$, and the transistor will thus stay out of saturation.

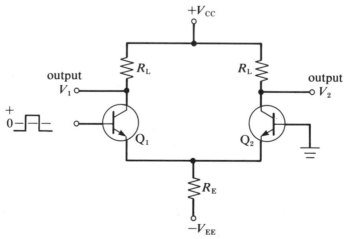

Fig. 3-26 Nonsaturating transistor switches.

The outputs V_1 and V_2 are complementary, with V_2 in phase with the input signal. Only a few tenths of a volt change of input signal are required to switch states.

Experiment 3-8 *Nonsaturating Switching Circuit*

A nonsaturating transistor switch circuit is constructed and its characteristics compared with a saturated switch.

3-3   Field-Effect Transistor Switches

A device which has the advantage of very high input impedance is the field-effect transistor (FET). This is a unipolar device as compared to the bipolar transistors already discussed. The designation "unipolar" indicates that it is a device that essentially contains only one type of current carrier, either holes or electrons, but not both as for the bipolar pnp and npn transistors. Whether holes or electrons are the current carriers will depend on the type of unipolar material chosen.

A semiconductor provides a current path whose resistance R to the flow of a specific carrier is varied by applying a transverse electric field. For example, a bar of silicon containing excess n impurity is illustrated in Fig. 3-27. It has a length L, width W, and thickness T. The resist-

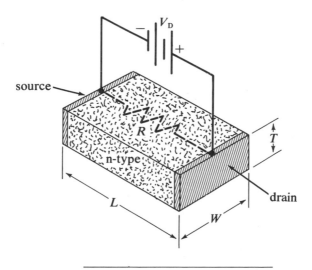

Fig. 3-27   Unipolar semiconductor bar.

ance of the bar is directly proportional to its length L and is inversely dependent on the width W, thickness T, and conductivity σ,

$$R = \frac{L}{WT\sigma} \qquad (3\text{-}29)$$

The conductivity σ depends on the number, mobility, and charge of each type of carrier; so

$$\sigma = q(n\mu_n + p\mu_p)$$

where q is the electron charge, n the electron density, p the hole density, and μ_n and μ_p the mobilities of electrons and holes, respectively. If it is assumed that the electron density n is much greater than hole density p (n $\gg$ p), then $\sigma \approx qn\mu$ and the conductance G is

$$G = \frac{1}{R} = \frac{\sigma WT}{L} \approx \frac{q\mu nWT}{L} \qquad (3\text{-}30)$$

From Eq. (3-30) it is obvious that with q, μ, W, and L constant, the conductance G could be varied only by changing n or T. The operation of the unipolar FET as a variable resistance depends on the formation of a "channel" of variable thickness through which the electrons must pass. The channel is formed by introducing p-type impurities into opposite sides of the bar, as illustrated in Fig. 3-28. A space charge is developed

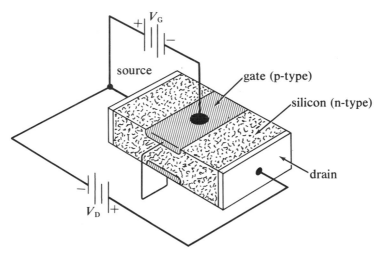

Fig. 3-28 Unipolar FET with pn junctions reverse biased.

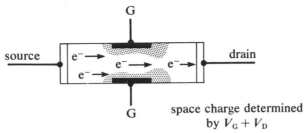

Fig. 3-29   Illustration of channel space charge.

at the p-type "gate" regions which provides a certain effective channel thickness through which electrons can flow. Now, by applying a reverse bias between the source and the gate diodes, the space charge is increased, and the effective channel thickness for electron flow is decreased, as illustrated in Fig. 3-29.

When the voltage V_D, is applied as shown in Fig. 3-28, the drain is made positive with respect to the source and electrons flow from the source through the channel to the drain. The gate voltage V_G controls the current by controlling the effective channel thickness. As would be expected, there is a critical drain voltage above which there is no increase in drain current. This is called the "pinch-off voltage" V_p and is dependent on the mobility of the electrons in the silicon bar. As shown in Fig. 3-30 a plot of drain current vs drain voltage provides a family of characteristic curves when the gate voltage V_G is changed.

The typical symbols for junction FET's are shown in Fig. 3-31.

When the FET is used as a switch the gate bias V_G is made sufficiently large to turn OFF the carrier flow. The OFF resistance is very high, typically greater than 10^9 Ω. The ON resistance for zero gate bias can be estimated at worst to be $R_{ON} = V_p/I_{DSS}$, where I_{DSS} is the current determined from where the pinch-off voltage V_p crosses the characteristic

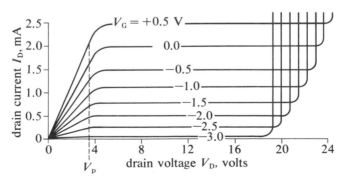

Fig. 3-30   Characteristic curves for junction FET.

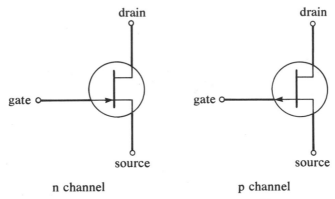

Fig. 3-31   Circuit symbols for FET's.

curve in Fig. 3-28.   Typical ON resistance values are 100–500 Ω.   A junction FET switch circuit is given in Experiment 3-9.

Insulated Gate Field-Effect Transistor (IGFET)

The IGFET can be constructed so that essentially no drain current flows under zero bias conditions, as illustrated in Fig. 3-32.   The pn junctions in the substrate very effectively isolate the source from the drain under zero bias on the gate.   The drain-source voltage reverse biases the drain junction so that drain current is essentially zero.   Typical resistances are 10^{14} Ω or more.

A thin layer of metal oxide (e.g., SiO_2) is deposited on the surface over the channel and this is covered with a thin layer of deposited metal. This forms a capacitor between the gate and substrate.   The application of a positive voltage between gate and source will induce a negative

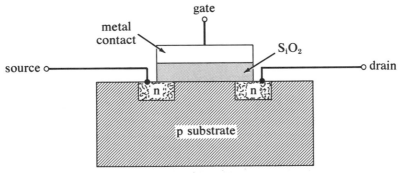

Fig. 3-32   Representation of an IGFET.

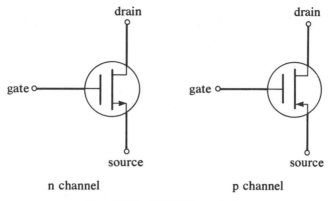

Fig. 3-33   IGFET symbols.

charge in the channel so that it becomes in effect an n-type material and current flows between source and drain.   The gate acts to *enhance* the number of current carriers in the channel.   It is from the use of the *metal oxide semiconductor* that the name MOSFET originates, which is frequently used to identify the device.

The IGFET can also be fabricated so that with $V_G = 0$ there is a conductive path between source and drain.   A negative voltage applied to the gate will then induce a positive charge in the channel which causes a *depletion* of carriers and can turn OFF the IGFET.

The symbols for the IGFET are shown in Fig. 3-33.

Experiment 3-9   *Field Effect Transistor (FET) Switch*

The characteristic curves for a field-effect transistor are obtained from the curve tracer and the load line and ON–OFF characteristics are measured.   A junction FET switch is constructed and investigated.

3-4   Silicon-Controlled Rectifier and Other pnpn Switches

There are several pnpn four-layer semiconductor devices (often called *thyristors*) that are of considerable importance as versatile switches for controlling ac and dc power.   In this family of pnpn devices are the silicon-controlled rectifier (SCR), the gate turn-off switch (GTO), the silicon-controlled switch (SCS), the Shockley or 4-layer diode, and the triac (essentially an integrated double SCR).   The characteristics and general mechanisms of switching pnpn devices are first considered. Then the specific use of the SCR for the discrete control of power in a load (such as a motor) is described.

Two-Transistor Representation of pnpn Devices

In Fig. 3-34a the pnpn device is shown as a four-layered pellet of alternating p- and n-type semiconductor material (usually silicon). The four-layer diode has leads connected only to the end layers of semiconductor. The SCR has three layers accessible by external leads, and the SCS has all four layers accessible as shown in Fig. 3-34b and c, respectively. Each layer interacts with its adjacent layers as for the transistor. It can be seen in Fig. 3-34 that starting from one end of the stack, three adjacent layers form a pnp transistor, and from the other end an npn transistor. This two-transistor concept is illustrated in Fig. 3-35 where it is apparent that the pnpn structure can be considered as a complementary (npn-pnp) transistor feedback pair. The collector of each transistor is direct-coupled to the base of the other, which provides a positive feedback loop. That is, a small change in base current of the npn transistor causes a larger change in its collector current which is the base current of the pnp transistor. The collector current of the pnp transistor also contributes to the base current of the npn transistor. Therefore, what begins as a small change in the base current of the npn transistor becomes a large change because of positive feedback.

If the cathode of the pnpn device is connected to the negative terminal and the anode to the positive terminal of the power supply, the center pn junction (J_2 of Fig. 3-35a) is reverse biased. Therefore, the device does not conduct unless the current gain around the feedback

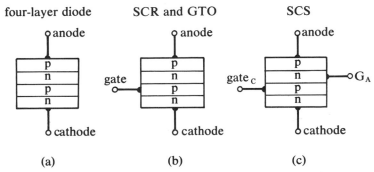

Fig. 3-34 **Illustration of external lead connection for various pnpn devices.**

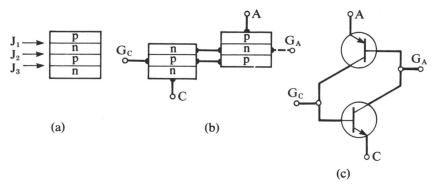

Fig. 3-35 Two-transistor representation of pnpn device: (a) junctions indicated; (b) hypothetical separation into two transistors; (c) schematic of two-transistor equivalent for pnpn device.

loop is approximately unity. This is seen more clearly by considering the equations expressing the total anode-to-cathode current I_A.

If the current gains of the npn and pnp transistors are h_{FE1} and h_{FE2}, respectively, then the current gain G_1 of the internal feedback loop is

$$G_1 = h_{FE1} \times h_{FE2} \tag{3-31}$$

Designating the base leakage current of the npn as I_{CO1} and that of the pnp as I_{CO2}, it follows that

$$I_{C1} \text{ (for the pnp)} = h_{FE1}(I_{C2} + I_{CO1}) + I_{CO1} \tag{3-32}$$

and

$$I_{C2} \text{ (for the npn)} = h_{FE2}(I_{C1} + I_{CO2}) + I_{CO2} \tag{3-33}$$

The total anode-to-cathode current I_A is

$$I_A = I_{C1} + I_{C2} \tag{3-34}$$

Combining Eqs. (3-32), (3-33) and (3-34), it follows that

$$I_A = \frac{(1 + h_{FE1})(1 + h_{FE2})(I_{CO1} + I_{CO2})}{1 - h_{FE1} \times h_{FE2}} \tag{3-35}$$

When the anode is positive with respect to the cathode the center section is reverse biased and h_{FE1} and h_{FE2} are very small compared to 1, so $I_A \approx I_{CO1} + I_{CO2}$, i.e., the sum of only the junction leakage currents. Therefore, the device is in its high impedance OFF state. However, it can be seen from Eq. (3-35) that as $G_I \rightarrow 1$, $I_A \rightarrow \infty$. In other words, as the loop gain approaches unity the circuit feedback is sufficient so that each transistor drives the other into saturation, and the device goes to its low impedance or ON state. When ON the anode current is limited only by the external circuit.

Mechanisms for Switching pnpn Devices

Any mechanism that increases h_{FE} so that $G_I \rightarrow 1$ will turn the device ON. These mechanisms are all based on the emitter-current dependence of h_{FE} as illustrated in Fig. 3-36. When the emitter current is very low the gain h_{FE} is low, but, if anything temporarily increases I_E sufficiently, then the pnpn device is rapidly switched ON.

Some important mechanisms for producing the necessary emitter current to switch the device ON include anode-to-cathode voltage, rate of voltage change, temperature, transistor action, and radiant energy.

When the applied voltage across the device gets high enough there is avalanche breakdown, and when the avalanche current causes $G_I \rightarrow 1$ the device is switched ON. This is the normal method used to switch the pnpn diodes into conduction. These diodes can be constructed to switch ON at certain selected anode-to-cathode voltages.

As previously discussed, any pn junction has capacitance. Therefore when a step function is impressed across the pnpn device a charging current $i = C \ dV/dt$ will flow from anode to cathode to charge the device

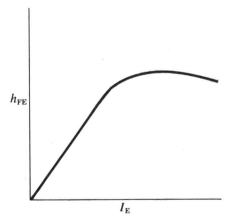

Fig. 3-36   Dependence of h_{FE} on emitter current in a silicon transistor.

capacitance.   If current i is sufficient for $G_I \rightarrow 1$ the device switches ON, and this mechanism is known as "the *dV/dt* effect."

Because the leakage current in a silicon pn junction doubles for every 8°C increase in temperature, it is possible for $G_I \rightarrow 1$ and the device to switch ON from high junction temperature.

A relatively small base current can turn a transistor ON.   Likewise for the pnpn device the injection of current carriers into either "gate" (base region) can switch the device ON.   This is the normal method of switching the SCR and SCS.

When photons of a suitable wavelength strike the silicon lattice many electron-hole pairs can be released.   If a sufficient number of current carriers are released per unit time so that $G_I \rightarrow 1$, the device is switched ON.   This mechanism is the basis for the light-activated SCR (LASCR), which is provided with a translucent window so that light can reach the silicon lattice.

The pnpn devices can be turned OFF by the removal of the anode-to-cathode dc supply voltage, or if the supply voltage is ac the device will turn OFF when the voltage is decreased sufficiently.   It is necessary to exceed a certain load current called the *holding current,* to maintain the device in conduction after a gate signal is removed.

The junctions in the SCR are all forward biased when it is conducting, and the base regions are heavily saturated with current carriers. Therefore, to turn OFF the SCR in a minimum time, it is necessary to apply a reverse voltage.   This causes the electrons and holes in the region of the two end junctions J_1 and J_3 in Fig. 3-35 to diffuse to these junctions so as to cause a reverse current in the external circuit.   Therefore, the voltage across the SCR will remain at about 0.7 V.   When the holes and electrons are removed from the region of the junctions the reverse current ceases and the junctions are again in the blocking state.   The reverse voltage across the SCR is now determined by the applied supply voltage.   However, the recovery of the SCR is not complete until the center junction is cleared of excess charge carriers, which depends primarily on the recombination process.   This process is quite independent of external applied voltage.   When the excess charge carriers at the center junction have been reduced to a low value it is again safe to apply a forward voltage to the SCR without danger of turn-on.   It usually requires about 10 μsec after cessation of the forward current flow before the forward voltage may be safely reapplied.   This delay time is known as the *turn-off time.*

The gate turn-off switch.   A device similar to the SCR is the gate turn-off switch (GTO) which has special gate construction characteristics that enable a small reverse gate current to turn OFF the device. It can also be turned OFF like a conventional SCR. For dc applications the GTO enables the turn-off circuitry to be simpler than with the SCR.

Circuit Symbols for pnpn Devices

The circuit symbols used for the pnpn diode, SCR, SCS, and triac are given in Fig. 3-37. One symbol commonly used for the four-layer diode is in the shape of a number four (4). The SCR symbols show its one gate lead which is attached to the semiconductor layer adjacent to the cathode, whereas the SCS symbols show leads adjacent to both the cathode and anode. The circles around the symbol are only for appearance in a diagram and are often eliminated in drawing circuit diagrams.

Power Control with pnpn Devices

Silicon controlled rectifier. The use of SCR's for the control of motor speed, light intensity, electrical heating, and similar power control applications is widespread. The reasons for using SCR's include the simplicity of introducing discrete packets of power at easily adjustable rates (i.e., digital control) and the high current and voltage ratings of relatively inexpensive SCR's.

An example of a circuit that can be used for control of power in a

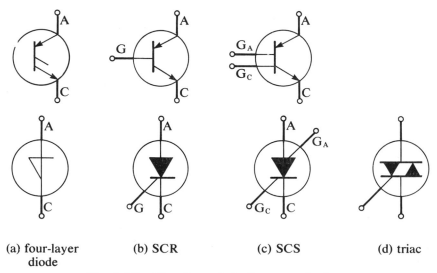

(a) four-layer diode   (b) SCR   (c) SCS   (d) triac

Fig. 3-37 Circuit symbols for pnpn devices.

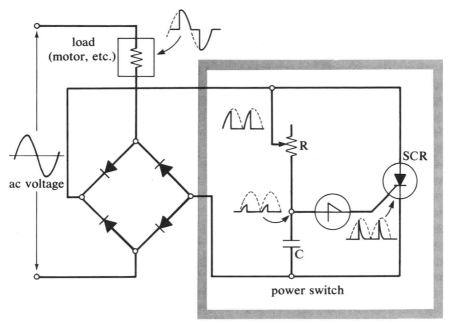

Fig. 3-38   Use of SCR switch to control power in a load.

load is shown in Fig. 3-38.   The section of the circuit enclosed in the
block acts as a switch to provide current through the load only during
a discrete selected fraction of each half-cycle.   The RC circuit determines the portion of each half-cycle required to charge the capacitor C
to a sufficient voltage for the four-layer diode to turn ON, which provides
the necessary gate signal for the SCR to turn ON.   Because the power
delivered to the load in each half-cycle depends on the portion of each
half-cycle required to charge the capacitor to the firing potential of the
diode and SCR, it is easy to change the power to the load by varying
resistor R or capacitor C.   The voltage waveforms shown in the various
portions of the circuit of Fig. 3-38 as related to the input supply voltage
illustrate the circuit action.   Note that because the effective resistance
of the ON SCR is very small, the circuit current is essentially determined
by the load resistance.

Silicon-controlled switch.   The silicon-controlled switch (SCS) can
be used like the SCR with only one gate control.   The second gate control can provide greater versatility for applications other than as a switch.
It is also possible to substitute an SCS for an SCR.   The SCS is thoroughly described in the *GE Transistor Manual*.[1]

[1] Seventh Edition (J. F. Cleary, ed.), General Electric Co., 1964.

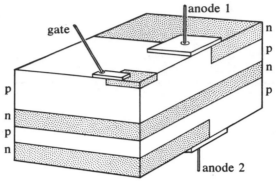

Fig. 3-39   Triac construction.

Triac. The triac (for "triode ac switch") is electrically equivalent to two SCR's on an integrated disk as illustrated in Fig. 3-39. The triac simplifies control modules which require full-wave power and is, thus, frequently preferred over the SCR circuits.

Experiment 3-10   *Control of Power in a Load with SCR Circuit*

A silicon-controlled rectifier circuit for the control of power in a load is constructed and its characteristics investigated.

3-5   Unijunction Transistor Switch

The unijunction transistor (UJT) has some unique features as compared to the two-junction npn and pnp transistors. It is a 3-terminal device with only one pn junction. The inherent stability of the characteristic parameters provide advantages for certain circuit applications.

Unijunction Transistor Characteristics

The UJT is conveniently considered as a bar or strip of silicon to which a pn junction is made, as illustrated in Fig. 3-40a, although present structures of the UJT deviate from the bar structure for simplicity and economy of production. The leads from the two ends are called *base 1* and *base 2*. The rectifying contact is called the *emitter*, and it is represented as being between base 1 and base 2 on the bar of silicon. The

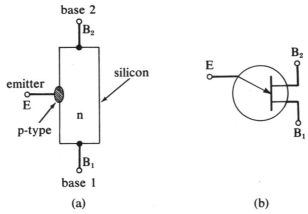

(a)   (b)

Fig. 3-40   Unijunction transistor: (a) structure representation; (b) circuit symbol.

schematic circuit symbol is shown in Fig. 3-40b.   The operation of the UJT is readily illustrated by consideration of the equivalent circuit in Fig. 3-41.

Biasing the unijunction transistor.   By applying a voltage V_{BB} so that base 2 is positive, a current will flow between base 2 and base 1 as deter-

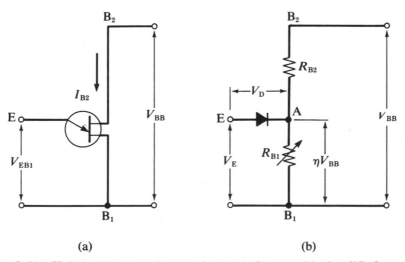

(a)   (b)

Fig. 3-41   Unijunction transistor: (a) nomenclature; (b) simplified equivalent circuit.

mined by the interbase resistance R_{BB} (typically 7 kΩ) of the silicon bar, so that

$$I_{B2} = \frac{V_{BB}}{R_{BB}} \qquad (3\text{-}36)$$

The resistance R_{BB} can be considered as consisting of the resistances R_{B1} and R_{B2} in series, which act as a voltage divider. A fraction η (typically 0.6) of the applied voltage (i.e., ηV_{BB}) is divided between the emitter contact and base 1. Therefore, the pn junction at the emitter is reverse biased and only a small leakage current normally flows in the emitter lead.

When the voltage V_E that is applied between the emitter and base 1 is increased, a voltage is reached where V_E is equal to the sum of the voltage across R_{B1} and the forward voltage drop of the pn junction. This voltage is known as the peak-point voltage V_p, so

$$V_p = V_D + \eta V_{BB} \qquad (3\text{-}37)$$

For emitter-base voltages greater than V_p the pn junction is forward biased so that holes are injected from the emitter into the silicon bar. Because base 1 is negative with respect to the emitter the electric field is such that most holes move toward the base 1 terminal. An equal number of electrons are injected from base 1 to maintain electrical neutrality in the bar. The increase in current carried in the silicon bar decreases the value of R_{B1}. This causes the fraction of voltage between point A and base 1 to decrease which causes a further increase of emitter current I_E and a lower resistance for R_{B1}.

A plot of V_E vs I_E is shown in Fig. 3-42 which shows the region of *negative resistance* where the voltage across R_{B1} decreases as I_E increases. The region of negative resistance exists between the peak-point voltage V_p and the so-called valley-point voltage V_v. At the valley-point voltage and higher voltages the density of charge carriers is so high that the lifetime of the carriers is decreased which counteracts the effect of new carriers being generated, and the emitter voltage V_E increases gradually at currents above I_v. The region to the right of the valley point is known as the saturation region, where the dynamic resistance is positive.

Static emitter characteristic curves. The emitter characteristic curve given in Fig. 3-42 is not drawn to scale in order to show clearly the various operating regions. A typical emitter characteristic curve drawn to scale is shown in Fig. 3-43, where only a small portion of the cutoff region is

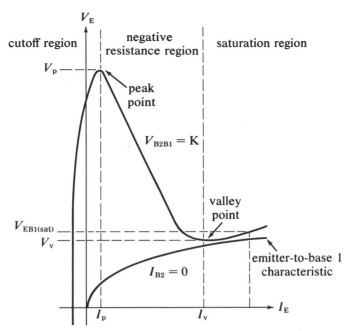

Fig. 3-42   Static emitter characteristic curves for unijunction transistor. (Courtesy General Electric Co.)

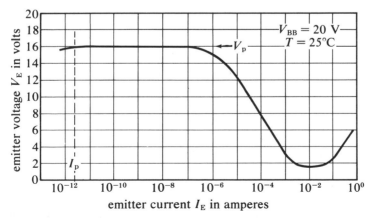

Fig. 3-43   Static emitter characteristics for unijunction transistor. (Courtesy Motorola Inc.)

shown. The peak-point voltage is about 16 V and is reached at a forward emitter current of about 10 pA. The voltage then remains constant until about 0.1 μA before it starts to decrease. The valley voltage of 1.6 V is reached at a valley current of about 8 mA. The saturation resistance can be determined by the slope of *I-V* curve in the saturation region and is seen to be about 5 Ω.

Peak-point voltage. The peak-point voltage V_p is the most important characteristic of the unijunction transistor because it determines the trigger or switch point in its circuit applications. The value of V_p is expressed in Eq. 3-37 as a function of the intrinsic standoff ratio η.

Intrinsic standoff ratio. The rearrangement of Eq. (3-37) provides the definition of the intrinsic standoff ratio,

$$\eta = \frac{V_p - V_D}{V_{BB}} = \frac{R_{B1}}{R_{B2}} \tag{3-38}$$

This ratio is not completely independent of temperature but the temperature coefficient is quite small, about 0.06%/°C. The intrinsic standoff ratio is also slightly dependent on the supply voltage V_{BB}, but the variation is generally negligible for practical purposes.

Unijunction Time Delay Circuit

A typical application of the unijunction transistor is shown in Fig. 3-44. The UJT is used to provide a precise time-delayed trigger pulse to turn ON the SCR power control circuit. The emitter-base 1 voltage for the UJT is supplied by the voltage across the capacitor C. At the start C is shorted by switch S_1. By opening S_1 the capacitor C charges through R from the stable 18-V Zener regulated supply. The voltage across C is determined by the *RC* time constant and Zener voltage source. When the capacitor charges to the UJT peak-point voltage V_p the UJT fires. This provides a pulse across R_2 which triggers the SCR. The full supply voltage (less the small voltage drop across the SCR) then is applied across the load terminals.

The voltage across the UJT drops to less than 2 V when the SCR triggers so that a low voltage is maintained on capacitor C, but this varies as a function of time.

The upper limit on the time delay depends on the specified accuracy, the V_p of the UJT, leakage current of capacitor C, and the temperature.

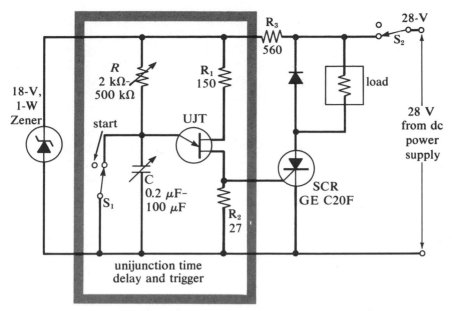

Fig. 3-44   Precision solid-state time delay circuit.

The upper limit for *R* is determined by the requirement that the current to the UJT emitter be large enough for it to fire (larger than I_p).

   To reset the circuit for another timing cycle the dc supply must be removed from the SCR by opening switch S_2 and the capacitor C should be shorted by closing S_1.

Experiment 3-11   *Unijunction Transistor Trigger Circuit*

A unijunction transistor time delay circuit is to be constructed and its characteristics investigated and then used to trigger an SCR power control circuit.

3-6   Semiconductor Optoelectronic Switches

   The combination of a light source with a photosensitive device inside an opaque enclosure can perform electronic functions that are difficult by other means.   The information from the light source and associated devices can be transferred to the photodetector without any electrical coupling.   Light is the coupling link.   Because the light source terminals and photodetector terminals are completely isolated the pair forms an ideal 4-terminal network where there is no interaction between input and output circuits. There are many types of photodetectors, but only a few of the most used devices are considered here. The basic characteristics are presented and several applications are described to indicate how these devices can be advantageously utilized in many instruments.

Light Sources

The emitted radiation from the light source as "seen" by the photo-detector depends on many variables, including the relative sensitivity of each as a function of wavelength, the geometry, use of light filters and light pipes, and others. It is important that the source have a long stable life and where many are to be used the cost is important. Where distance between the two must be long the intensity and focusing are especially important. Incandescent lamps, neon bulbs, electroluminescent devices, light-emitting injection diodes, ruby and gas lasers, and others have been used.

Incandescent lamp. The incandescent light is inexpensive and has a wide spectral output which is similar to blackbody radiation as shown in Fig. 3-45. However, the rise time of the output light is poor (typically

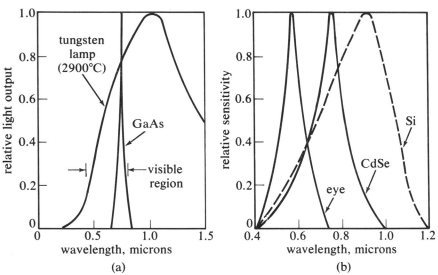

Fig. 3-45 **Relative output or sensitivity of sources and detectors vs wavelength. (a) Sources—tungsten lamp and gallium arsenide diode emitter. (b) Detectors—human eye, cadmium selenide photoconductive cell, and silicon photodiode.**

100 msec), and tungsten bulbs are susceptible to vibration breakage and burnout. They are primarily used with dc systems.

Neon lamp. The neon lamp is very low cost, has long life, and the required input power is much less than for the incandescent bulb. Response time is in the microsecond range. Most of the light output is limited to two orange spectral lines, and the maximum available output is quite limited. This limits the flexibility of the optical link and a detector must be chosen that is sensitive to the orange lines.

Semiconductor sources. Light-emitting diodes and junction lasers are dependent on semiconductor pn junctions which emit photons as a result of hole-electron recombinations in the structure. In silicon and germanium diodes the energy from recombinations is delivered to the crystal primarily as heat, but for junction diodes made from gallium arsenide and gallium phosphide the recombination energy is largely in the form of emitted photons. This light is either incoherent and concentrated in a narrow bandwidth or it can be coherent and concentrated primarily in a narrow line. When emitting coherent light it is called a *junction laser.* These diodes have very fast response times in the nanosecond range and have good dynamic range and linearity. Present diode devices have their light output typically in the infrared region (about 9000 Å) so that a detector must be used that is sensitive in this range.

Photodetectors

The characteristics of input light source and output detector should, of course, be matched. It would usually not be wise to use a nanosecond response source with a millisecond response detector. If the light source must be positioned far from the detector it would not be efficient to use a wide-angle light source such as a neon bulb without using some type of light pipe or focusing arrangement to direct the light to the light-sensitive surface of the detector.

The wavelength response curve for the detector must be in the wavelength region of emission from the source. The relative sensitivities vs wavelength for the human eye, the cadmium selenide photoconductive cell, and the silicon photodiode detector are shown in Fig. 3-45b.

Photoconductors. The cadmium sulfide (CdS) and cadmium selenide (CdSe) detectors are *photoconductive* devices which are used as light-sensitive resistors that decrease in resistance for an increase in light intensity and vice versa. They are used in hundreds of switching applications. The response time of the CdSe cell is in the millisecond range and the CdS response is somewhat slower. The detectors can have large surface response areas at low cost and can control relatively high powers. The ON-OFF resistances can be selected over wide ranges.

Photovoltaic cells. The *photovoltaic* cell is unique in that it does not require an additional power supply to obtain a light-sensitive output signal. That is, an output voltage is generated that is dependent on the incident light intensity. These are often called "solar cells" because they provide a source of electrical power from exposure to the sun. However, the efficiency is low and their usefulness is limited in compact optical links.

Semiconductor photodetectors. The *photodiode* operated with re-verse-biased pn junctions can provide currents in the microampere range which are linear to the incident light. Frequency response is very high, in the gigacycle region. *Phototransistors* are photodiodes with a gain of 10–100 built in.

Some typical optical links matching source-detector pairs are shown in Fig. 3-46.

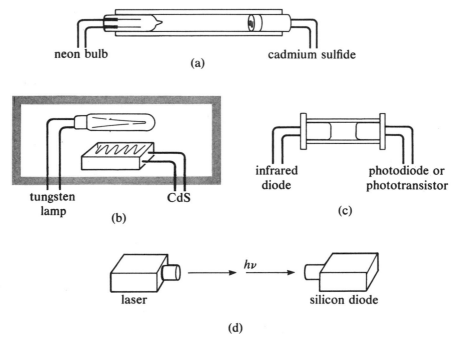

Fig. 3-46   Source-detector pairs for optoelectronic devices.

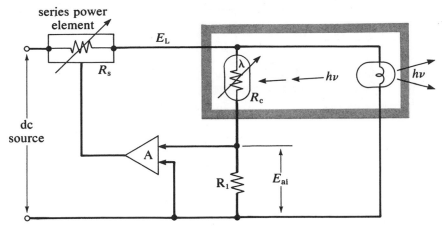

Fig. 3-47   Illumination control using a tungsten source and CdS detector.

Applications

The applications of semiconductor optoelectronic devices could be classified into two groups — one group that depends on the device acting as a continuously variable resistor (rheostat), and another group that depends only on the low ON resistance and high OFF resistance for switching applications.   A few examples are presented to illustrate some of the many uses of source-detector pairs.

The use of the rheostat characteristics of a CdS detector together with a tungsten source is illustrated in Fig. 3-47 for an illumination control.   A portion of the light radiation is made to fall on the CdS cell which is in series with resistor R_1.   The voltage drop across R_1 is fed to an amplifier which provides an output signal to a series power element.   The effective resistance R_s of the series element depends on the input to the amplifier, which depends on the light incident on R_c.   The amplifier and power series element are connected so that an increase in lamp intensity for any reason will cause a decrease in R_c.   This causes an increase in the amplifier input voltage $E_{ai} = E_L R_1/(R_1 + R_c)$, which causes the resistance R_s of the series element to increase, thus decreasing the lamp voltage E_L.   The resulting optoelectronic feedback, therefore, brings the light illumination down toward its original value, and tends to hold illumination constant, as "seen" by the CdS detector.

Another possible application of the rheostat characteristics of a source-detector pair is shown in Fig. 3-48 for a voltage-to-frequency converter.   The radiant power output P_L from the light source should be directly proportional or a known function of the input voltage or current

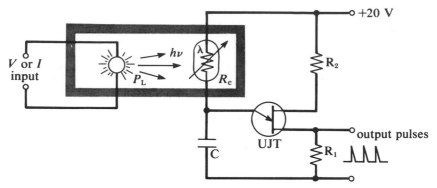

Fig. 3-48 Voltage-to-frequency converter.

signal. The resistance R_c of the photodetector changes the charging time for the capacitor and, therefore, the firing rate for the unijunction transistor. Because an increase in incident light decreases R_c, the number of output pulses per unit time will increase. Consequently, the number of pulses at the output is related to the input signal amplitude.

A widely used switch application of the neon source-CdSe detector is as a chopper (i.e., a modulator to convert dc to ac voltage). This is illustrated in Fig. 3-49. Lights L_1 and L_2 are alternately turned ON and OFF so that when L_1 is ON, L_2 is OFF. Therefore, when R_1 is LOW (e.g., 1 kΩ), R_2 is HIGH (e.g., 1 MΩ) and amplifier input $E_{ai} = E_{in}R_2/(R_1 + R_2) \approx E_{in}$; and when R_1 is HIGH R_2 is LOW, so that $E_{ai} \approx 0$. The amplifier

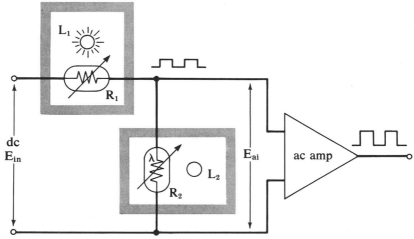

Fig. 3-49 Neon source and cadmium selenide detector used as a chopper.

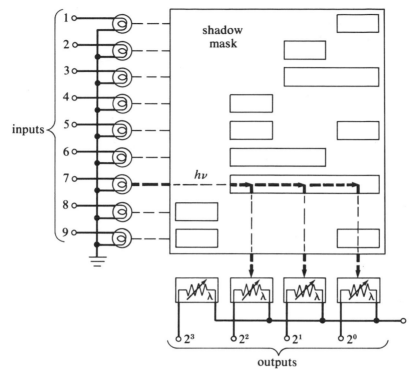

Fig. 3-50   Decimal-to-binary encoder.

input voltage is therefore a chopped signal proportional in amplitude to the input dc signal E_{in}.   This chopped signal can now be advantageously amplified or manipulated as an ac signal.

The reading, converting, or translating of digital information from a punched card or tape by optoelectronic devices is illustrated in Fig. 3-50.   In Figs. 3-51 and 3-52 the use of source-detector pairs for remote and relay switching are illustrated.   From the previous discussions of the optoelectronic and associated devices the operation of the illustrated circuits should be obvious.

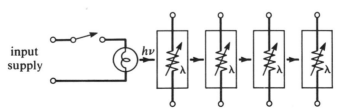

Fig. 3-51   Four-pole single-throw switch.

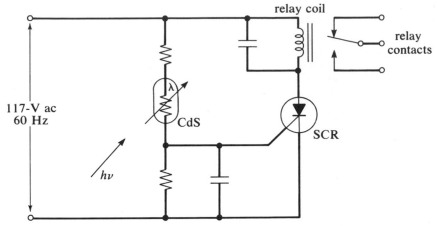

Fig. 3-52   Light-operated relay.   The SCR is triggered and relay pulls in when the cadmium sulfide cell is illuminated.

Experiment 3-12   *Optoelectronic Switch Circuit*

A photoconductive light chopper is investigated as a modulator for converting a dc signal to an ac signal for subsequent amplification.

3-7   Relay Switches

Relays are widely used in electronic circuits as remotely controlled mechanical switches used to turn ON or OFF a sequence of events.   They are not so fast as transistor or optoelectronic switches but they have higher open-circuit resistance, lower contact resistance, and can generally switch higher loads.   The electromagnetic relay is a 4-terminal device where the actuating terminals are electrically isolated from the switched signal terminals.   Reed relays can be activated by a permanent magnet or a coil.   The types of relays, relay contacts, and the characteristic contact bounce and operate and transfer times of relays are discussed in the following sections.

Types of Relays

Electromagnetic relays utilize a current through a coil to provide a magnetic field that moves the switch contacts, as illustrated in Fig. 3-53 for an *armature relay*.

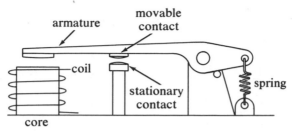

Fig. 3-53 A single-pole, single-throw, normally open (SPST NO) relay.

An *armature relay* operates by energizing an electromagnet (with a suitable current) which attracts a pivoted lever of magnetic material to a fixed pole. The pivoted lever is called the *armature.* A switch contact moves with the armature to provide a *movable contact.* If the current in the coil (and, therefore, the magnetic force) is sufficient, the armature moves the movable contact until it touches the *stationary contact.* A spring holds the movable contact (and armature) in the open position when the coil is not energized.

There are many types of electromagnetic relays. With a *plunger relay,* movable contacts are attached to a plunger that moves within a tubular magnetic coil (solenoid). *Rotary relays* often utilize the rotation of a motor shaft to move switch contacts.

Relays are constructed according to many designs to perform a wide range of functions. Interlock, stepping, sequencing, time delay, latch-in, polarized, differential, and general purpose are some of the types based on function.

The *reed relay* is a relatively new type of switch. Two or more metal reeds are enclosed in a hermetically sealed glass capsule. A normally open SPST reed relay is shown in Fig. 3-54. The overlapping reeds can be closed or opened by positioning a permanent magnet near to or away from the reed contacts. Also the reed contacts can be switched by actuating an electromagnet. Some typical arrangements for the use of the reed relay are illustrated in Fig. 3-55.

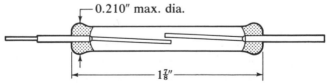

Fig. 3-54   Reed relay, with silver alloy contacts. (Courtesy Hamlin, Inc.)

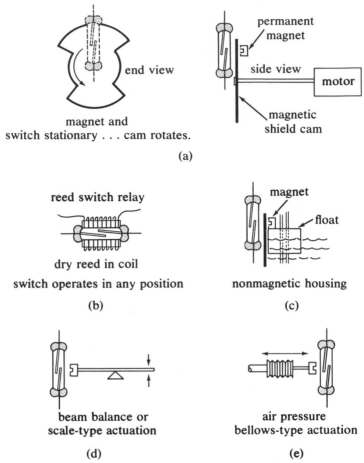

end view

magnet and
switch stationary . . . cam rotates.

permanent
magnet

side view

motor

magnetic
shield cam

(a)

reed switch relay

dry reed in coil

switch operates in any position

(b)

magnet

float

nonmagnetic housing

(c)

beam balance or
scale-type actuation

(d)

air pressure
bellows-type actuation

(e)

**Fig. 3-55 Typical reed switch arrangements. (a) Switch operates when-
ever cut-out portion of shielding metal is between magnet and
switch. (b) Switch located in cylindrical coil. (c) Solution level
control. (d) Beam position indicator. (e) Magnet positioned
by air-driven bellows. (Courtesy Hamlin, Inc.)**

Relay Contacts and Characteristics

The number of relay contacts and contact arrangements are determined by the application. The nomenclature[2] for the basic contact forms
that are used to make up specific contact arrangements is illustrated
in Fig. 3-56 for four of the most common forms. Many other contact

[2] Accepted by the National Association of Relay Manufacturers.

forms are available. The normally closed contacts are often abbreviated NC, and the normally open contacts NO.

A combination of two stationary contacts and *one movable contact* which engages one stationary contact when the coil is energized and the other stationary contact when the coil is not energized is called a single-pole double-throw (SPDT) relay. It is one of the most common contact arrangements and is illustrated in Fig. 3-56, form C.

Pull-in and drop-out currents. The current in the relay coil must exceed a certain minimum value for the armature to "pull in" and close the NO contacts. At a somewhat lower current the armature will "drop out" and the NO contacts will open. In switching circuits it is common to design to exceed the pull-in current by several times the minimum so as to insure operation of the relay.

Operate, transfer, bounce, and release time. Relays do not operate instantaneously. In fact, most relays require at least a few to many milliseconds to complete their contact function, although the reed relays have operate times of about 1 msec.

The *operate time* is the time interval from the instant of coil-power application until completion of the last contact function. The *release time* is the time interval from the instant of coil-power cutoff until the completion of the last contact function.

The measurement of the operate and release times gives an appreciation for the limitations of relays as switches, and a simple circuit for obtaining this information is shown in Fig. 3-57.

The circuit in Fig. 3-57 provides a voltage V (5 V in the example)

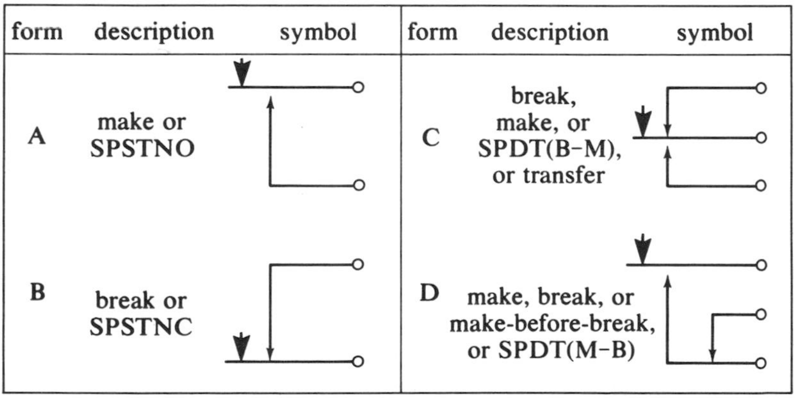

form	description	symbol	form	description	symbol
A	make or SPSTNO		C	break, make, or SPDT(B–M), or transfer	
B	break or SPSTNC		D	make, break, or make-before-break, or SPDT(M–B)	

Fig. 3-56 Four common forms of relay contacts with designations.

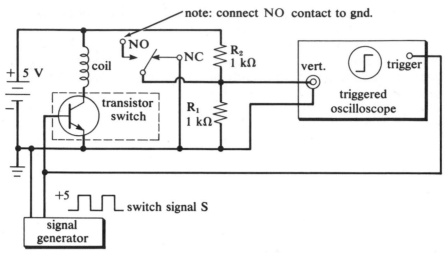

Fig. 3-57 Circuit for measuring operate, transfer, bounce, and release times.

across a voltage divider consisting of two 1-kΩ resistors. The voltage across the 1-kΩ resistor, R_1, is fed to the input of the oscilloscope. However, at the start the normally closed contacts are connected across the 1-kΩ resistor, thereby shorting the input resistor, R_1, and providing zero voltage input to the scope vertical amplifier. The switch signal S operates the coil of the relay and is also connected to the trigger input of the scope. That is, the +5 V signal triggers the sweep and applies the drive current to the relay coil. There is then a finite time period (while the scope beam is sweeping across the horizontal axis at a known preset velocity) before the movable contact breaks away from the stationary contact. When the contacts do break away one-half of the voltage V appears across the input to the scope vertical amplifier, and the trace on the scope is deflected by the input voltage, and the magnitude depends on the sensitivity setting of the vertical amplifier.

If it is the function of the relay in a specific circuit to utilize the breaking of the normally closed contacts then the time from start of trace to the voltage jump is the *operate time*.

After the normally closed contacts *break* there is a finite *transfer time* before the movable contact reaches the normally open stationary contact. It is possible to measure this transfer time by connecting the normally open contact, as shown in Experiment 3-13. Note that the transfer time should not include the *bounce time* which is very obvious on the scope trace and of considerable duration.

When the contacts *break* or *make* there is always some *contact bounce*. Although this bounce and its duration can be serious in operating digital circuits, it is *not* included in the operate and release times. The bounce time is also a characteristic of manual switches. Circuits for blanking the contact bounce will be discussed in subsequent chapters.

Experiment 3-13 *Relay Characteristics*

The operate and release and contact-bounce times can be critical for proper operation of some electronic circuits, and it is the purpose of this experiment to measure these values. Also, the pull-in and drop-out currents are determined for one of the relays.

---

chapter four

Switching Logic and Logic Gates

The apparently complex logical operations performed by even the most sophisticated digital instruments are based on a relatively simple system of symbolic switching logic (Boolean algebra) and implemented by repeated use of a few basic logic circuits. The three basic logic circuits, AND, OR, and NOT are illustrated in the subsequent sections. Also, the NAND and NOR inverting gates, various forms of integrated-circuit logic gates, and some representative applications are described.

4-1   Basic Logic Concepts

AND Concept

If two switches are wired in series as in Fig. 4-1, the circuit is completed only when switch A is CLOSED and switch B is CLOSED. These two switches are said to perform the AND operation with respect to transmitting current through the circuit by connecting the source to the load. The necessary condition for transmission can be expressed by the following equation in Boolean algebra:

$$A \cdot B = T \tag{4-1}$$

which reads, "If A is CLOSED AND B is CLOSED, then transmission will occur." The symbol A stands for the CLOSED state of switch A.

The symbols A, B, C, etc., could each represent the truth of one of

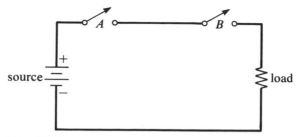

Fig. 4-1   AND operation implemented by manual switches.

the following specific statements: "the error in the resistance is less than 1%," "the straightness of the shaft is within specifications," "the temperature is above 24.6°C," "the weight of the sample is equal to 10.2 ± 0.1 g," "the radioactivity is at a dangerous level," etc.

In more general terms Eq. (4-1) could read, "If one specific statement (represented by the symbol A) is TRUE AND another specific statement (represented by the symbol B) is TRUE at the same time, then it can be logically concluded that a specific result (represented by the symbol T) is TRUE as a direct consequence of the specified A AND B statements being TRUE simultaneously.

Since the CLOSED or ON state of a switch can be used to represent the truth of a specific statement (A, B, C, etc.) and the OPEN or OFF state of the same switch to represent the falsity of the same specific statement, it is possible to use a switch to represent the state (TRUE or FALSE) of each logic statement, and a switching circuit to represent the AND function— whereby A AND B AND C AND . . . , etc., must all be TRUE statements simultaneously for a specific result T to be TRUE. As seen in Fig. 4-1, the AND function can be implemented by manual or relay switches operating in series.

It is customary to represent conditions or statements that are TRUE at a particular time by a **1**, and FALSE statements by a **0**. The equation $A =$ **1** means that the A statement or condition exists (is TRUE) at a predetermined time, and $A =$ **0** means that the A condition does not exist at the selected time.

Truth table. A "truth table" provides an outline of all possible combinations of states for all variables at a specific time and the results of each combination. A truth table is given in Table 4-1 for the AND operation with variables A and B, and result T. At some selected time the variables A and B can both be TRUE **(1)**, or FALSE **(0)**, or A TRUE **(1)** and B FALSE **(0)** (or vice versa)—thereby making four possible combina-

Table 4-1 Truth Table for A · B = T

Inputs		Output
A	*B*	*T*
0	0	0
0	1	0
1	0	0
1	1	1

tions. For the AND operation all specified conditions must exist simultaneously for the result to be TRUE. Therefore, for the AND truth table (Table 4-1), the *T* column will have a 1 when, and only when, $A = 1$ AND $B = 1$. This table can be verified by considering a 0 as an OPEN switch and a 1 as a CLOSED switch in Fig. 4-1, or by wiring the circuit of Fig. 4-1 and trying all combinations of *A* and *B*, observing the conditions under which transmission occurs.

Experiment 4-1 *Relay AND Circuit*

The AND circuit of Fig. 4-1 is wired using relay switches and a light driver for the load. The truth table for $A \cdot B = T$ is verified.

For the case of performing the AND operation with three variables (*A*, *B*, and *C*) the truth table would be as shown in Table 4-2.

Table 4-2 Truth Table for A · B · C = T

Inputs			Output
A	*B*	*C*	*T*
0	0	0	0
0	0	1	0
0	1	0	0
0	1	1	0
1	0	0	0
1	0	1	0
1	1	0	0
1	1	1	1

To be certain that all possible combinations of conditions are considered in an orderly fashion when constructing any truth table, it is simplest to count in binary numbers, starting with **0** for all variables in the top row and proceeding sequentially with the binary number **1** in the second row, **10** in the third row, **11** in the fourth row, etc., until in the final row each condition column (A, B, C, etc.) is filled with **1**, as in Table 4-2. For three variables the number of combinations is $2^3 = 8$; for n variables the number of condition combinations is 2^n.

OR Concept

If two switches are put in parallel as in Fig. 4-2, the circuit is completed when $A = 1$ OR $B = 1$ OR both $= 1$. This circuit is said to perform the OR operation. The Boolean algebra equation is written

$$A + B = T \qquad (4\text{-}2)$$

where the $+$ sign is a logic symbol that is read as OR. Table 4-3 gives the truth table for Eq. (4-2).

In general, the result T will exist only when either condition A OR B OR both conditions exist.

Table 4-3   Truth Table for $A + B = T$

Inputs		Output
A	B	T
0	0	0
0	1	1
1	0	1
1	1	1

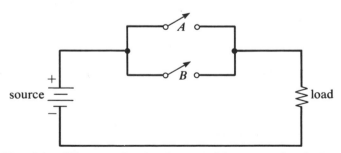

Fig. 4-2   OR operation implemented by manual switches.

Experiment 4-2 *Relay OR Circuit*

The truth table for the OR function given in Table 4-3 is verified using relay switches in the circuit of Fig. 4-2. An indicator light driver is used as the load.

NOT Concept

In general, a NOT means simply the "other state" of a logic condition — that is, the opposite or inverse of a stated condition. If the condition is given the symbol A, then the opposite of that condition is given the symbol $\overline{A}$ or A', and is read NOT A. For example, if condition A is "the temperature is above 24.6°C," then $\overline{A}$ is the inverse of this condition, i.e., "the temperature is NOT above 24.6°C."

The NOT function can be implemented by a single-pole double-throw relay switch as illustrated in Fig. 4-3. There is a closed contact when the relay is not activated so that when $A = 0$, then $\overline{A} = 1$, and vice versa. Note that by using the $\overline{A}$ contact and the $\overline{B}$ contact in series with a source and lamp, the lamp T would light when A and B were both FALSE, i.e., $\overline{A} = 1$ AND $\overline{B} = 1$. The Boolean equation would then be $\overline{A} \cdot \overline{B} = T$ and the truth table would be as shown in Table 4-4.

Table 4-4 Truth Table for $\overline{A} \cdot \overline{B} = T$

Inputs		Output
A	B	T
0	0	1
0	1	0
1	0	0
1	1	0

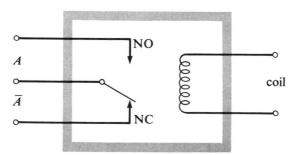

Fig. 4-3   SPDT relay for implementing the NOT operation.

Experiment 4-3   *Relay Logic NOT Circuits*

The truth table for $\overline{A} \cdot \overline{B} = T$ given in Table 4-4 is verified using relay switches and indicator light driver. Circuits can also be wired to light the light when A is TRUE AND B is FALSE or when A is FALSE AND B is TRUE to prove that there is only one 1 in the T column of the truth table for any AND circuit.

4-2   Basic Theorems of Boolean Algebra

Boolean algebra provides a simple mathematical technique for the design, simplification, and analysis of switching systems as well as a symbolism for the description of switching and logic functions. Some of the basic theorems are readily understood in terms of their switching circuit equivalents. Then, once understood, these theorems can be used to aid the understanding of switching circuits too complex for intuitive or visual analysis of the circuit.

Theorems in 1 and 0

There are only two states that a 2-terminal switch system can be in no matter how complex it is — OPEN or CLOSED. For transmission through the network, a CLOSED switch or network is in the 1 state and an OPEN switch or network is in the 0 state. Combinations of 0's and 1's and their switching circuit equivalents are shown in Table 4-5.

Theorems in One Variable

The table of theorems (Table 4-5) is illustrated with combinations of switches that are fixed OPEN or CLOSED. Other combinations could include variables such as A, where A can be either 1 or 0 depending on whether the statement for A is TRUE or FALSE at that moment. Table 4-6 lists Boolean theorems for combinations including one variable, and illustrates the equivalent switching circuit.

Theorems with More Than One Variable

Table 4-7 gives some Boolean theorems in more than one variable.

The commutation theorems indicate that the order of the conditions in the statement is not significant. The association theorems show that the indicated AND or OR operations may be performed in any order.

Table 4-5   Boolean Theorems in 0 and 1 with Equivalent Switching Circuits

Table 4-6   Boolean Theorems in One Variable with Equivalent Switching Circuits

Table 4-7   Some Theorems of Boolean Algebra in More Than One Variable

Commutation theorems:	$A + B = B + A$ $AB = BA$
Absorption theorems:	$A + AB = A$ $A(A + B) = A$
DeMorgan's theorems:	$\overline{A + B} = \overline{A} \cdot \overline{B}$ $\overline{AB} = \overline{A} + \overline{B}$
Association theorems:	$A + (B + C) = (A + B) + C$ $A(BC) = (AB)C$
Distribution theorems:	$A + BC = (A + B)(A + C)$ $A(B + C) = AB + AC$

Redundant terms in an expression can be eliminated by the absorption theorems. The expression $A + AB$ depends only on the state of A. If $A = 0, 0 + 0 \cdot B = 0$ and if A is $1, 1 + 1 \cdot B = 1$ so that $A + AB = A$. Similarly $A(A + B) = A$. Inspection of the equivalent switch network would also show this to be true. The distribution theorems indicate that combinations of variables with AND and OR operations may be factored or distributed according to the rules of ordinary algebra. For instance, when $(A + B)(A + C)$ is "multiplied out,"

$$(A + B)(A + C) = AA + AC + AB + BC$$

From Table 4-6, $AA = A$

$$(A + B)(A + C) = A + AC + AB + BC$$

From the absorption theorem, $A + AB + AC = A$, so

$$(A + B)(A + C) = A + BC$$

DeMorgan's theorems, dealing with the NOT function on entire terms, are especially interesting and useful. They can be proven by the truth table method; that is, to make a truth table for all four combinations of A and B for $\overline{A + B}$ and another for $\overline{A} \cdot \overline{B}$. If the T column is identical in the two tables, then the two functions are proven to be logically equivalent.

4-3 Manipulation of Logic Statements

Any logic statement can be implemented by a combination of AND, OR, and NOT circuits. However, the use of the Boolean theorems to change the form of the logic statement can lead to a simplification of the statement and the necessary circuit. It is frequently necessary to design a circuit to carry out a given logic operation taking the limitations of available circuitry into account. In this case, an equivalent equation may be derived which, though not necessarily simpler, is more practical in that case. For instance, systems of logic gates frequently provide only AND and NOT functions (or OR and NOT) in the basic gate packages. Here DeMorgan's theorems are often helpful showing how an expression using AND operations can be converted into an equivalent expression using OR operations and vice versa.

In general, any logic expression can be inverted or "complemented" simply by inverting every term so that an A becomes $\overline{A}$, B becomes $\overline{B}$, and every AND becomes OR, and vice versa; or as stated by DeMorgan:

$$\overline{AB} = \overline{A} + \overline{B} \tag{4-3}$$

and

$$\overline{A + B} = \overline{A} \cdot \overline{B} \tag{4-4}$$

Where more complex logic statements are made, each term should be considered as an entity. For example, follow the steps in deriving various equivalent expressions for the inverse of the statement

$$AB + BC + CA$$

Consider AB as one term and BC and CA as other terms. Invert and apply DeMorgan's theorem,

$$\overline{AB + BC + CA} = \overline{AB} \cdot \overline{BC} \cdot \overline{CA} \tag{4-5}$$

Again, each term on the right

$$= (\overline{A} + \overline{B})(\overline{B} + \overline{C})(\overline{C} + \overline{A}) \tag{4-6}$$

Applying the distribution theorem to the second two terms

$$= (\overline{A} + \overline{B})(\overline{B}\overline{C} + \overline{A}\overline{B} + \overline{C} + \overline{A}\overline{C})$$

Since $\overline{C} + \overline{B}\overline{C} + \overline{A}\overline{C} = \overline{C}$ from the absorption theorem

$$= (\overline{A} + \overline{B})(\overline{A}\overline{B} + \overline{C})$$

Applying the distribution theorem once more

$$= \overline{A}\overline{B} + \overline{B}\overline{C} + \overline{A}\overline{C} \tag{4-7}$$

Then applying DeMorgan's theorems twice more, first to each term and then to the entire expression

$$= \overline{(A + B)} + \overline{(B + C)} + \overline{(A + C)} \tag{4-8}$$

$$\overline{AB + BC + CA} = \overline{(A + B)(B + C)(A + C)} \tag{4-9}$$

Thus also proving that

$$AB + BC + CA = (A + B)(B + C)(A + C) \tag{4-10}$$

The proof of the following identities is suggested as an exercise for the reader:

$$A(\overline{A} + B) = AB$$
$$A + \overline{A}B = A + B$$
$$(A + B)(A + \overline{B}) = A$$
$$ABC + AB\overline{C} = AB$$
$$A(\overline{B} + C) = \overline{A} + B\overline{C}$$
$$AB + CD = (A + C)(A + D)(B + C)(B + D)$$

Binary Addition

Any operation that can be restated in logic terms can be performed by an appropriate logic circuit. A good example is that of the binary addition of digits A and B. The binary addition table can be stated logically as follows:

$$0 + 0 = 0,\ 0 + 1 = 1,\ 1 + 0 = 1,\ \text{and}\ 1 + 1 = 0\ \text{and carry 1}$$

The addition table is summarized in Table 4-8, where S is the sum and C is the carry. Note that throughout this text, binary numerals are written 0 and 1 and should not be confused with digital logic levels **0** and **1** used in truth tables, logic expressions, and digital circuit descriptions.

Table 4-8   Binary Addition

Inputs		Outputs	
A	B	S	C
0	0	0	0
0	1	1	0
1	0	1	0
1	1	0	1

Table 4-8 shows that $S = 1$ when $A = 0$ AND $B = 1$ OR when $A = 1$ AND $B = 0$ and that $C = 1$ when A AND B are 1. This can be written

$$S = \bar{A}B + A\bar{B} \tag{4-11}$$

and

$$C = AB \tag{4-12}$$

thus converting the binary addition table into two logic expressions. The equivalent switching circuit for these equations is shown in Fig. 4-4. The circuit is simplified somewhat by recognizing that condition A can be obtained between the A and $\bar{B}$ contacts rather than using another A contact for the C output.

EXCLUSIVE-OR

The function $\bar{A}B + A\bar{B}$ is called the EXCLUSIVE-OR function. It can be stated as A OR B but NOT A AND B. The EXCLUSIVE-OR is used so frequently that it has a separate symbol $\oplus$. Thus

$$A \oplus B = \bar{A}B + A\bar{B} \tag{4-13}$$

The S output of Fig. 4-4 is an implementation of the EXCLUSIVE-OR

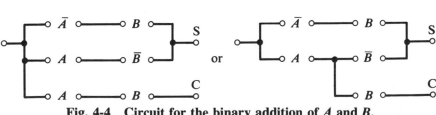

Fig. 4-4   Circuit for the binary addition of A and B.

function.　Equivalent circuits could be obtained by deriving equivalent expressions.　For instance, from the distribution theorems,

$$\bar{A}B + A\bar{B} = (\bar{A}B + A)(\bar{A}B + \bar{B})$$

Again applying the distribution theorems to each term in parentheses,

$$\bar{A}B + A\bar{B} = (A + \bar{A})(A + B)(\bar{A} + \bar{B})(B + \bar{B})$$

From Table 4-6, $(A + \bar{A}) = 1$, thus

$$A \oplus B = \bar{A}B + A\bar{B} = (A + B)(\bar{A} + \bar{B}) \tag{4-14}$$

Equation (4-14) shows that $A \oplus B$ can also be implemented by parallel contacts A and B in series with parallel contacts $\bar{A}$ and $\bar{B}$.

Equality Comparator (Coincidence Circuit)

Sometimes the EXCLUSIVE-OR expression $A\bar{B} + B\bar{A}$ is referred to as an "inequality comparator" because it states that the result is TRUE if A and B are *not equal* to each other, i.e., if one is 1 and the other is 0. If the expression for the EXCLUSIVE-OR is inverted and manipulated using DeMorgan's inversion theorem,

$$\begin{aligned}
\overline{A\bar{B} + B\bar{A}} &= (\bar{A} + B)(\bar{B} + A) \\
&= \bar{A}\bar{B} + A\bar{A} + B\bar{B} + BA \\
&= AB + \bar{A}\bar{B}
\end{aligned} \tag{4-15}$$

This states that the inverse of the inequality comparator is an "equality comparator"; the output is true when $A = 1$, AND $B = 1$ OR when $A = 0$ AND $B = 0$; that is, when A and B are equal.

The implementation of the equality comparator is very similar to that of S in Fig. 4-4 except that the series combination of A and B is in parallel with the series combination $\bar{A}$ and $\bar{B}$.

Experiment 4-4　*Relay Comparator or EXCLUSIVE-OR Circuit*

A relay circuit to provide the function $T = A \cdot B + \bar{A} \cdot \bar{B}$ is wired and tested to verify the expected response.　An EXCLUSIVE-OR circuit $T = \bar{A} \cdot B + A \cdot \bar{B}$ may be wired instead.　Either circuit, it will be noted, is an implementation of the upstairs-downstairs light switch problem.

4-4 Binary Information Gates

Boolean algebra provides a symbolism and a convenient means of manipulating and simplifying complex combinations of 2-state quantities or devices. The usefulness of Boolean algebra applies as well to the processing of binary (2-state) information as it does to the development of 2-state (ON-OFF) switching circuits. The implementation of the algebra for binary signals is generally different than for switches though there are some areas of overlap and similarity.

Binary Signals

A binary information signal has, at any instant, only two possible states: logic level **1**, e.g., about +4 V and logic level **0**, e.g., about 0 V. Because the output level of each binary signal source can indicate no more than the truth of a single statement at any instant, a combination of several such statements (or several thousand) may have to be analyzed logically to yield the desired quantity. This logical analysis is accomplished by logic circuits.

A logic circuit produces an output logic level that is a logical function of the input logic levels. The exact input and output voltage levels are not critical so long as they are clearly identifiable by the succeeding circuit as logic **0** or logic **1**. Therefore, small changes in voltage level because of nonideal switching elements can cause no loss of accuracy unless they accumulate to the point of interfering with logic level recognition. Circuits that perform the simple logic functions on binary signals are called "gates." Note that logic gates cannot act to transmit or connect an analog signal directly from the input to the output as switching circuits can do.

AND Gate

An AND gate is a circuit with input and output logic levels that correspond to the truth table for the AND function as shown in Table 4-2. The AND gate has two or more input terminals, A, B, C, etc., and one output terminal. Only when *all* the signals connected to the input terminals are simultaneously at the logic **1** level will the output be at logic **1**. Under all other conditions the output logic level will be **0**. One form of the AND gate which uses relay switches is shown in Fig. 4-5. A logic

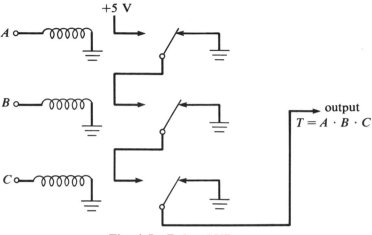

Fig. 4-5   Relay AND gate.

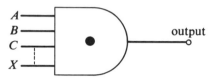

Fig. 4-6   AND gate symbol.

level 1 signal (~+5 V) at the *A*, *B*, or *C* inputs is sufficient to actuate the relay, but a **0** level signal (~0 V) will not. Only when all three relays are actuated (*A* = *B* = *C* = 1) will the signal from the +5 V source be connected to the output. At all other times the output is grounded. Frequently a transistor switch is used to drive the relay coil so that the entire coil current does not have to be supplied by the binary input signals.

When drawing circuits using logic gates, it is awkward to draw the complete circuit for each gate. Thus symbols are used to represent the entire gate circuit and/or function. The symbol for an AND gate is shown in Fig. 4-6. This symbol is used regardless of the kind of switches used in the circuit.

OR Gate

The output of an OR gate circuit will have a logic 1 level when any one or more of the inputs is a logic 1. A relay OR circuit is shown in Fig. 4-7. An examination of Fig. 4-7 will verify that the input and output

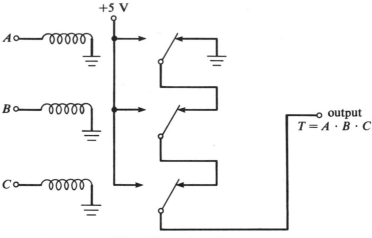

Fig. 4-7   Relay OR gate.

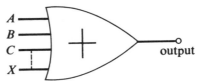

Fig. 4-8   OR gate symbol.

levels of the circuit correspond to the OR truth table, Table 4-3.   The general symbol for an OR gate is shown in Fig. 4-8.

The Inverter

A logic inverting circuit completes the list of basic logic operations. An inverter "inverts" the logic level of the input signal as shown in Table 4-9.   A relay inverting circuit is shown in Fig. 4-9.   Note that when the relay coil is not energized (input = **0**) the output is **1** and vice

Table 4-9   Truth Table for Inverter

Input	Output
1	0
0	1

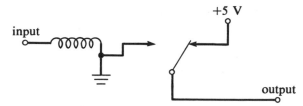

Fig. 4-9   Relay inverter circuit.

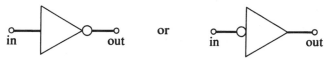

Fig. 4-10   Inverter logic symbols.

versa according to Table 4-9.   The inverter performs a NOT operation
on the input function.   If the input function is A, the output is $\bar{A}$; if the
input is $A + B$, the output is $\overline{A + B}$, and so forth.   The logic symbol for
an inverter is shown in Fig. 4-10.   It is the triangular symbol for a noninverting amplifier with a circle at the input or output connection to indicate inversion.

Logic Gate Applications

EXCLUSIVE-OR Gate.   Any simple expression in Boolean algebra
can be implemented with the use of AND, OR, and inverter gates.   The
design of the gate circuit follows the algebraic expression directly, using
the appropriate gates to perform each indicated operation.   For example,
consider the EXCLUSIVE-OR expression, Eq. (4-13), discussed in
Section 4-3, $A \oplus B = A\bar{B} + B\bar{A}$.   This can be implemented by gates performing the AND operation on A and $\bar{B}$, and on B and $\bar{A}$, and then the
OR operation on $A\bar{B}$ and $B\bar{A}$, as shown in Fig. 4-11.

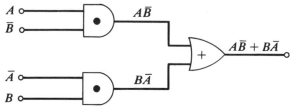

Fig. 4-11   Gate diagram for $A\bar{B} + B\bar{A} = A \oplus B$.

Half-adder. The implementation of Fig. 4-11 assumes that the input functions $\bar{A}$ and $\bar{B}$ are conveniently available. If this is not the case it would be necessary to use inverters to generate $\bar{A}$ and $\bar{B}$ signals from A and B. If the EXCLUSIVE-OR is to be used for binary addition, the carry function $C = A \cdot B$ will also be needed. A circuit which provides the sum and carry outputs for A and B inputs is called a "half-adder." A full-adder will also accept the carry signal from the column to the right. A logic gate half-adder is shown in Fig. 4-12. Using the theorems of Boolean algebra, equivalent expressions can be derived which can be implemented in the same way and will perform the same function. Frequently equivalent implementations, though identical in function, will require a different number of gate circuits and, therefore, may vary in economy and/or efficiency. The process of finding the optimum circuit to perform a given function is called "minimization." This process is of primary interest in the design of computing circuits which will be repeated many times over. An example of this is the half-adder of Fig. 4-12. An equivalent may be found for the sum which has $A \cdot B$ as one of the terms. Begin with

$$S = A\bar{B} + B\bar{A}$$

Expand each term

$$S = \bar{B}(A + B) + \bar{A}(A + B)$$

Factor $A + B$,

$$S = (\bar{B} + \bar{A})(A + B)$$

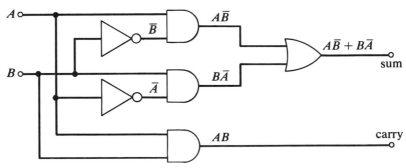

Fig. 4-12   Logic gate half-adder.

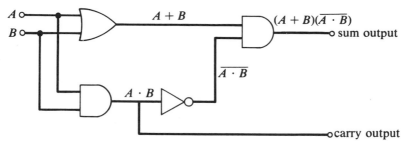

Fig. 4-13   A minimized half-adder circuit.

Use DeMorgan's theorem for $\overline{B} + \overline{A}$,

$$S = \overline{A \cdot B}(A + B)$$

and implement as shown in Fig. 4-13.

Full-adder. When two binary numbers are added, any carry resulting from adding the first column must be added to the two digits in the second column. This is illustrated by adding 1011 to 0101. Starting with the 2^0 column, $1 + 1 = 0$, carry 1. In the 2^1 column, $0 + 1 + $ carry $1 = 0$, carry 1. In the 2^2 column, $1 + 1 + $ carry $1 = 1$, carry 1 and in the 2^3 column, $1 + 0 + $ carry $1 = 0$ carry 1.

	2^3	2^2	2^1	2^0	
	1	1	0	1	A
+	0	1	1	1	B
1	0	1	0	0	

The final carry is placed in the 2^4 column. This is the binary equivalent of the decimal addition $13 + 7 = 20$. A circuit which will perform the addition for columns 2^1, 2^2, etc., must add digits A, B, and C_i and produce outputs S and C_o, where C_i and C_o are the carry in and carry out, respectively.

Following the laws of binary addition, a truth table or table of combinations can be made up for all combinations of input states and for both S and C_o outputs. This table, Table 4-10, is made up by considering what S and C_o should be for each input combination. From this truth table Boolean expressions for S and C_o are obtained as shown in Eqs. (4-16)–(4-20).

Table 4-10   Truth Table for Full-Adder

Inputs			Outputs	
C_i	A	B	S	C_o
0	0	0	0	0
0	0	1	1	0
0	1	0	1	0
0	1	1	0	1
1	0	0	1	0
1	0	1	0	1
1	1	0	0	1
1	1	1	1	1

$$S = \overline{C_i}\overline{A}B + \overline{C_i}A\overline{B} + C_i\overline{A}\overline{B} + C_iA \cdot B \qquad (4\text{-}16)$$
$$= \overline{C_i}(\overline{A}B + A\overline{B}) + C_i(\overline{A}\overline{B} + AB)$$

Equation (4-15) has shown that $(\overline{A}\overline{B} + AB) = (\overline{\overline{A}B + A\overline{B}})$,

$$S = \overline{C_i}(\overline{A}B + A\overline{B}) + C_i(\overline{\overline{A}B + A\overline{B}}) \qquad (4\text{-}17)$$
$$= C_i \oplus (A \oplus B) \qquad (4\text{-}18)$$

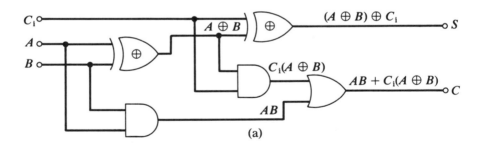

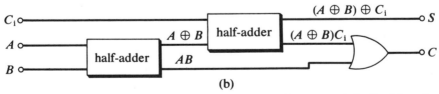

(b)

Fig. 4-14   **Full-adders using (a) EXCLUSIVE-OR and (b) half-adder circuits.**

and

$$C_o = \bar{C_i}AB + C_i\bar{A}B + C_iA\bar{B} + C_iAB$$
$$= AB + C_i(\bar{A}B + A\bar{B}) \qquad (4\text{-}19)$$
$$= AB + C_i(A \oplus B) \qquad (4\text{-}20)$$

The quantities S and C_o can be generated with AND and OR gates from Eq. (4-16) and (4-19) or by EXCLUSIVE-OR gates and AND gates using Eq. (4-18) and (4-20) or by two half-adders as shown in Fig. 4-14. By using one of these circuits for each digit, binary numbers of any practical length can be added. Other circuits for the half- and full-adders and the technique for adding more than two numbers together will be discussed in later sections.

4-5 Diode Logic Gates

Electronic switches have obvious advantages over mechanical switches in many applications. They are faster, smaller, and often more economical. On the other hand, electronic switches are generally not as "ideal" as mechanical switches and only rarely achieve as good isolation between the actuating and switched circuits as the relay. One application where the limitations of the simplest electronic switches can be tolerated is in the logic circuits which deal with binary information signals.

Diode AND Gate

A diode circuit which performs the AND function is shown in Fig. 4-15. The logic level of the signal sources is +4.0 V and 0 V for **1** and **0**, respectively. (A 0-V signal source is a connection to the common through the output impedance of the generator; it is *not* an open circuit.) Prior to the first pulse $A = B = C = 0$. All three diodes may conduct current from the +5-V source through R and the signal sources to the common. If R is very much larger than the output impedance of the sources and the forward resistance of the diodes, the output voltage will be V_γ for the diode (about 0.6 V for a Si diode). This is close enough to 0 V to be called a logic **0** output.

During the first pulse, all three inputs are at +4.0 V. The diodes are still forward biased between the +5-V supply and the +4-V signals. The output voltage is 4.0 V + V_γ, a logic **1**. Since there is no pulse at C during the second pulse at A and B, the diode at input C will continue to conduct to a 0-V source, the output remains at logic **0**, and the diodes at

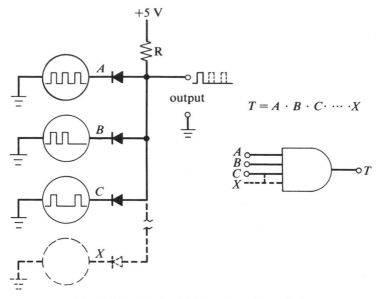

Fig. 4-15 Diode AND gate and symbol.

inputs A and B are reverse biased. During the third pulse, the **0** input at B keeps the output at logic **0**. Only when all three inputs are **1** will the output be **1**. This can be stated, "If A AND B AND C are **1**, then, and only then, the output T will be **1**." Using Boolean notation, $T = A \cdot B \cdot C$, the AND function can be extended to include additional signal sources by simply adding more diodes as shown. If an unused input is left open, the diode cannot conduct in any case and, therefore, cannot influence the output level. If all the inputs are unused and open, what will the output level be? If one input is permanently grounded what will the output level be?

It is generally true that only one diode of the AND gate will be conducting at one time. Whichever one is connected to the least positive source will conduct and determine the output voltage. All the other diodes will be reverse biased. Even when two or more inputs are grounded, the diode which is the best conductor or has the least V_γ will conduct the majority of the current.

This simple diode resistor AND circuit is basic to a majority of the circuits in digital instruments and computers. The reason it is often called a "gate" is shown in the following example. Consider a two-input circuit. When one input is held at logic level **1**, the output will follow the level changes at the other input, i.e., the gate is OPEN. On the other hand, if one of the inputs is held at **0**, the output will be **0** regardless of the other input signal, i.e., the gate is CLOSED. This circuit for which the gating action has been shown as the diode switch of Fig. 2-44, could be used as the gate circuit in most of the block diagram instruments in Chapter 1.

Experiment 4-5   *Diode AND Gate*

A three-input diode AND gate as shown in Fig. 4-15 is wired. Logic signals are applied to the inputs and a light driver is connected to the output to verify the AND function of the circuit.

Diode OR Gate

If the diodes and the power supply polarity of the AND gate are reversed, the gate will perform the OR function. The circuit is shown in Fig. 4-16. Whichever diode has the highest input voltage will be forward biased transmitting that input voltage to the output (reduced by V_γ). The other diodes will be reverse biased. Therefore, if any combination of the inputs have a logic level of **1**, the output will be **1**. Only when all inputs are **0** will the output be **0**. The output level thus corresponds to the OR

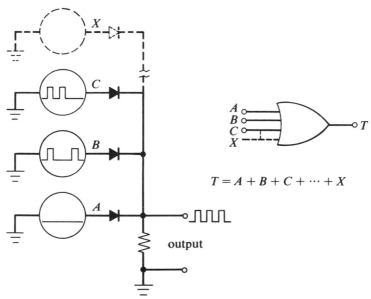

$$T = A + B + C + \cdots + X$$

Fig. 4-16   Diode OR gate and symbol.

function of the input levels. Additional inputs may be accommodated by adding more diodes as shown. Unused diodes will not affect the operation of the gate.

All circuits performing the AND or OR function are called "gates." The gating action of the OR circuit is accomplished by holding one of the inputs at logic 1. During this period, variations in logic level at the other inputs cannot appear at the output. However, the most common application of the OR gate is preventing binary signal sources which need to be connected to one point in a circuit from interfering with each other. Since only one diode is conducting at one time, all sources are isolated from each other by reverse-biased diodes.

Experiment 4-6 *Diode OR Gate*

The diode OR circuit shown in Fig. 4-16 is wired. The OR function is verified using an output level indicator and logic level input signals.

Transistor Inverter

To complete the list of basic electronic logic circuits, an electronic inverting circuit is required. The transistor switch shown in Fig. 4-17 is used for this purpose. The input resistance is chosen so that when the source is at **0** logic level, the base current is low enough to keep the transistor near the cut-off region resulting in about +5 V at the output. However, when the source signal is at logic **1**, the base current will drive

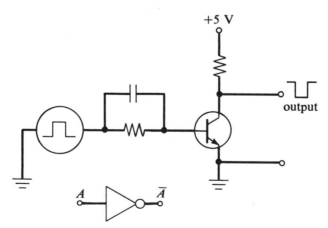

Fig. 4-17 Transistor inverter and symbol.

the transistor to saturation resulting in a **0** output level. Thus the input function is inverted at the output. Because of the gain of the transistor amplifier, the input signal need not be as large as the output signal. Therefore, the inverter can be used as a logic level restorer for signals of which the logic levels have deteriorated and are not as far separated as desired for the system. For level restoration with no inversion, two inverters are required.

Emitter Follower

The emitter-follower amplifier described in Chapter 3 is sometimes used in logic circuits. Its high input impedance and low output imped-ance make it a useful buffer amplifier in many applications. However, in this configuration the transistor is acting as a linear amplifier of less than unity gain and not as a switch. Thus, the emitter-follower amplifier degrades the logic level separation of the input signal somewhat and should not be used without considering this effect.

Experiment 4-7   *The Inverter*

The inverter circuit of Figure 4-17 is wired and its NOT function is verified.

Combinations of Diode AND and OR Gates

Diode gates are very convenient for performing a single logic opera-tion. The limitations of the diode gates become clearer when more complex logic functions requiring several gates are considered. When one AND gate follows another, the logic **0** increases by V_γ for each gate. For only two silicon diode gates in succession, the logic **0** output would be about 1.2 V. Similarly, the logic **1** output decreases by V_γ for each OR gate in succession.

Input and output impedances are another problem. Consider, for instance, a logic diagram in which the output of an OR gate is connected to the input of an AND gate. To maintain the desired logic level at a gate output the load impedance connected to the output must be very large compared to the output impedance of the gate. The output im-pedance of the OR gate is equal to the gate resistor for a **0** output level and to the highest input signal source impedance for a **1** output level. The input impedance for an AND gate is equal to the gate resistor for a **0** input signal. When a **0** output from an OR gate is connected to an

AND gate, the output resistance of the OR gate is equal to the AND gate input resistance (assuming equal-valued gate resistors). Since the OR gate resistor is grounded and the AND gate resistor is connected to +5 V, the **0** level output voltage of the OR gate is about +2.5 V.

Other combinations such as an AND gate followed by an OR gate work out much better. The example proves, however, that with simple diode gates, it is important to analyze each gate's output considering its load condition to assure that reasonable logic levels will be maintained. General purpose diode gates used for logic patching often have a transistor inverter or emitter follower at each gate output to assure a low output impedance for any reasonable load.

Experiment 4-8 *Combinations of Gates*

The circuits wired in Experiments 4-5 through 4-7 are used with an emitter follower and another diode AND circuit to test combinations of gates. Input-output compatibility and expected logic functions are checked. The AND-OR, OR-AND, OR-FOLLOWER-AND, AND-AND, and finally the half-adder circuit of Fig. 4-13 are wired and tested.

Choice of Logic Levels

In the previous discussions of binary information signals and logic gates, the **1** logic level has been taken as +5 V and the **0** logic level as 0 V. This choice of logic levels is completely arbitrary. It happens to correspond to the current trend, but is by no means universal. None of the arguments or circuits advanced in previous sections change if some other positive voltage, e.g., +1 V, +3 V, or +10 V is chosen as the **1** logic level. However, it is interesting to consider the result of reversing our previous choice of logic levels, i.e., make 0 V the **1** state and +5 V the **0** state. This is called "negative logic" because the more negative level is the **1** state. The Boolean algebra does not change, of course, since its laws are entirely independent of the method of implementation. However, the gates do not perform the same operation for negative logic as they do for positive logic.

Consider the diode AND gate of Fig. 4-15. The output voltage will be 0 V (logic **1**) when any one or more of the inputs is 0 V (logic **1**). The output can only be logic **0** (+5 V) when all the inputs are +5 V (logic **0**) simultaneously. The truth table for this gate now exactly corresponds to the OR function. This is to say that a positive logic AND gate is an

OR gate when negative logic is used. A similar consideration of the positive logic OR gate (Fig. 4-16) will show that it is a negative logic AND gate. An inverter can either be thought of as logic level inverter (the NOT operation previously discussed) or as a logic inverter since a positive logic 1 at the input becomes a negative logic 1 at the output.

As we shall see later, with certain types of gates, it is helpful to be able to think of the inverter as changing the signal from positive to negative logic and vice versa. Suppose one were given a logic problem to work out and had only positive OR gates and inverters to do it with. The above arguments suggest that the AND function could be performed by inverting the logic level of each input signal, using a positive OR gate (negative AND) and then returning to positive logic again by using another inverter at the gate output. This is shown in Fig. 4-18. Boolean algebra also shows this to be valid since the inverters at the input change A to $\overline{A}$ and B to $\overline{B}$. After the $+$OR gate, we have $\overline{A} + \overline{B}$. DeMorgan's theorem says that $\overline{A} + \overline{B} = \overline{A \cdot B}$ which, after inversion becomes $A \cdot B$.

Still other choices of logic levels could be made if the supply voltage were negative instead of positive. Figure 4-19 shows the circuits of diode AND and OR gates with a negative supply voltage. The operation is as shown in the figure for negative logic, i.e., -5 V is a **1**, and 0 V is a

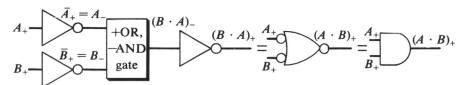

Fig. 4-18   +AND gate using +OR gate and inverters.

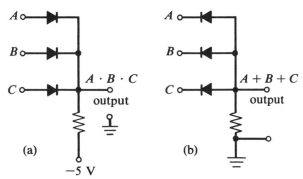

Fig. 4-19   Diode gates for negative supply: (a) −AND; (b) −OR.

0. It is also clear that for positive logic (0 V = **1**, −5 V = **0**), Fig. 4-19a is a +OR gate and Fig. 4-19b is a +AND gate. In terms of logic capability, there is no reason to choose any one set of logic levels over the others. The choice is generally made on the basis of simplicity and economy of implementation of the active elements in the gating circuits.

4-6 Transistor Gates

The logic gate is the basic circuit element in all modern digital circuits. The characteristics of a computer or instrument are determined, in large part, by the speed, versatility, noise immunity, durability, etc., of the gate circuit which is used hundreds of times over in the simplest digital instruments. Improvements in gate design have come frequently in recent years resulting in currently available gates which are very nearly the ideal universal gate.

DTL Gates

A natural development from the diode gates discussed in the previous section was the addition of an amplifier to the gate output to reduce the output impedance as shown in Fig. 4-20. This is a diode-transistor logic or DTL gate. The input diodes and R_1 are the AND circuit. When the AND output is near 0 V, the transistor is cut off and the output is +5 V — a logic **1**. A **1** output from the AND gate is sufficient to saturate the transistor making the output very near 0 V. The gate function is now $T = \overline{A \cdot B \cdot C}$, called NOT-AND or NAND. The symbol for a

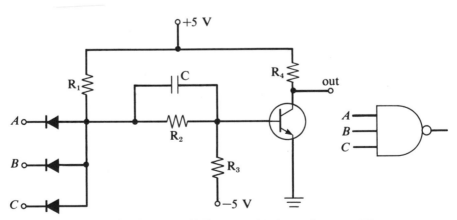

Fig. 4-20   Positive AND gate with inverting amplifier.

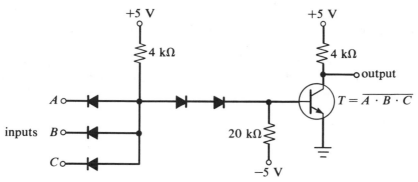

Fig. 4-21   Improved DTL NAND gate.

NAND gate is showing in Fig. 4-20. For negative logic levels, it is an
OR gate and inverter or NOR gate. The inverting amplifier was chosen
over the emitter follower to avoid the problem of deterioration of the logic
level voltages through successive gates. The impedance mismatch problems are also greatly improved. When a **0** level signal source is connected to one of the inputs, a current of $5/R_1$ amperes must pass to ground
through the signal source without raising the input voltage beyond the
upper limit of the **0** level. Therefore a signal source for the gate of Fig.
4-20 must have a very low output impedance when its output level is **0**.
When the output of the NAND gate shown is **0**, the transistor is saturated
and conducting heavily. Its output impedance is so low that it can be
connected to many gate inputs. The number of gate inputs which can
be connected to a gate output without jeopardizing the logic level is
called the "fan-out," an important gate characteristic. Very little current is required of a signal in the **1** state and the amplifier is easily designed to supply enough current at the **1** logic level to equal the fan-out
limitation set by the **0** state conditions.

A similar gate using a positive OR gate and inverters could be made,
but this is generally not done. As demonstrated in the previous section,
all logic can be done with just OR gates and inverters, or just AND gates
and inverters. Users of the DTL system employ only AND gates, or
rather NAND gates, and inverters. In fact, when an inverter is needed,
a NAND gate is used employing just one input. Even though this may
seem to waste some components, it is very convenient to use just one
gate circuit. The speed, fan-out, noise immunity, and other characteristics of that gate are known and are the same throughout the circuit or
digital system.

An improved DTL gate is shown in Fig. 4-21. The coupling capacitor and resistor have been replaced by two diodes in series. The diodes improve the speed and the noise immunity. The noise immunity of a gate is the smallest **1** level output voltage minus the minimum effective **1** level input voltage, or the minimum effective **1** level input voltage minus the maximum **0** level output, whichever is smaller under the worst case conditions. A noise pulse exceeding this value has a chance of affecting the state of the gate. The DTL gate shown has a noise immunity of about 1.0 V and a propagation delay of 50–100 nsec.

Experiment 4-9   *The DTL NAND Gate*

The DTL NAND gate of Fig. 4-21 is wired. The NAND function of the circuit is verified and the input signal requirements and fan-out are measured.

RTL and RCTL Gates

Another gate circuit employing transistors is simply a number of transistor switches in parallel, as shown in Fig. 4-22. When any one or more of the transistors is ON (a positive logic **1** input) the output will

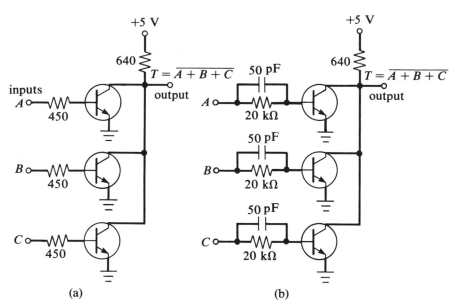

Fig. 4-22   Logic gates: (a) RTL; (b) RCTL.

be **0**. The output will only be **1** if all the inputs are **0**. This is the inverse of the positive OR function so the gate is a positive NOR or negative NAND gate. The NOR gate symbol is an OR gate with a circle at the output connection. The RTL gate requires forward-biased base current from a **1** source and virtually no current from a **0** source. Since the output impedance of the gate shown for a **1** level is 640 Ω, the fan-out may only be three–five gate inputs, about half that of the DTL. The noise immunity and speed are also inferior to the DTL, but the RTL is simple and economical.

Fan-out and speed are improved somewhat by increasing the base resistance and providing a charge-compensating capacitor as shown in Fig. 4-22b.

Gates such as the RTL and RCTL which require current from the signal source to activate the gate input device are in a class of circuits called *current-sourcing* logic. This is because the gate output acts as a current source for all the gate inputs connected to it. The DTL circuits, on the other hand, require the gate output to absorb or "sink" current from a gate input in order to hold that input in the **0** state. Gates of this type are called *current-sinking* logic. Because of the different output and input requirements, these two types of logic circuits are not usually mixed in a single system without careful consideration or special interfacing circuits.

Experiment 4-10 *The RTL NOR Gate*

The RTL NOR gate circuit shown in Fig. 4-22a is wired and its NOR function is verified. The **1** and **0** logic level regions and noise immunity are measured.

4-7 Integrated Circuit Gates

Many of the currently practiced techniques of semiconductor fabrication involve a series of diffusion of dopants and chemical etching on selected areas of a silicon crystal "chip." These techniques allow for hundreds of transistors and/or diodes to be made on a single chip. Many transistors and diodes can then be interconnected right on the chip to form more complex circuit units called "integrated circuits." Resistors can also be made by careful control of doping area and depth. The doping and etching operations are carried out first to make the desired array of transistors, diodes, and resistors. Then an oxide insulating layer is put over the chip. This is followed by selective etching through the insulation

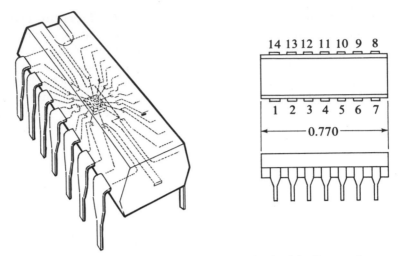

Fig. 4-23　An integrated circuit, plastic dual-in-line package.

where the connections to the components are to be made. Next aluminum is deposited on the chip in the form of the required interconnections. The chip is then mounted in a metal or plastic body and connections are welded from the connector pins of the body to aluminum pads on the interconnection pattern where the external connections are to be made. Finally, the body is sealed and the device is given a final check. An example of an integrated circuit is shown in Fig. 4-23.

Integrated circuits offer many advantages: (1) substantial reduction in size and often cost over equivalent discrete component circuits; (2) higher speed circuitry due to much shorter leads; (3) many fewer, externally wired connections and components; (4) improved circuitry due to the ability to use more transistors per function than is practical with discrete component circuits. In principle, there is almost no limit to the complexity of circuit which can be made as an integrated circuit. In practice, the complexity is limited by the number of elements that can be combined and still give a reasonable yield of operating devices and the number of external connections that have to be made.

As semiconductor technology has improved the number of active devices and components which can be economically put on a single silicon chip has increased greatly. Several complete gates of considerable complexity are now available in a single tiny package as are many prewired combinations of gates that perform special functions. Indeed, it is the availability of such versatile circuits in convenient and inexpensive form which has sparked a new revolution in scientific instrumentation and many other areas of electronics.

TTL Gates

Available in integrated circuit form are the DTL, RTL, and RCTL gates described in the previous section in many variations. Along with the development of these circuits also came the realization that some circuit improvements were possible and economical with IC's which were not practical with discrete components. One such example is the TTL gate shown in Fig. 4-24. Q_1 is a multiple-emitter transistor. Grounding any one or more inputs forward biases transistor Q_1 which puts the collector of Q_1 at a low potential and turns Q_2 OFF. This, in turn, turns Q_4 ON and Q_3 OFF and results in a logic 1 output. If, however, all inputs are HIGH (or unused) the base-collector junction of Q_1 will conduct, forward-biasing the base-emitter junction of Q_2. When Q_2 is ON, Q_3 is ON and Q_4 is OFF resulting in a 0 at the output. This corresponds to the inverted AND or NAND function.

The circuit is actually very similar to the DTL gate with Q_1 replacing the gate and coupling diodes. Q_4 is a dynamic collector resistor for Q_3 resulting in a larger current capability in the 1 state, and Q_2 provides gain for increased speed and noise immunity. Since the connecting leads inside the integrated circuit (IC) gate are extremely short, lead capacitance and inductance are minimized and very high speed is possible. The standard TTL gate has only a 13-nsec propagation delay with still

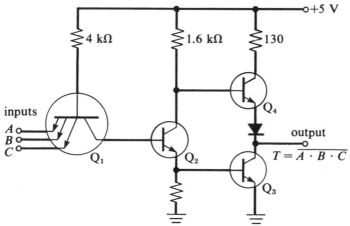

Fig. 4-24   TTL positive NAND gate.   (Courtesy Texas Instruments.)

faster TTL gates available for special purposes. The fan-out is about 10 and the noise immunity is at least 1.0 V, making them a very convenient general purpose gate.

ECL Gates

In some applications, the switching speed or propagation delay of the gate is a very important factor. There are always needs for counters that count faster and computers that compute faster. One of the limitations on the maximum switching speed of all the transistor gates described thus far is the storage time of the saturated transistor. Several gate designs have been made which eliminate this delay by not allowing the transistor to saturate. One of the most successful of these circuits, the emitter-coupled logic or ECL, is made in integrated circuit form. The circuit of a representative ECL gate is shown in Fig. 4-25.

Transistors Q_3 and Q_4 are in the nonsaturating current-steering switch circuit of Fig. 3-26. The potential at the base of Q_4 is -1.175 V. If inputs A, B, and C are all at logic **0** (-1.5 V) Q_1, Q_2, and Q_3 will be cut off. Q_4 then acts as an emitter follower producing $-1.175 - V_{BE} \approx -1.9$ V at the emitter. The current through Q_4 and R_2 is nearly equal to the current through R_3. Therefore, the voltage across R_2 is (300/

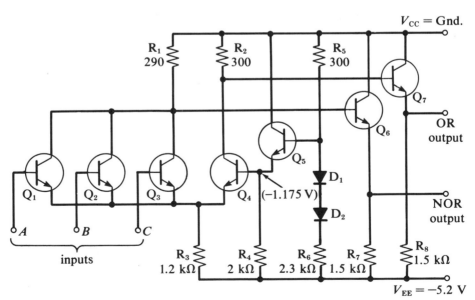

Fig. 4-25   ECL gate circuit.   (Courtesy Motorola.)

1.2 kΩ)(5.2–1.9) = 0.82 V. Thus the voltage at the base of emitter follower Q_7 is -0.82 V and the OR output voltage is -0.82 V $- V_{BE} \approx -1.5$ V (a logic **0**). Since Q_1, Q_2, and Q_3 are cut off, there is almost no current through R_1 and the voltage on the base of emitter follower Q_6 is nearly 0 V. Thus, the NOR output voltage is $0 - V_{BE} \approx -0.75$ V (a logic **1**). The -1.175-V fixed bias voltage at the base of Q_4 is obtained from the voltage divider R_5 and R_6 and the emitter-follower amplifier Q_5. Diodes D_1 and D_2 are to compensate for the temperature dependence of the base-emitter junction potentials of Q_5 and Q_4.

If one or more of the inputs A, B, or C is raised to a logic **1** (-0.75 V) the resulting increase in current through R_3 raises the emitter potential of Q_4 cutting that transistor off. The emitters are now at $-0.75 - V_{BE} \approx -1.5$ V. The forward bias on Q_4 is not sufficient for it to conduct appreciably resulting in 0 voltage drop across R_2 and ≈ -0.75 V (logic **1**) at the OR output. The IR drop across R_1 is (290/1.2 kΩ)(5.2–1.5) = 0.77 V resulting in ≈ -1.5 V (logic **0**) at the NOR output. Thus the outputs are the stated logic function of the input levels. The great speed of the ECL gate is due to the elimination of the saturated state for the transistors and to the relatively small logic level swing required to go between **1** and **0** (~ 1.5 V).

The current-steering switch naturally provides differential or "complementary" outputs. The gate designers have taken advantage of this to provide both OR and NOR outputs. Gates of this type are also sometimes called "current-steering" logic (CSL) and "current-mode" logic (CML).

FET Gates

Gates made with field-effect transistors offer the advantages of high input impedance, low power consumption, greater circuit simplicity, and further reduced size. With these advantages goes a disadvantage — significantly slower speed due to high input capacitance and large logic level swings. One version of an FET gate is shown in Fig. 4-26. The FET's used are insulated-gate, p-channel, enhancement mode FET's. No other components are required. Transistor Q_4 acts as a dynamic load impedance. A negative voltage at A or B or C will cause Q_1 or Q_2 or Q_3 to conduct making the output near 0 V. Only if all three inputs are 0 V will the output be negative. Thus, this gate acts as a NOR gate for negative logic.

FET logic circuits have been manufactured in more complex integrated circuits than other logic forms. Integrated circuits containing

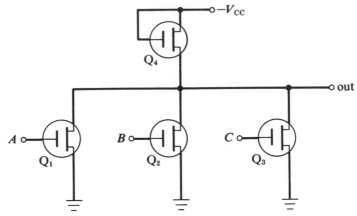

Fig. 4-26 MOSFET NOR gate.

hundreds of gates are called LSI for "large-scale integration." The term "MOS gates" is also frequently used for these circuits after the type of FET's used.

4-8 Applications of Logic Gates

This section contains some information which is useful in the selection and application of logic gates. The logic functions and gate types and characteristics are summarized. Then a short section is devoted to the use of NAND/NOR Gates. A few examples of gate circuits are presented at the end of this chapter. Of course, many more applications of this most basic circuit will be found throughout the remainder of the book.

Summary of Logic Functions and Gate Types

Logic functions. Seven basic logic functions have been described in this chapter: AND, OR, INVERT, NAND, NOR, EXCLUSIVE-OR, and equality comparator. The truth table for all these functions (except INVERT) is summarized in Table 4-11. The interrelationships among the functions can been seen by comparing the output columns. The NAND is obtained by inverting the AND output. The OR function is the NAND column upside down which can be obtained by inverting the inputs A and B. Of course, the NOR is the inverse of the OR output. The AND function would result if the inputs to the NOR gate were in-

Table 4-11 Comparison of Six Basic Logic Functions

Inputs		Outputs					
A	B	AND $A \cdot B$	NAND $\overline{A \cdot B}$	OR $A + B$	NOR $\overline{A + B}$	EXCL.-OR $A \oplus B$	Eq. comp. $\overline{A \oplus B}$
0	0	0	1	0	1	0	1
0	1	0	1	1	0	1	0
1	0	0	1	1	0	1	0
1	1	1	0	1	0	0	1

verted, and so on. The EXCLUSIVE-OR is not obtainable from the AND or OR functions by simple inversion of inputs or outputs. It is in a different family from the AND/OR group as its uniquely symmetrical truth table shows. Some people consider the EXCLUSIVE-OR to be the third basic logic gate function (after AND/OR and INVERT). It has been proven that the equality comparison gate is the inverse of the EXCLUSIVE-OR. However, it is interesting to note here that the equality can also be obtained from the EXCLUSIVE-OR by inverting either one of the inputs. This can be seen in the truth tables or proven algebraically.

Gate types. The currently available gate types and their typical characteristics are summarized in Table 4-12. Each of the currently popular types has some advantages over the others in terms of cost, speed, power requirements, noise immunity, etc. It is not possible to state categorically, then, that one type is better than another without specifying the application requirements. In larger instruments or systems, it may be desirable to use different types in different parts of the instrument. For instance, in a counter one may want to use ECL in the first decades of counting for highest speed, then change to DTL for economy or to TTL for the convenience of the integrated circuit DCU packages.

Gate designs having the same logic levels and interconnection types are electrically compatible and may be interconnected directly. One example of compatible types is the DTL and TTL gates which are both of the current-sinking type. (The gate output must "sink" current from gate inputs to maintain a **0** level input.) When gates of different interconnection type or logic levels are to be used together, special interfacing circuits are generally required. In a system where a number of functions are to be performed, gate types may be chosen on their versatility and compatibility to avoid the problems and circuitry of interfacing

Table 4-12   Gate Types and Characteristics

Characteristics	Gate Types				
	RTL	DTL	TTL	ECL	FET
Basic logic function	+NOR	+NAND	+NAND	+OR, NOR	−NOR
Interconnection type	Current sourcing	Current sinking	Current sinking	Current mode	Voltage
Logic levels	0.2 V = 0 1.6 V = 1	0.2 V = 0 3.0 V = 1	0.2 V = 0 3.3 V = 1	−1.6 V = 0 −0.75 V = 1	−2 V = 0 −10 V = 1
Fan-out	5	8	10	25	5
Propagation delay	25 nsec	25 nsec	13 nsec	3 nsec	~0.5 μsec
Noise immunity, mV	~200	750	1000	~200	
Supply voltage, V	+3.6	+5	+5	−5.2	−20
Primary advantages	Low cost	Moderate cost Package variety	High speed Noise immunity High fan-out, Versatile Package variety	Highest speed Package variety	Low powers Large-scale integration

disparate gate types. The DTL/TTL group would appear to have the edge in this regard at the present time, and the FET is developing very rapidly.

NAND/NOR Gate Implementation

Direct substitutions. The use of the NAND and NOR gates can be just as easy as the AND, OR, and INVERT gates discussed earlier. A table of logic functions using NAND and NOR gates such as Table 4-13

Table 4-13   Logic Functions Using NAND and NOR Gates

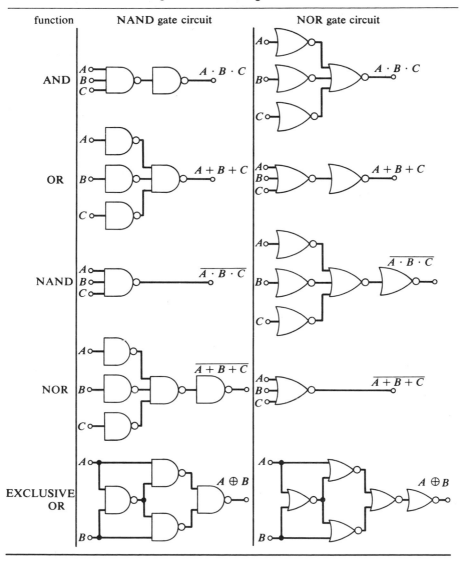

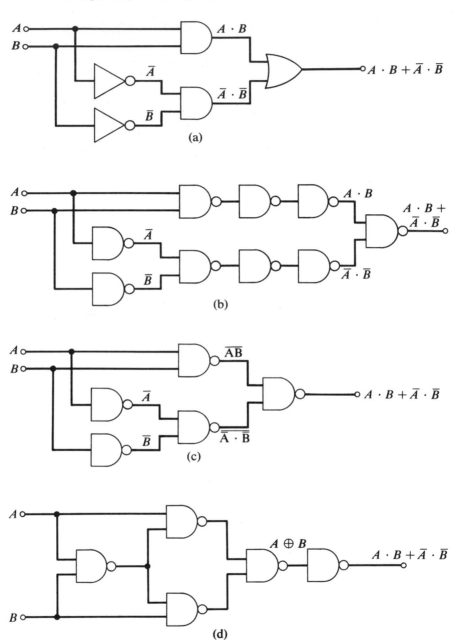

Fig. 4-27 Implementations of $T = A \cdot B + \bar{A} \cdot \bar{B}$. (a) AND, OR gates. (b) NAND gates, direct translation. (c) NAND gates, simplified. (d) EXCLUSIVE-OR, inverted.

is often helpful. These circuits can be used directly in place of the AND and OR gates in a conventional logic diagram. Consider the implementation of the coincidence or equality expression $T = A \cdot B + \bar{A} \cdot \bar{B}$. This is shown with AND and OR gates in Fig. 4-27a. A direct translation into NAND gates using Table 4-13 results in the circuit of Fig. 4-27b. Inspection of this circuit reveals two inverters in a row at the AND outputs and OR inputs. Inverting twice clearly results in no change so those inverters can be eliminated as shown in Fig. 4-27c. Another approach is to invert the output of the EXCLUSIVE-OR function to obtain the equality function as shown in Fig. 4-27d. The NAND and NOR gate implementations of the EXCLUSIVE-OR function will be described later.

Experiment 4-11   *TTL NAND Gate*

The input and output logic level regions are determined for an integrated circuit TTL NAND gate. The logic function is verified and a gate delay time measurement is suggested.

Equivalent symbols. It is also sometimes helpful to draw logic diagrams with the DeMorgan's theorem equivalent gates. These equivalents are shown for the NAND and NOR gates in Fig. 4-28. Since $\overline{ABC} = \bar{A} + \bar{B} + \bar{C}$ according to DeMorgan's theorem, the NAND gate which produces $\overline{A \cdot B \cdot C}$ from A, B, and C at the input is exactly equivalent to an OR gate with an inverter at each input. A similar argument holds for the NOR equivalent. Thus when an OR function is to be implemented with NAND gates, the equivalent OR gate symbol can be used and additional inverters inserted where necessary.

To implement the function $T = AB + \bar{A}\bar{B}$ with NAND gates by this

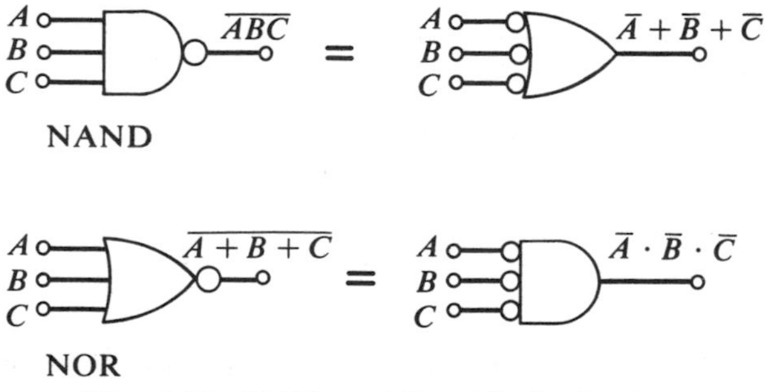

Fig. 4-28   DeMorgan's equivalent gates.

approach, use NAND gates to perform the indicated AND operations, then draw the NAND-equivalent OR symbol to perform the OR operation. This is shown in Fig. 4-29a. Now the inversions must be checked. The inversions at the outputs of gates 1 and 2 are reinverted at the inputs to gate 3. Thus the AND function AB from gate 1 is ORed at gate 3 as desired. If inversion were desired, an inverter would be added to the circuit.

The function $AB + \bar{A}\bar{B}$ can be implemented with NOR gates in the same way. It is convenient to convert the AND-then-OR function $AB + \bar{A}\bar{B}$ to the equivalent OR-then-AND expression $(A + \bar{B})(\bar{A} + B)$. The OR functions are drawn with NOR gates symbols and the AND with the NOR-equivalent AND symbol as shown in Fig. 4-29b. Again, the inversions must be checked.

Equivalent symbols for the NOR and NAND gates are also frequently useful in analyzing gate circuits. The NAND implementation of the EXCLUSIVE-OR circuit of Table 4-13 will be used as an example. Applying the NAND function to each gate input as the circuit is drawn, the output is $(\overline{\overline{AB}A})(\overline{\overline{AB}B})$ which will require some algebraic manipulations before being recognized as equivalent to $A \oplus B$. Now the circuit is redrawn using some NAND-equivalent OR symbols as shown in Fig. 4-30. Gates 1 and 4 are drawn as OR gates to minimize the number of inversions between gates, that is, to try to use a non-inverting output to connect to a noninverting input and an inverting output to connect to an inverting input. Where this can be done, the inversions cancel and can be ignored. Gate 1 is an OR gate operating on $\bar{A}$ and $\bar{B}$ to produce $\bar{A} + \bar{B}$. Gates 2, 3, and 4 can be treated as ordinary AND and OR gates yielding $A(\bar{A} + \bar{B}) + B(\bar{A} + \bar{B})$ at the output. A one-step application of the absorption theorems yields $A\bar{B} + \bar{A}B$, the EXCLUSIVE-OR function.

When this circuit is used as a half-adder, the output of gate 1 is $\overline{AB}$ or NOT CARRY. It is suggested that the reader show that the full-

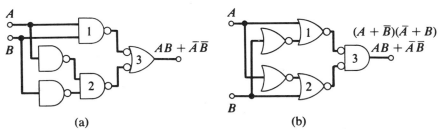

(a)   (b)

Fig. 4-29 Implementation of $AB + \bar{A}\bar{B}$ using equivalent gate symbols. (a) NAND gates. (b) NOR gates.

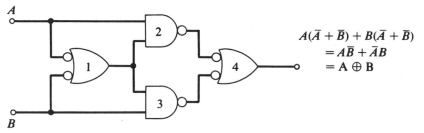

$$A(\bar{A} + \bar{B}) + B(\bar{A} + \bar{B})$$
$$= A\bar{B} + \bar{A}B$$
$$= A \oplus B$$

Fig. 4-30   EXCLUSIVE-OR using NAND gates.

adder of Fig. 4-14b can be wired using two of the NAND gate EXCLU-SIVE-OR circuits described above and *one* additional NAND gate. The NAND-equivalent OR gate concept is useful in this application.

Experiment 4-12   *Binary Full-Adder*

Two half-adder circuits are wired, tested, and combined to make a full-adder. A third half-adder can be wired, if desired, to make an operating two-bit binary adding circuit.

Signal Selector Gate

It has been shown in this chapter that the logic gate acts as a logic signal-actuated switch for logic level signals. The switching action is analogous to that of a single-pole, single-throw (SPST) switch as shown in Fig. 4-31.[1] If several logic signals are to be gated ON and OFF simultaneously (analogous to the multiple pole switch), several different gates are connected to the same control signal.

The input of a logic circuit such as a counter might have to be switched from one signal source to another to accommodate various desired applications. (The need for such a switching capability is clearly demonstrated by Figs. 1-13 to 1-18 and Table 1-2 for the general purpose counter-timer.) This requires the gate analog of a multiple-throw or *selector* switch as shown in Fig. 4-32. The outputs of the four signal-switching gates are combined in an OR operation to provide a single output. This is drawn using NAND gates in Fig. 4-32. Notice the convenience of using the NAND-equivalent OR symbol in this diagram.

[1] A difference in terminology for switch and gate circuits should be noted. An OPEN switch *prevents* the passage of the signal whereas an OPEN gate *allows* the signal to pass. Similarly, the term CLOSED has opposite meanings for switch and gate circuits.

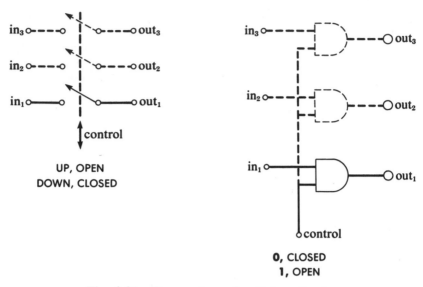

Fig. 4-31   Comparison of switch and gate.

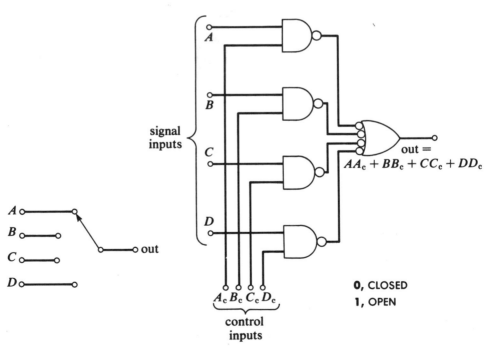

Fig. 4-32   NAND gate circuit for signal selection.

According to the output function, the state of the control inputs A_c, B_c, C_c, and D_c determine which of the input signals A, B, C, and D will appear at the output. In practice, only one control input will be **1** at a time. For instance, if $A_c = $ **1** and B_c, C_c, and D_c are all **0**, the output function is A regardless of the states of the signal inputs. Thus the input signal that appears at the output depends on which of the control inputs is **1**.

This circuit can be extended to any number of inputs and can also be implemented with NOR gates or with the AND-OR-INVERT gate package as shown in Fig. 4-33. For the NOR gate circuit, a **0** at a control input is required to OPEN a gate. If A_c, B_c, and D_c are **1** and C_c is **0**, the output will be C. The AND-OR-INVERT gate circuit of Fig. 4-33b is exactly the same as the NAND gate circuit of Fig. 4-32 except for the inverted output. If B_c is the only control input at logic **1**, the output will be $\bar{B}$. The B signal is still gated through, but it comes out inverted.

Binary Decoder

In the four-input selection gate applications above, there were only four allowable combinations of logic levels at the control inputs, **1000,**

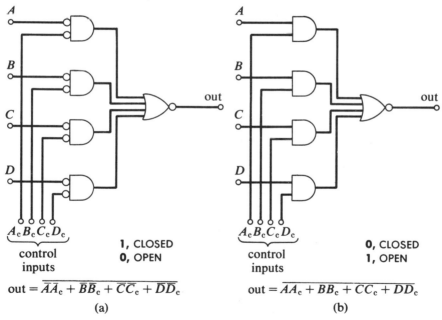

out $= \overline{A\bar{A}_c + B\bar{B}_c + C\bar{C}_c + D\bar{D}_c}$

(a)

out $= \overline{AA_c + BB_c + CC_c + DD_c}$

(b)

Fig. 4-33   Signal selection gate circuits.   (a) NOR gates.   (b) AND-OR-IN-VERT gate.

0100, 0010, and **0001.** Of course, only two logic level signals (bits) are required to give four distinguishable states, **00, 01, 10,** and **11.** It is a simple matter to design a logic circuit that will convert the four states of a two-wire input into the four signals required by the control inputs of the signal selection gates. Such a circuit is shown in Fig. 4-34. Four logic gates are connected so that one of them will have a **1** output for each of the four possible input states of signals X and Y. The use of this circuit to operate the control inputs of Figs. 4-32 and 4-33 reduces the number of control connections and eliminates the possibility of more than one control input being **1** at once.

The P's at the gate outputs of Fig. 4-34 are a short-hand notation for the four possible states of X and Y. It is obtained by considering X and Y to be two digits of a binary number, where X is read as **1** and $\overline{X}$ is read as **0** and similarly for Y. Thus $\overline{X}\overline{Y}$ is 00 = 0 and is symbolized by P_0. Similarly, $X\overline{Y} = (10)_2 = (2)_{10} = P_2$. This circuit thus "decodes" the binary input into a separate output for each number. If there were three inputs, there would be eight combinations. Thus, eight three-input gates would be required for complete decoding, and eight outputs (P_0 through P_7) would be produced. Other decoders, including an eight-level decoder, are described in Chapter 6.

It is suggested that the reader design a circuit which will encode from four inputs (only one of which can be **1** at a time) to two binary outputs.

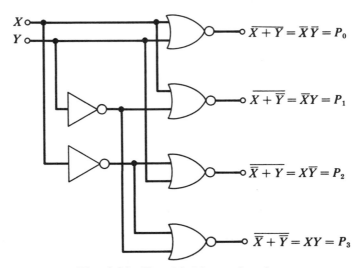

Fig. 4-34   Two-bit binary decoder.

P and S notation. The P notation introduced above is used for logic quantities combined in the AND function. There are also four possible combinations of X and Y with the OR function: $\overline{X} + \overline{Y}$, $\overline{X} + Y$, $X + \overline{Y}$, and $X + Y$. These are called S_3, S_2, S_1, and S_0, respectively. Note that for the S terms, $\overline{Y} = 1$, and $Y = 0$. It can be seen from DeMorgan's theorem, $\overline{AB} = \overline{A} + \overline{B}$, that $P_0 = \overline{S}_0$, or in general terms,

$$P_i = \overline{S}_i \qquad (4\text{-}21)$$

The P and S notation is often convenient in dealing with complex functions. A very helpful theorem for quantities that can be written with this notation states that if a function is in the form $P_i + P_j + \cdots$ (sums of products form), it can also be written in the form $S_a \cdot S_b \cdots$ (products of sums form) where the S terms are all those that did *not* appear in the equation with P's. Taking the EXCLUSIVE-OR for an example again,

$$X \oplus Y = \overline{X}Y + X\overline{Y} = P_1 + P_2 \qquad (4\text{-}22)$$

There are four possible P terms and four possible S terms for two variables. The sum of products expression contains P_1 and P_2. According to the above theorem, the equivalent product of sums expression contains S_0 and S_3. Thus,

$$\overline{X}Y + X\overline{Y} = P_1 + P_2 = S_0 S_3 = (A + B)(\overline{A} + \overline{B}) \qquad (4\text{-}23)$$

which provides an easy way of transforming equations from one form to the other.

Equality Detector

The equality circuit of Fig. 4-27 gives a true output if the A and B inputs satisfy the equation $AB + \overline{A}\overline{B}$, that is, $A = B$. This circuit can be thought of as an equality detector for the one-bit binary signal sources A and B. In computing and digital instrumentation, it is often necessary to compare binary signal sources of more than one bit to detect equality. The circuit which will accomplish this simply determines whether equality is established for each of the separate bits simultaneously, that is $A_1 = B_1$ AND $A_2 = B_2$ AND $A_3 = B_3$, and so on. An AND circuit combines the outputs of all one-bit equality detectors as shown in Fig. 4-35.

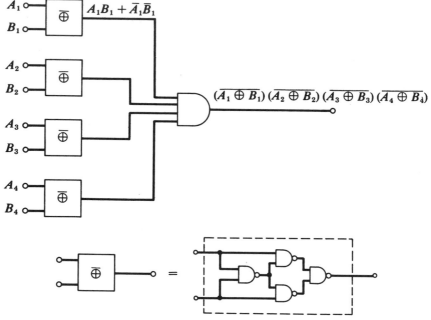

Fig. 4-35   Equality detector.

chapter five

Flip-Flops and
Multivibrators

The logic gates described in the previous chapter are the "decision-making" elements of a digital instrument or computer. The response to every command or event requires a decision. In a counting operation, for example, the circuit will make a decision if an event occurs, e.g., if the current count is six, change the count to seven. Or in an adding circuit; if one input is **1**, and the other input is **0**, and if the carry from adding the previous column is **0**, then make the sum output **1** and the carry output **0**. Decision-making elements are thus used to give the desired results for every possible combination of input states.

In both of the above examples, more than the input information was required for the decision. The availability of information about the results of previous counts or additions implies the existence of memory or storage units to keep this information until needed. The storage unit is a device with two stable states and at least two connections, one to put it in the desired state, and the other to read out the stored information. The basic electronic digital storage unit is the bistable multivibrator. This circuit can be made simply with two transistors as shown in Section 5-1, or in more complex forms with inverting logic gates as shown in Sections 5-2 and 5-3. The monostable and astable multivibrators, which are simple variations of the basic multivibrator circuit are described in Sections 5-4 and 5-5. Finally, another very important bistable circuit, the Schmitt trigger is described in Section 5-6.

These circuits complete the list of basic digital circuits. Subsequent chapters are devoted to the applications of these circuits in digital instrumentation and computation.

5-1 The Transistor Bistable Multivibrator

The ideal bistable electronic storage circuit has two stable, easily distinguishable, electronic states and can be easily, but not accidentally, switched from one state to the other. It is the electronic analog of the spring-loaded toggle switch. Bistable memory devices have been made of virtually every type of 3- and 4-terminal switching device described in Chapter 3. The most common circuit uses the transistor switch and is described in this section.

A Transistor Bistable Switch

The two most easily recognized states of the transistor switching circuit are, of course, ON and OFF. The common-emitter transistor switching circuit described in Chapter 3 is shown below in Fig. 5-1. Recall that for the switch to be ON (saturated, ≈ 0 V output) the input must be positive, and to be OFF (cutoff, $\approx +V_{CC}$ output) the input must be 0 V or negative. Some additional circuitry should be added to this simple switch to make it stable in these two states and unstable in all states between cutoff and saturation. Such a circuit would sense the switch output level and apply a signal to the switch input which will keep it in that state. That is, when the output is 0 V, the input should be positive and vice versa. The switching circuit itself has this inverting characteristic as described in Chapter 3. Thus a second switching circuit (Q_2) can be used to make the first transistor (Q_1) a bistable switch as shown in Fig. 5-2. When Q_1 is saturated, its 0-V output level is sensed by Q_2 which is then cut off. The $+V_{CC}$ level at the Q_2 output is connected to the Q_1 input assuring that Q_1 will remain saturated. Similarly, when Q_1 is OFF, its posi-

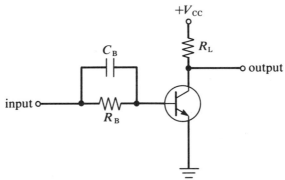

Fig. 5-1   Transistor switching circuit.

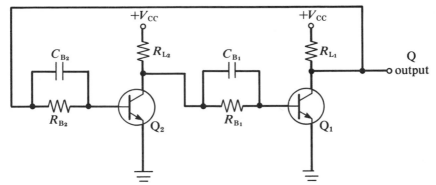

Fig. 5-2 **Transistor bistable switch.**

tive output will turn Q_2 ON, which will apply 0 V to the Q_1 input to keep Q_1 OFF. That the circuit is not stable between cutoff and saturation can be seen by postulating that both Q_1 and Q_2 are in a partially conducting state (in the amplification region of the transistor characteristics). An infinitesimally small change in potential anywhere in the circuit, e.g., at the collector of Q_2 will be amplified and inverted by Q_1, then further amplified and inverted again by Q_2 causing a greater change in the same direction. This change, thus, continues and increases until the transistors are out of the amplification region, i.e., saturated or cutoff. Note that in the two stable states of the circuit of Fig. 5-2, the transistors are always in opposite states of conductance.

A Bistable Memory Circuit

The transistor bistable switch is redrawn in Fig. 5-3. Note that this circuit is basically the same as that of Fig. 5-2, except that inputs have been added so that it can function as a memory. The circuit has been drawn to demonstrate the symmetry of the Q_1 and Q_2 circuits and a second output at the Q_2 collector is shown. Since Q_1 and Q_2 are always in opposite states, the logic levels at the two outputs are always opposite; thus the Q_2 output is often called "NOT Q" and symbolized by $\bar{Q}$.

The dc voltages in the circuit of Fig. 5-3 are easily determined as follows. Assume that Q_2 is saturated making the potential at $\bar{Q}$ approximately 0.2 V. The 10-kΩ resistor between the collector of Q_2 and the base of Q_1 and the 3.3-kΩ resistor from the base of Q_1 to ground are a voltage divider determining the potential on the base of Q_1 which is 0.2 V $\times$ 3.3 kΩ/(10 kΩ + 3.3 kΩ) = 0.05 V. This base voltage is too low to turn on the base-emitter junction of Q_1 and assures that Q_1 is cut off. The base current of Q_1 in cutoff is very small and will not affect the previ-

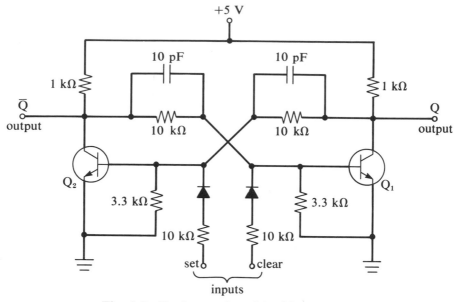

Fig. 5-3   Basic transistor bistable memory.

ous calculation of the base voltage. With Q_1 cut off, the only significant current through the collector resistor of Q_1 is through the voltage divider which determines the Q_2 base voltage. A first calculation of the Q_2 base voltage, neglecting Q_2 base current is [5 V × 3.3 kΩ/(3.3 kΩ + 10 kΩ + 1.0 kΩ)] = +1.2 V. Clearly the base-emitter junction of Q_2 would be forward biased and the base current of Q_2 cannot be neglected. The potential drop across a forward-biased silicon pn junction is about 0.6 V. Using this value as the potential on the base of Q_2, the potential on the collector of Q_1 can be calculated:

$$[(5 \text{ V}{-}0.6 \text{ V})(10 \text{ k}\Omega)/(1.0 \text{ k}\Omega + 10 \text{ k}\Omega)] + 0.6 \text{ V} = +4.6 \text{ V}.$$

For Q_2 to be in saturation, the base current must be equal to, or greater than I_C/h_{FE} (Eq. (3-14)). From the collector voltage and load resistance, $I_C = 4.8$ V/1 kΩ = 4.8 mA. From the Q_2 base voltage, the coupling resistor and the collector load resistor of Q_1, I_B of $Q_2 = (5.0 - 0.6)/10$ kΩ + 1.0 kΩ) − (0.6/3.3 kΩ) = 0.22 mA. Therefore, any transistor with an h_{FE} of 22 or more will be saturated with this circuit.

From the previous analysis and the symmetry of the circuit, it can be seen that, in a stable state, one transistor is saturated and the other is cut off and that the logic 1 output is +4.0 V and the logic 0 output is 0.2 V.

It is desirable for the base current to be from 1.4 to 2.5 times the minimum calculated from the h_{FE} as above,[1] so that the resistance values of Fig. 5-3 are optimum for a transistor with a h_{FE} of 30 to 55. To accommodate transistors with a higher h_{FE}, the collector-to-base coupling resistors could be increased or the base resistors could be connected to a negative supply instead of ground. The first approach reduces the current in the coupling circuit and increases the logic 1 output voltages. The second approach, which increases the fraction of the coupling circuit current diverted from the saturated base, is sometimes also used when it is desirable to put the OFF transistor further into the cut-off region.

Inputs to the bistable memory. For operation as a memory, the circuit of Fig. 5-3 is set in the desired state which can later be referred to or "read" at any time by noting the logic level at Q or $\overline{Q}$. The state of the binary memory is determined by the application of a logic 1 signal (+4.0 to +5.0 V) to the set or clear input. For instance, if a +4.0-V signal is applied to the clear input (a 10-kΩ resistor and diode connection to the Q_1 base), Q_1 will turn ON (or stay ON) and the Q output level will be **0**. This is called "clearing" the binary. Remember that when Q_1 turns ON, Q_2 must turn OFF. Similarly, the binary is "set" by applying a logic 1 to the set input which turns Q_2 ON and Q_1 OFF thus producing a 1 output at Q_1. After a set or clear signal has been applied, the binary will remain in that state until commanded to change by a signal at the opposite input. The simultaneous application of 1-level signals to set and clear inputs would make both outputs **0**. This situation would be avoided in normal operations. In summary, then, dc inputs to the memory circuits are usually made to the transistor bases and the applied positive signal causes the transistor with the OFF base to turn ON.

Experiment 5-1 *Basic Transistor Bistable Memory*

The circuit of Fig. 5-3 is wired and tested. Voltage measurements are made to confirm the dc analysis and the set and clear inputs are used to verify their effect. The Q and $\overline{Q}$ states are monitored with a meter or more conveniently with a light- and logic-actuated light driver circuit.

Of course, there are many other ways to change the state of the memory flip-flop. A negative signal could be applied to the ON base to turn it OFF. Set and clear signals are sometimes applied to the transistor collectors. One such circuit is shown in Fig. 5-4. When a positive

[1] For an excellent discussion of transistor bistable circuit design, see Louis Delhom, *Design and Application of Transistor Switching Circuits,* Chapter 12, McGraw-Hill, New York, 1968.

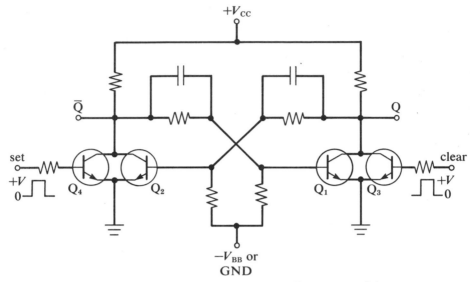

Fig. 5-4 Set-clear flip-flop with pull-over transistors.

voltage is applied to the set input the "pull-over" transistor Q_4 goes ON which brings the collector voltage of Q_2 down thus turning Q_1 OFF. With **0** at the set and clear inputs, Q_4 and Q_3 are cut off and have no effect on the state of the memory circuit. The pull-over transistors are used to provide the higher signal currents which are required to force the collector of Q_1 or Q_2 from one state to the other.

Alternating current inputs. In many applications, especially counting, it is desirable to have a binary memory SET or CLEAR as a result of an input signal transition, such as from logic **1** to logic **0,** rather than by a dc signal level. This can be accomplished by RC coupling to the set and clear inputs through a diode, as shown in Fig. 5-5.

The leading edge (0 to $+V$ transition) of the set input, for instance, is differentiated by C and R and the resulting positive pulse passes diode D_2 to turn Q_2 ON. The trailing edge has no effect since the coupling diode is reverse biased.

Clocked RS Flip-Flop

A classic flip-flop circuit is shown in Fig. 5-6. In this circuit, a clock pulse is diode-capacitor coupled to the transistor base through C_1 and D_1 to Q_1 and through C_2 and D_2 to Q_2. Whether the clock pulse

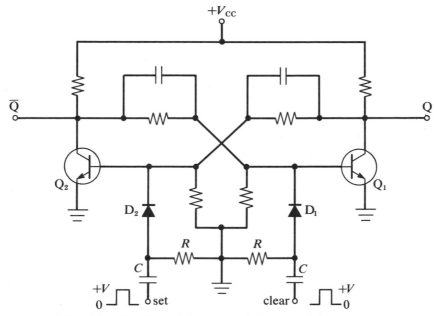

Fig. 5-5 Flip-flop with ac-coupled set and clear inputs.

is effective in turning Q_1 or Q_2 OFF depends upon the potential at points R and S. Figure 5-7 illustrates the effect of the logic level at the steering input R or S upon the clock pulse. In Fig. 5-7a the steering input is grounded, putting point P at ground potential. The clock pulse is differentiated by the small capacitor. The diode D_1 or D_2 blocks the positive pulse from the transistor base while D_3 or D_4 shunts it to ground. When the clock pulse falls, a 4-V negative pulse results which is passed by D_1 or D_2 to the transistor base turning the transistor OFF.

If the steering connection is connected to a logic **1** as shown in Figure 5-7a, point P is at logic **1** potential. Again, the positive pulse resulting from differentiating the clock waveform is blocked by $D_{1(2)}$ and clipped by $D_{3(4)}$, while point P remains at +4 to 5 V. When the clock pulse returns to **0,** the 4-V drop in potential at P cannot make P negative. Therefore D_1 or D_2 cannot conduct and the transistor base potential is unaffected. Thus, a **0** at the steering input prior to the clock pulse causes the clock pulse to turn the transistor OFF. The clock pulse will have no effect on a transistor with a **1** at the steering input.

Assume that R of Fig. 5-6 is at logic **1** and S is **0.** The trailing edge of the trigger signal is thus steered to turn OFF Q_1 by the **0** at S while leaving Q_2 unaffected because of the **1** at R. When Q_1 is OFF, Q is at logic **1** and the flip-flop is set.

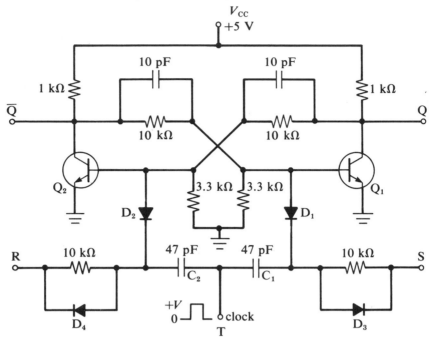

Fig. 5-6 Clocked RS flip-flop.

Similarly the flip-flop could be cleared by reversing the logic levels at S and R and directing the turn-off pulse to Q_2. If both S and R are at logic **1**, neither base will receive a turn-off pulse and the state of the flip-flop will remain unchanged. On the other hand, if both R and S are at **0**, the turn-off pulse is directed to both transistors and the result is uncertain or indeterminate. The four possible combinations of R and S levels are summarized in the truth table of Fig. 5-6 where t_n represents the logic

Truth Table for Fig. 5-6

Inputs, t_n		Output, t_{n+1}
R	S	Q
0	0	Indeterminate
0	1	0
1	0	1
1	1	Q_n (Q at t_n)

levels at R and S before the trailing edge of T, and t_{n+1} represents the logic levels of Q after the trailing edge of T. The notation Q_n means that Q at time t_{n+1} is the same as it was at time t_n.

The memory circuits of Figs. 5-3, 5-4, and 5-5 changed state when the appropriate pulses or transitions occurred at their set and clear inputs. There is an important timing difference in the action of the triggered flip-flop of Fig. 5-6. In this case the information directing the desired change is present at the R and S inputs before the change takes place. A change in state can only occur when the trailing edge of the T signal occurs. Thus the changes are "timed" or clocked by the T signal. The T input is sometimes referred to as the clock connection.

If the set and clear (reset) inputs of Figs. 5-3, 5-4, or 5-5 are combined with the T input of Fig. 5-6, the result is a complete R-S/T flip-flop. Of course, transition commands would not be simultaneously applied to more than one input at once.

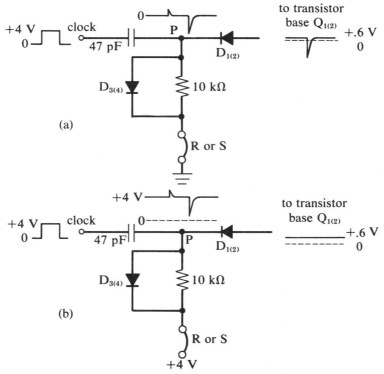

Fig. 5-7   Steering inputs of Fig. 5-6.

A "T" Binary

For simple binary counting and many other applications it is desirable to have a flip-flop that alternates or complements its output state upon each input trigger signal. The triggered flip-flop of Fig. 5-6 can accomplish this if the logic levels at R and S are opposite and alternated after each T pulse. For instance, if Q is 1, R and S should be 0 and 1, respectively, for Q to change on the next T pulse. After the transition, Q is 0 and R and S should change to 1 and 0 so that another reversal will occur on the next pulse. A convenient place to obtain the opposite and alternating signals for R and S is Q and $\overline{Q}$ of the flip-flop itself. If Q is connected to S and $\overline{Q}$ to R, the flip-flop will alternate state on each T input pulse. This action follows that of the "toggle-switch" analogy used in describing the counter circuits in Chapter 1. For this reason, this complementing action is sometimes called "toggling" and the T input in such cases is sometimes referred to as the "toggle input." When connected to an alternating signal source, the output frequency will be half that of the input signal source.

Many improvements have been made in this basic circuit to improve speed, versatility, and noise rejection. However, the best discrete component circuits are now no better than several types of integrated circuit flip-flops and are more expensive. Therefore it is no longer very significant to discuss these speed-up techniques.

Experiment 5-2   *The Triggered Flip-Flop*

The trigger input and steering networks of Fig. 5-6 are added to the memory circuit of the previous experiment. Q and $\overline{Q}$ are connected to logic-level light drivers to observe the flip-flop states. Inputs S and R are connected to logic-level switch outputs and T is connected to a low-frequency square-wave signal or a manually triggered pulse generator. The effect of the R and S levels on the outputs as stated in the truth table is confirmed. The toggling action is observed when S is connected to Q and R to $\overline{Q}$. The frequency and time relationship between the T signal and the Q output is studied using indicator lights or a dual-trace oscilloscope.

5-2 Logic Gate Memory Circuits

The Basic Memory

In the previous section, it was shown how a transistor bistable circuit was made from two cross-coupled inverter circuits. The circuit was then further embellished by additional inputs. Integrated circuit gates were shown in the previous chapter to be inverting circuits with multiple

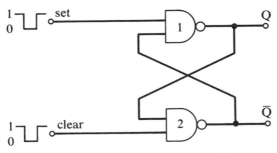

Fig. 5-8   Memory circuit using NAND gates.

inputs.   A bistable circuit can be made very simply, then, by cross-coupling two inverting gates as shown in Fig. 5-8.   Either NAND or NOR gates could be used.

Recall that with the NAND gate, the output will be **1** if any input is **0,** and that the output can be **0** only when all inputs are at logic **1** or unused.   Assume the clear and set inputs are **1.**   For either gate, then, a **1** at the cross-coupled input would produce a **0** at the output or vice versa. Since the output of one gate is the input of the other, the two outputs must be in opposite states, i.e., $Q = 1$ while $\overline{Q} = 0$ or $Q = 0$ while $\overline{Q} = 1$. Follow the logic levels through on the diagram to show that either of the above are stable states, and that Q and $\overline{Q}$ cannot both be **0** or **1** when clear and set are **1.**

If a **0** is applied to the clear input and a **1** to the set input, the output $\overline{Q}$ must become a **1,** and Q a **0.**   If the set input is **0,** and the clear input is **1,** then $Q = 1$ and $\overline{Q} = 0$.   By applying a momentary **0** level to the clear or set input, the output will become **1** or **0** as desired.   Since the output will remain in this state until a **0** is applied to the alternate input, the circuit "remembers" which input had the latest momentary **0** applied to it. In a typical application, the memory will be "cleared" by applying a momentary **0** to the clear input producing a **0** at the Q.   Then some se-

Truth Table for Fig. 5-8

Inputs		Outputs	
Clear	Set	Q	Q̄
0	**1**	**0**	**1**
1	**0**	**1**	**0**
0	**0**	**1**	**1**
1 (Last **0**)	**1**	**0**	**1**
1	**1** (Last **0**)	**1**	**0**

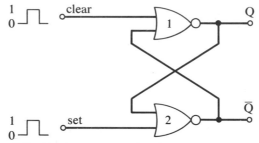

Fig. 5-9 Basic memory circuit using NOR gates.

Truth Table for Fig. 5-9

Inputs		Outputs	
Clear	Set	Q	$\overline{Q}$
0	1	1	0
1	0	0	1
1	1	0	0
0	0 (Last 1)	1	0
0 (Last 1)	0	0	1

quence of events will take place which might produce a **0** pulse at the set input to make Q = **1**. Later the Q output will be "read" to see if the events produced a set pulse, or not.

If **0** levels are applied to both set and clear inputs, both Q and $\overline{Q}$ outputs must be **1**. In a true bistable circuit this condition would be avoided.

NOR memory. Cross-coupled NOR gates form the basic memory circuit of Fig. 5-9. The NOR gate has a **0** output level if any input is at logic **1**. For this reason the truth table is very similar to the NAND gate memory except that the set and clear inputs are reversed and the Q and $\overline{Q}$ output levels are opposite. Notice that the NOR gate inputs are normally **0** between set and clear pulses.

Experiment 5-3 *Logic Gate Memory*

The basic NAND memory circuit of Fig. 5-8 or the NOR circuit of Fig. 5-9 is wired to verify the truth table. It is noted with indicator lights that the outputs are always in different states. Momentary **0** pushbutton outputs are connected to the set and clear inputs (momentary **1** outputs for the NOR circuit) to observe the alternation of memory state when the set and clear buttons are pushed alternately.

Gated Memory

The output of the basic cross-coupled NAND gate memory circuit indicates which input had a momentary **0** level last. By adding gates to the inputs of this circuit the memory can be made to respond to input levels only during a specified time interval, as shown in Fig. 5-10. As long as the clock input is at **0**, gates 3 and 4 have **1** at their outputs regardless of the set and clear input levels. Recall that with a **1** level at both inputs to the basic NAND gate memory, the output level depends on which input was **0** last. To allow new information to reach the basic memory, the clock input goes to a **1** level. A **1** at the set input and a **0** at the clear input will now result in a **0** input to gate 1, and a **1** input to gate 2, and Q will be **1**. If clear were **1** and set were **0**, Q would be **0**. If both set and clear are **0**, the inputs to the basic memory remain **1**'s and the output does not change from its previous state. When the desired input level sampling interval is over, the clock returns to **0** and further changes in set and clear levels will have no effect until the next clock pulse.

This memory circuit is said to have a "clocked" input because the sampling interval can be "timed" to coincide with the appearance of the

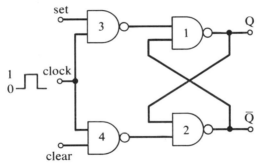

Fig. 5-10   Gated memory circuit.

Truth Table for Fig. 5-10

Inputs, t_n		Output, t_{n+1}
Clear	Set	Q
0	0	Q_n (Q at t_n)
0	1	1
1	0	0
1	1	Indeterminate

desired information at the set and clear inputs. The truth table for a clocked circuit is usually written as shown in Fig. 5-10 when the output state after the clock pulse depends on its previous state.

The set and clear input signals to the gated memory do not have to be pulses or momentary level changes as the basic memory required. In this case the input pulsing is accomplished by the clock signal. This is a very useful feature when set or clear inputs are obtained from other logic circuits.

Data latch. A frequently used variation of the gated memory circuit is the data latch circuit shown in Fig. 5-11. This circuit has a single data input and an inverting gate with the gated memory to apply the data level to the set input and its complement to the clear input. This confines the

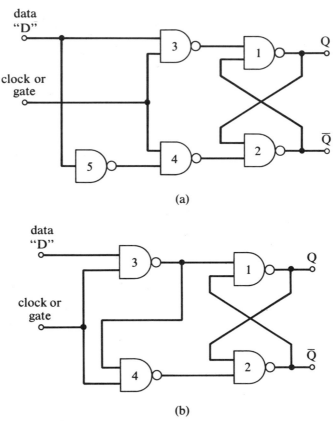

Fig. 5-11   Data latch: (a) Inverted input. (b) Minimized.

Truth Table for Fig. 5-11

Input, t_n	Output, t_{n+1}
D	Q
0	0
1	1

operation of the data latch to the middle two entries in the truth table for Fig. 5-10. Thus the data latch stores in the bistable memory and presents at Q the data logic level during the most recent clock pulse. The data latch is frequently used for storing output information in counters and computers until the readout has taken place. The use of this circuit allows the counting or arithmetic circuits to begin their next cycle while the readout from the previous cycle is still taking place.

The data latch circuit can be reduced by one gate as shown in Fig. 5-11b. This is accomplished simply by obtaining the $\overline{D}$ information required at the gate 4 input from the gate 3 output instead of the added inverter, gate 5. Gate 3 acts as an inverter for D when the clock is high which is the only time the $\overline{D}$ is needed at gate 4.

Gated NOR memory. The gated memory and data latch circuits are also frequently made from combinations of AND and NOR gates as shown in Fig. 5-12. The use of AND instead of NAND gates for the gating function makes the NAND and NOR circuits logically equivalent as shown by comparing the truth tables of Figs. 5-10, 5-11, and 5-12. The NOR clocked memory is as easy to implement as the NAND because of the availability of the AND/NOR or AND-OR-INVERT gates in a single integrated circuit package.

Experiment 5-4 *Gated Memory and Data Latch*

The gated memory and data latch circuits of Figs. 5-10 and 5-11 or 5-12 are wired and their truth tables are verified. A repetitive clock signal is used with the data latch to observe the timing of the data transfer.

Clocked RS Flip-Flop

Another technique for clocking the simple memory circuit is shown in Fig. 5-13. The clock signal is ac coupled to both set and clear inputs

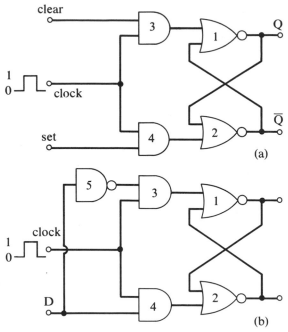

Fig. 5-12   AND-NOR gates as (a) gated memory and (b) data latch.

Truth Table for Fig. 5-12a

Inputs, t_n		Output, t_{n+1}
Clear	Set	Q
0	0	Q at t_n
0	1	1
1	0	0
1	1	Indeterminate

Truth Table for Fig. 5-12b

Input, t_n	Output, t_{n+1}
D	Q
0	0
1	1

of the basic memory circuit using the same diode steering network as the triggered flip-flop of Fig. 5-6. When S is **0** the trailing edge of the T signal applies a negative pulse to the upper gate making $Q = 1$. If S is **1**, the zero-going edge of T cannot pull the gate input down far enough to change the output level. From the truth table, it is seen that this circuit is logically identical to the T flip-flop of Fig. 5-6 and, because of the integrated circuit, it is simpler to build, faster, and has better output load characteristics. Direct set and clear inputs may be added to the circuit by using additional inputs to the NAND gates as shown.

A "T" Flip-Flop

A complementing flip-flop can readily be made by connecting S to Q and R to $\overline{Q}$ as in the earlier circuit.

It is worthwhile to take a more careful look at the timing in the toggle application of the triggered memory of Fig. 5-13. While the T

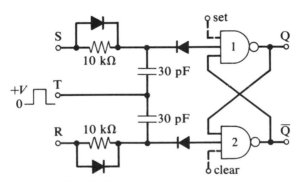

Fig. 5-13   Triggered memory.

Truth Table for Fig. 5-13

Inputs, t_n		Output, t_{n+1}
S	R	Q
0	0	Indeterminate
0	1	1
1	0	0
1	1	Q at t_n

input is at logic **1**, the steering input voltage is developed across the input capacitors; no charge for a **1** input, 4–5 V with positive side at T for a **0** input. When T returns to **0**, the capacitor for which the steering input was **0**, dumps its charge into the gate input. If R or S were to change logic level at that same instant, it would make no difference in the trigger steering since it is the capacitor that provides the charge from the previous information. In fact, in the toggle mode of operation R and S do change logic level as soon as the trigger signal changes the gate output levels. Since the delay in the gate may be 10–25 nsec or less, the new levels appear at R and S very shortly after the trigger pulse.

However, the ac-coupled trigger circuit is dependent on a very sharp trailing edge of the trigger pulse. A direct-coupled gate input has the advantage of triggering on a logic level change of any reasonable shape.

The gated flip-flop (Fig. 5-10) cannot be easily made to complement by connecting Q and $\overline{Q}$ to the set and clear inputs, because the set and clear inputs remain coupled to the memory circuit for the duration of the clock or gate pulse. If this pulse is not shorter than the transition time of the memory bistable, new information is fed to the gates to return them to their earlier state, and so on.[2] Although it is possible to make such short trigger pulses and gates that can react to them, other techniques are generally used. These techniques will be discussed in the next section.

Experiment 5-5 *Triggered Memory*

The ac clocked memory of Fig. 5-13 is wired. The truth table and timing of data transfer are verified. The circuit is wired to toggle and the alternation of the output is observed. Time relationships between high-frequency clock and output signals are observed on a dual-beam oscilloscope, and the frequency division is noted.

5-3 Logic Gate Flip-Flops

The basic cross-coupled gate memory circuits in the previous section could not be made to toggle except with ac-coupled inputs because the inputs are coupled to the bistable gate circuit all the time the clock pulse is high.

---

[2] For a thorough analysis of time stability in logic circuits, see G. Maley and J. Earle, *The Logic Design of Transistor Digital Computers*, Prentice-Hall, Englewood Cliffs, New Jersey, 1963, and M. P. Marcus, *Switching Circuits for Engineers*, Prentice-Hall, Englewood Cliffs, New Jersey, 1967.

It was desired, of course, to overcome the timing problem without giving up the advantages of dc logic level coupling. Consequently, circuits were developed that effectively prevent the output transitions (or other transitions which occur during the clock pulse) from altering the effect of the input signals present when the leading edge of the clock pulse occurs. The circuit is usually two gated memory circuits; one to hold the output state and the other to hold the input information present at the beginning of the clock pulse for later transfer to the output memory. One such combination of memories is called a *master-slave* flip-flop and is shown in Fig. 5-14.

RS Master-Slave Flip-Flop

The RS master-slave flip-flop follows the same truth table as the transistor RS flip-flop of Fig. 5-6. The timing problem is solved by using two basic memory circuits. The operation of the RS flip-flop of Fig. 5-14 through one clock pulse is described below. The clock input is normally **0** which keeps the outputs of gates 7 and 8 at **1** and which prevents the R and S inputs from having any effect on the circuit. With a **1** at each input, the memory bistable made up of gates 5 and 6 could be in either state. The slave flip-flop will be recognized as the gated memory circuit with the master flip-flop supplying the input signals on the inverted clock signal. When the clock input is **0**, the output of gate 9 is **1** so gates 3 and 4 to the slave flip-flop are OPEN. Therefore the gates 1-2 flip-flop will be in the same state as the gates 5-6 flip-flop.

The **0** level current at the gate 9 input is sunk to the clock source through a 220-Ω resistor. This makes gate 9 a little closer to the **1** input state than gates 7 and 8. When the clock pulse is applied, a four-step sequence occurs as outlined in Fig. 5-14. First, as the clock goes positive, because of the 220-Ω resistor mentioned above, gate 9 reaches the **1** input state before gates 7 or 8. A **1** at the gate 9 input results in a **0** at the output which CLOSES gates 3 and 4 and isolates the slave flip-flop from the master. This isolation occurs before any change in the state of the master could take place. Thus the state of the master flip-flop is stored at the slave outputs Q and $\overline{Q}$.

Second, gates 7 and 8 are OPENED by the clock signal and the information at the S and R inputs determines the state of the master flip-flop according to the gated memory truth table. Third, the clock pulse begins to fall closing gates 7 and 8 which isolate the master flip-flop from the S and R inputs. Fourth, gate 9 achieves a **0** input which OPENS gates 3 and 4. At this time, the master flip-flop outputs are transferred to the

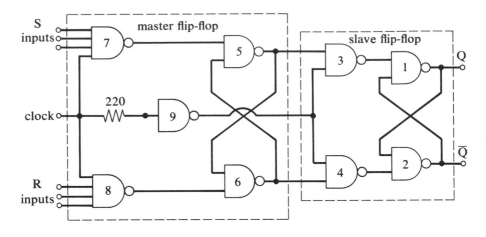

1. gates 3 and 4 close isolating slave from master

2. gates 7 and 8 open connecting master to inputs

3. gates 7 and 8 close isolating master from inputs

4. gates 3 and 4 open connecting master to slave

Fig. 5-14   RS master-slave flip-flop from NAND gates.

Truth Table for Fig. 5-14

Inputs, t_n		Output, t_{n+1}
S = $(S_1 \cdot S_2 \cdot S_3)$	R = $(R_1 \cdot R_2 \cdot R_3)$	Q
0	0	Q at t_n
0	1	0
1	0	1
1	1	Indeterminate

slave flip-flop and appear at the circuit output terminals. In this way, the output change does not occur until the clock pulse is over and, therefore, the effects of output changes cannot appear at the input terminals during the clock pulse.

There are several types of master-slave flip-flops designed on this basic circuit. They are generally named for the type of inputs which are provided.   The circuit of Fig. 5-14 is called an RS master-slave flip-flop because the inputs simply RESET (CLEAR) or SET the output levels as shown in the truth table.   Additional inputs on the master flip-flop input gates are sometimes useful.   As seen from the circuit and the truth table, a **0** at any S input results in Q = **0**, and a **0** at any R input results in Q = **1**. A **0** at both S and R results in a **1** output level for gates 7 and 8 which leaves the master flip-flop unchanged from its previous state.   The action of the R and S inputs in this circuit is very similar to that of the R and S inputs of the triggered memory, Fig. 5-13, and the RS flip-flop of Fig. 5-6.

Experiment 5-6   *RS Master-Slave Flip-Flop*

The RS master-slave flip-flop of Fig. 5-14 is wired and its operation is verified.   It is noted that gates 7 and 8 can accept information at any time during the clock pulse, the state of the master flip-flop being determined by whichever input was **1** last.

JK Master-Slave Flip-Flop

Master-slave flip-flops offer great versatility, good freedom from timing problems, and the waveshape independence of dc coupling.   However, their widespread use was not possible until integrated circuits brought this complex circuit down in size and cost.   One of the most popular forms of the master-slave flip-flop is the JK flip-flop.   As Fig. 5-15 shows, the JK is identical to the RS except that the outputs are connected to the inputs to obtain a toggle operation.   According to the truth table of Fig. 5-14, if Q is **1** and should go to **0** on the next clock pulse, connect one of the S inputs to **0** ($\bar{Q}$).   Similarly, to return Q to **1** on the pulse after that, one of the R inputs should be connected to Q.

The remaining inputs are called J and K instead of R and S and the truth table is changed to show the alternation of output state when J and K inputs are both **1**.   Thus the JK flip-flop can be used as a simple toggle flip-flop or the J and K inputs can be used to establish particular output states as required.

Experiment 5-7   *JK Flip-Flop*

The RS flip-flop of Experiment 5-6 is modified to make the JK flip-flop of Fig. 5-15.   The truth table is verified and the toggling operation and frequency division are observed on a dual-trace or double-beam oscilloscope.

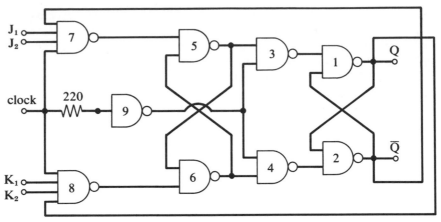

Fig. 5-15   JK master-slave flip-flop from NAND gates.

Truth Table for Fig. 5-15

Inputs, t_n		Output, t_{n+1}
$J_1 \cdot J_2$	$K_1 \cdot K_2$	Q
0	0	Q at t_n
0	1	0
1	0	1
1	1	$\overline{Q}$ at t_n

Integrated circuit JK flip-flop. When a master-slave flip-flop is fabricated in one integrated circuit, several special design techniques are
used which result in an improved circuit. Figure 5-16 shows the functional block diagram and the pin connections for an SN7472 TTL JK
master-slave flip-flop. The master flip-flop is a pair of cross-coupled
AND-OR-INVERT gates and the slave uses NAND gates, cross-
coupled. Multiple input AND gates provide the inputs to the master
while transistor switches are used to connect the information between
the master and the slave.

The signal levels at various points in the block diagram will be followed for typical operational sequence. Suppose that the master is in
the state with a **1** at the output of gate 3. The clock (T) input is at **0**.
The output of the J and K input gates 7 and 8 are **0** regardless of the other
inputs because of the **0** clock input. The inputs to gate 4 are **1** and **0**
resulting in a **0** at the gate 4 output. Since the gate 3 output is at **1**, the

connecting transistor Q_1 is ON, applying a **0** to gate 1 of the slave flip-flop. This forces a **1** at Q. Since the output of gate 4 is **0,** the transistor switch Q_2 is OFF and all **1**'s are applied to the slave gate 2 resulting in a **0** at Q. Thus the output of the master is imposed on the slave.

Now suppose that all J and K inputs are **1** or unconnected and the clock input begins to rise toward **1**. The JK flip-flop goes through the same four-step sequence as the RS flip-flop. First the transistor coupling switches Q_1 and Q_2 are turned OFF because the emitters are no longer at 0 V. (Step 1: isolate slave from master.) When the T input reaches true **1** level, the outputs of the J and K input gates 7 and 8 will be determined by the other inputs. In this case all will be **1**'s except the **0** level $\bar{Q}$ signal fed back to the J input gate 8. Thus the J gate (8) output is still **0,** but the K gate (7) output is now **1**. This makes the master gate 3 output a **0** which, in turn, makes the gate 4 output a **1**. (Step 2: Enter information from input gates to master.)

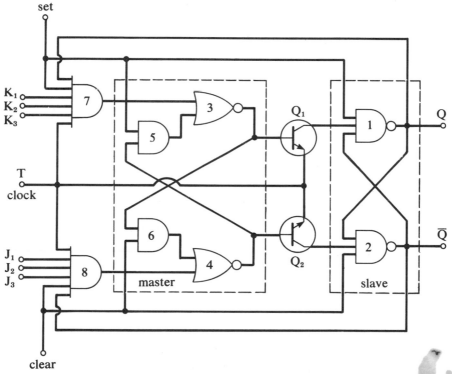

Fig. 5-16   Functional block diagram of an integrated circuit JK master-slave flip-flop.

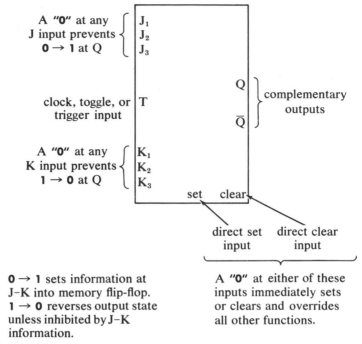

A "0" at any
J input prevents
0 → 1 at Q
J_1
J_2
J_3

clock, toggle, or trigger input T

Q
$\overline{Q}$
complementary outputs

A "0" at any
K input prevents
1 → 0 at Q
K_1
K_2
K_3

set   clear

direct set   direct clear
input   input

0 → 1 sets information at J–K into memory flip-flop. **1 → 0** reverses output state unless inhibited by J–K information.

A "**0**" at either of these inputs immediately sets or clears and overrides all other functions.

Fig. 5-17 Summary of relationships between input and output functions.

As the T input level begins to drop, it becomes a **0** at the J and K input gates bringing both gate outputs to **0**. (Step 3: Disable the input gates.) When the T input level approaches 0 V, the connector transistor switches can be turned ON. In this case the gate 4 master output is a **1** which will turn transistor switch Q_2 ON and force $\overline{Q}$ to be a **1**. (Step 4: Transfer information from master to slave.) Thus the output levels were reversed as expected from the last line in the truth table, Fig. 5-15. By similar reasoning, the response of the JK flip-flop to the other three com-

Truth Table for Fig. 5-17

Inputs, t_n		Output, t_{n+1}
J	K	Q
0	0	Q at t_n
0	1	0
1	0	1
1	1	$\overline{Q}$ at t_n

binations of J and K input information is shown to correspond to the JK truth table.

The action of the direct set and clear inputs is readily seen in Fig. 5-16. A **0** at the set input forces an immediate **1** at the master gate 3 output through both input AND gates 5 and 7. The **0** applied to the slave gate 1 also forces a **1** at Q. Thus the K gate 7 is disabled and both flip-flops are SET. The clear input operates the same way on the lower half of the circuit. The direct set and clear inputs act immediately regardless of the state of the clock input and they override all other input functions. Figure 5-17 and the accompanying truth table summarize the relationships among the input and output functions.

Experiment 5-8 *Three-Bit Binary Counter*

Two integrated circuit JK master-slave flip-flops are used in conjunction with the NAND gate JK flip-flop wired in Experiment 5-7 to make a simple three-bit binary counter. The counting action is observed with indicator lights and the frequency division by 2, 4, and 8 is observed on the dual-beam oscilloscope.

"D" Master-Slave Flip-Flop

The third type of master-slave flip-flop is the "D" flip-flop which is a "data" input modification of the RS flip-flop of the same type used to obtain the data latch from the gated memory circuit. The resulting circuit is shown in Fig. 5-18. The data level is applied to the S input and its complement is applied through gate 10 to the R input. When the clock signal goes high, the data level appears at the master flip-flop output (gates 5 and 6) and when the clock pulse is over, the data appears at Q.

The D flip-flop is very useful when the flip-flop output must respond to a single input level. By connecting the D input to $\bar{Q}$ the input information is always opposite to the state of the output so the flip-flop will toggle on successive clock pulses.

Experiment 5-9 *D Flip-Flop*

The RS flip-flop from Experiment 5-6 is modified to make the D flip-flop of Fig. 5-18. The data transfer is observed. It is noted that the D input remains active during the clock pulse. When $\bar{Q}$ is connected to D the toggling action is observed.

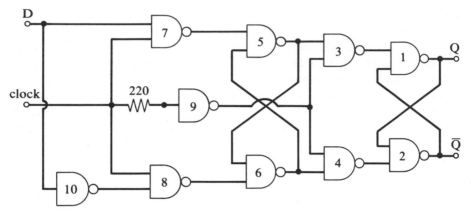

Fig. 5-18   D-type master-slave flip-flop.

Truth Table for Fig. 5-18

Input, t_n	Output, t_{n+1}
D	Q
0	0
1	1

Edge-Triggered Flip-Flops

Another type of flip-flop circuit which overcomes the timing problem is the edge-triggered flip-flop. This type reads the input information and makes the output transition in response to the leading or trailing edge of the clock pulse. There is an advantage in not having the inputs active during the clock pulse in system design, speed of operation, and clock width control. However, they are generally more complex.

Edge-triggered D flip-flop. An example of an edge-triggered flip-flop design is the D flip-flop of Fig. 5-19. With this circuit the data input and transfer occur on the leading edge of the clock pulse. Gates 1 and 2 comprise the bistable circuit with the set and clear functions. The output levels of gates 3 and 4 determine the state of the cross-coupled output gates. The remaining gates 5 and 6 determine whether the output of gate 3 and 4 (but not both) will become **0** in response to a trigger signal applied to the clock input.

A description of the operation of this circuit follows below. A positive transition (from **0** to **1**) triggers the circuit. Therefore, prior to the trigger the clock input is at **0**. Assume also that the set and clear are both at **1**. The level at the D input determines which transition of the bistable circuit will be allowed. To begin, assume that D is **1**. Since the clock input is **0**, the outputs of gates 3 and 4 must be **1**. Now all inputs to gate 6 are **1** so its output is **0**. This forces the output of gate 5 to a **1**. With all **1**'s at the inputs to gates 1 and 2, the bistable circuit could be in either state. Now when the clock input goes to **1**, all inputs to gate 3 are **1** causing its output to become **0**. However, the output of gate 4 stays at **1** because of the **0** output of gate 6 and the new **0** level from gate 3. The **0** at the gate 1 input will cause Q to become **1**. If Q had been **1** prior to the trigger, no output transition would occur.

Now consider the circuit again, but with a **0** level applied to the D input. Prior to the **0–1** transition at the clock input the outputs of gates 3, 4, and 6 are **1**; 5 is **0**, and the bistable may be in either state. When

Truth Table for Fig. 5-19

Input, t_n	Output, t_{n+1}
D	Q
0	0
1	1

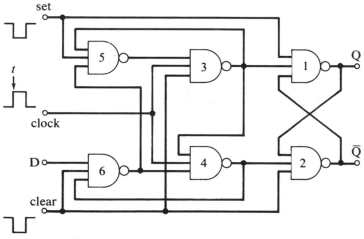

Fig. 5-19   D-type edge-triggered flip-flop.

the clock input goes from **0** to **1,** all inputs to gate 4 are **1** so its output becomes **0** and the output of gate 3 cannot change. In this case, the $\overline{Q}$ output will become **1** or remain **1** if it had been in that state prior to the trigger. Therefore, the action of the triggering circuit is to cause Q to become whatever level was applied to the D input just prior to the trigger signal. Note that any change in the level at D which occurs after the trigger transition has set the outputs of gates 3 and 4 cannot affect the outputs of gates 1, 2, 3, or 4 until the next triggering edge.

Edge-triggered JK flip-flop. An edge-triggered JK flip-flop is shown in Fig. 5-20. The input information is stored in gates 5 and 6 and is transferred to gates 3 and 4 on the leading edge of the clock signal. This information is fed back to the input gates to hold their output steady and is fed onto the output gates 1 and 2. By the time the new information is fed back to the input gates, they are locked out. When the clock returns to **0,** gates 5, 6, 7, and 8 will again respond to the J and K inputs.

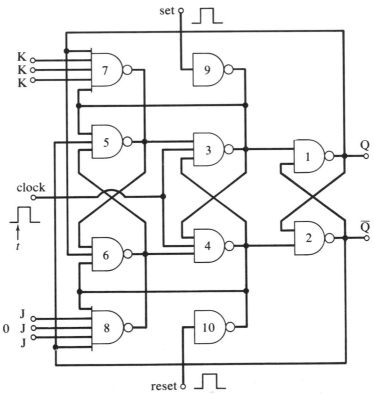

Fig. 5-20   Edge-triggered JK flip-flop (Motorola MTTL III MC3050).

In this diagram there are several instances of gate outputs being connected together. This can generally be done with DTL gates, but not with TTL gates. In the integrated circuit provision was made for this type of connection. The **0** output prevails in establishing the gate output logic level.

The logic gate flip-flops described in this section show how logic gates can be combined to perform virtually any specific desired operation that can be defined by a set of logic statements. New integrated circuit packages combining many gates into more complex and/or convenient circuits are under continuous development which greatly benefits the systems designer.

5-4   The Monostable Multivibrator

As the name implies, the monostable multivibrator is stable in just one state. It can be triggered into its other "semistable" state where it will remain for a time and then spontaneously return to its initial stable state. The time spent in the semistable state is not determined by the triggering signal but by the choice of component values in the monostable circuit itself. This circuit can be thought of as a triggered, adjustable width pulse generator. It is used as a pulse shaper to provide uniform-width pulses from a variable-width input pulse train. Its other principal application is as a delay element since it provides a logic output transition at a fixed time after the trigger signal.

The Transistor Monostable Circuit

A circuit of a monostable multivibrator is shown in Fig. 5-21. This circuit differs from the basic transistor bistable of Fig. 5-3 only in the mode of coupling from the collector of Q_2 to the base of Q_1. In the monostable circuit, Q_1 is normally ON because its base is biased to a positive voltage and only capacitor-coupled to the collector of Q_2. The action of this circuit is as follows: When the circuit is triggered, Q_1 turns OFF. The +4 V, now at the collector of Q_1, is coupled to the base of Q_2, causing that transistor to turn ON. The resulting 4.8-V drop in potential at the collector of Q_2 is transmitted through C to the base of Q_1. The potential at the base of Q_1, had been about 0.7 V ($V_{BE(ON)}$) but is now 4.8 V less than that or ~ -4.1 V. This negative voltage keeps Q_1 turned OFF until C discharges (through R) enough to allow Q_1 to turn ON again. This sequence is illustrated in the waveform diagram Fig. 5-22.

To understand the waveshapes in better detail, the sequence will be

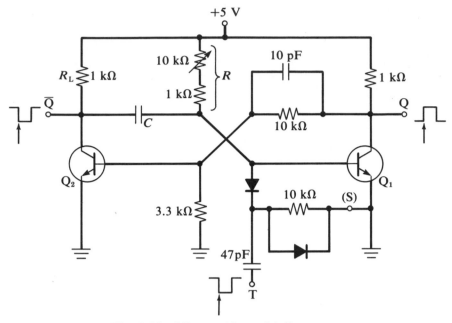

Fig. 5-21   Monostable multivibrator.

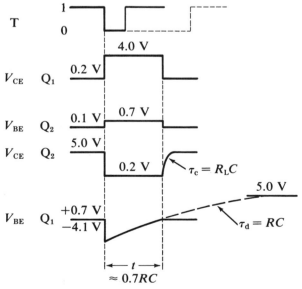

Fig. 5-22   Monostable waveforms.

reviewed again paying particular attention to the charging and discharging of capacitor C. In the stable state, Q_2 is cut off and Q_1 is saturated. The potential at the collector of Q_2 is $+5$ V since only leakage current passes through R_L. The potential at the base of Q_1 is the turn-on potential of Q_1 or about 0.7 V. The potential across C is the difference between these two potentials or about 4.3 V. When Q_1 is triggered OFF, turning Q_2 ON, the potential on the collector of Q_2 immediately drops 4.8 V to about 0.2 V. Since the potential across a capacitor cannot change instantly, the potential at the base of Q_1 also drops 4.8 V to -4.1 V. There is now a 9.1-V potential difference across R, and C begins to discharge through Q_2 and R, as shown in Fig. 5-23a. Since the resistance of Q_2 is generally much less than R, the discharge time constant τ_d is approximately equal to RC. The potential at the base of Q_1 charges toward 5.0 V, but upon reaching 0.7 V, the base-emitter junction of Q_1 becomes forward biased preventing any further increase in potential. At this time Q_1 turns ON terminating the output pulse. During the discharge time, the capacitor is discharged 4.8 V of a 9.1-V initial potential difference or 53%. This discharge requires about seven-tenths of a time constant so the duration of the pulse t is equal to about $0.7RC$ secs.

When Q_1 turns back ON, its base potential is fixed by the forward drop across the base-emitter junction at about 0.7 V. At the same time Q_2 turns OFF. For the potential at the collector of Q_2 to rise, C must charge through R_L. The capacitor charging path is shown in Fig. 5-23b. Since R_L is generally much larger than the forward resistance of the Q_1 base-emitter junction, the time constant for charging τ_c is approximately $R_L C$. This results in the curvature of the $\overline{Q}$ output shown in Fig. 5-22.

Base-emitter breakdown. When Q_2 turns ON, a reverse voltage almost equal to V_{CC} is applied to the base-emitter junction of Q_1. If V_{CC} is greater than the base-emitter breakdown voltage rating BV_{EBO}, the base-emitter junction will act as a Zener diode limiting the negative base voltage to the breakdown value. This will not damage the transistor, but it should be taken into account when calculating the pulse width since C will begin to discharge from the breakdown voltage. Two techniques have been devised to make the peak reverse base voltage more reproducible from one circuit to another. One is to put a diode with known, reproducible, and lower breakdown voltage in the base circuit and, the other, is to put a higher breakdown diode in series with the emitter to avoid breakdown altogether. These techniques are illustrated in Fig. 5-24. If V_{CC} is 10 V and the minimum BV_{EBO} of Q_1 is 5.0 V, a 3.7-V breakdown diode is used for D_1. Added to the forward drop across D_2, the conduction voltage through D_1 and D_2 is about -4.3 V. The silicon diode D_2 prevents D_1

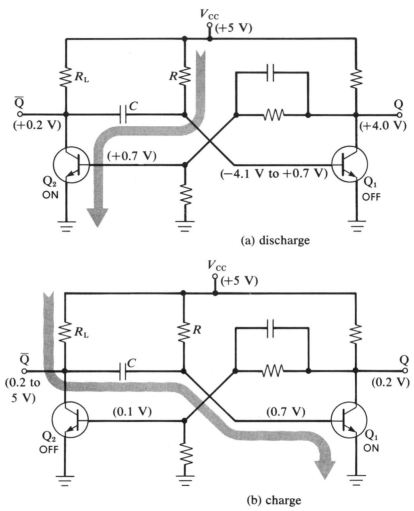

$V_{\rm CC}$
(+5 V)

$\overline{Q}$
(+0.2 V)

$R_{\rm L}$

C

R

(+0.7 V)

Q_2
ON

(−4.1 V to +0.7 V)

Q
(+4.0 V)

Q_1
OFF

(a) discharge

$V_{\rm CC}$
(+5 V)

$\overline{Q}$
(0.2 to
5 V)

$R_{\rm L}$

C

R

(0.1 V)

Q_2
OFF

(0.7 V)

Q
(0.2 V)

Q_1
ON

(b) charge

Fig. 5-23   Charging and discharging paths in monostable multivibrator.

from conducting when the base of Q_1 is forward biased. In Fig. 5-24b, D_3 has a breakdown voltage greater than $V_{\rm CC}$. The base-emitter junction of Q_1 is thus prevented from breaking down. The first approach has the disadvantage of shortening the discharge time; the second that the LOW state Q output voltage is increased by the forward drop across D_3.

Triggering. The triggering of the monostable may be accomplished by turning Q_1 OFF or Q_2 ON momentarily. Any of the triggering tech-

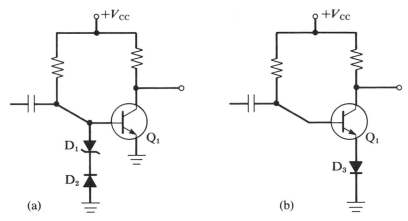

Fig. 5-24 Circuits to avoid base-emitter breakdown of Q_1.

niques discussed for the transistor bistable circuits in Section 5-1 applies equally well to this circuit. Of course, only one transition needs to be triggered since the circuit returns to the initial state automatically after time t. The triggering circuit of Fig. 5-21 turns Q_1 OFF on the negative transition of the trigger pulse. It is "half" the trigger circuit shown in the clocked RS flip-flop of Fig. 5-6. In that circuit a **0** logic level applied to the R or S connections steered the turn-off pulse to that respective base. Here the S connection of the diode resistor "steering" network is grounded so that every T pulse is effective in turning Q_1 OFF. If it were desired to lock out T pulses during a particular period, a logic **1** signal could be applied to point S during that period.

Another popular triggering circuit is the turn-on trigger of Fig. 5-25. The leading edge (**0–1** transition) of the T input applies a positive pulse to the base of Q_2 to turn it ON. When the Q_2 collector voltage drops, Q_1 turns OFF and the cycle begins. The trigger signal could be dc connected, omitting the coupling capacitor if the trigger pulse is shorter than the monostable output pulse. It would not generally be desirable for the input signal to be holding Q_2 ON after C had discharged allowing Q_1 to come back ON. For this reason, ac-coupled trigger inputs are far more common. The RC differentiator at the input can be designed to provide a trigger pulse shorter than the monostable pulse duration.

Duty cycle. The duty cycle of the monostable circuit is the fraction of the time it spends in its semistable state. For all monostable circuits there is a maximum duty cycle. In the circuit of Fig. 5-21 the duty cycle is limited primarily by the rate of charging C when the circuit returns

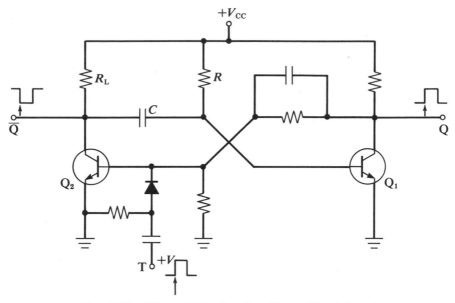

Fig. 5-25   Monostable circuit with positive trigger.

to the stable state.   If a new trigger pulse occurs before C is fully charged, t would be shorter because of the reduced charge on C.   If the trigger signal occurs very soon after the previous pulse is over, the charge on C may be insufficient to turn Q_1 OFF or keep Q_1 OFF, and no output pulse would result.   To reduce the pulse duration error to a minimum, the trigger pulse should not occur until about 3 charging time constants after the previous pulse.   Thus the minimum stable-state time is about $2R_LC$. The pulse width is about $0.7RC$.   The maximum duty cycle is the time in the semistable state over the minimum time of one complete cycle. In this case,

$$\text{max. duty cycle} \approx \frac{0.7RC}{0.7RC + 3R_LC} \approx \frac{R}{R + 4.3R_L}$$

Experiment 5-10   *Transistor Monostable*

The transistor bistable circuit of Experiment 5-1 is modified to make the monostable circuit of Fig. 5-21.   The effect of C and R on pulse width is measured and compared with the predicted relationship. The maximum duty cycle is determined and related to the rounding of the waveform at the $\bar{Q}$ output.   The pulse rise and fall times are measured and the maximum usable range of pulse widths is determined.

Logic Gate Monostable

Just as in the case of the bistable multivibrator, it is possible to make monostable multivibrator circuits with basic integrated circuit logic gates. A very simple design is shown in Fig. 5-26. The trigger signal T is applied to inputs of gates 1 and 2. At the output of gate 2, the T signal is inverted and delayed by the gate propagation time to give the signal $\overline{T}_d$. The T and $\overline{T}_d$ pulses thus appear at gate 1 slightly displaced in time as shown in the waveforms. The output will be a momentary **0** resulting from the overlapping logic **1** levels of T and $\overline{T}_d$ according to the NAND operation $\overline{T \cdot \overline{T}_d}$. The output appears delayed by one gate delay time from the leading edge of the T signal due to the propagation delay in gate 1.

RC delay monostable. The basic circuit of Fig. 5-26 has two limitations: (1) the output pulse duration is extremely short (generally about 10^{-8} sec with fast gates) and not adjustable, and (2) the T pulse must be longer than the output pulse. The first problem is solved by adding more signal delay in the gate 2 output circuit such as the low-pass filter circuit of Fig. 5-27. Before the trigger pulse, the T input is at logic **0**, and the output of gate 2 is logic **1**. The capacitor C is charged through R to this level. When T goes to logic **1**, the gate 2 output goes to logic **0**, and the capacitor discharges toward **0** through R and the gate 2 output impedance. The potential at $\overline{T}_d$ decreases and crosses the **1–0** logic threshold for the gate 1 input. At this time the logic **0** at $\overline{T}_d$ brings the gate 1 output back to logic **1**. The pulse duration t thus depends on R, C, and the logic threshold level and is generally of the same magnitude as τ_{RC}.

For most gate types, R in Fig. 5-27 is limited to small values. With current-sinking gates such as the TTL and DTL, the **0** level input current must pass through R without the IR drop bringing the gate **1** input potential too near the **1–0** threshold. Approximately 220 Ω is considered a

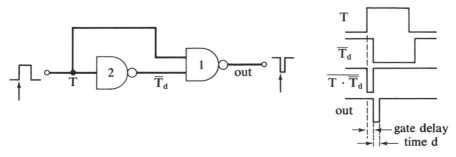

Fig. 5-26   Gate delay monostable.

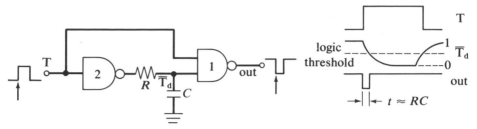

Fig. 5-27   RC delay monostable.

safe upper-limit input resistance for the standard TTL gate. This low R limits the practical pulse duration from such a circuit, for even if a 1-μF capacitor is used the pulse will only be about 0.25 msec long. However, for applications such as clearing flip-flop registers, triggering, and clocking, pulses of a microsecond or less are frequently used. This circuit is a convenient source of such pulses.

Gated RC delay monostable.   The circuit of Fig. 5-27 requires that the T pulse be longer than the output pulse. Because of the direct connection between T and gate 1, whenever T is **0**, the output is **1**. Thus if the T signal returns to **0** before the $\bar{T}_d$ signal, the output pulse will be terminated prematurely. This problem is solved by gating the T signal and using the output pulse to control the gate as shown in Fig. 5-28. Because of the inverting gate 3, the circuit now triggers on the **1−0** transition of T. When T goes to **0**, the monostable is triggered and the **0** at the output is applied to gate 3 to keep the gate 3 output at **1** as long as the output is **0**. The T pulse only needs to be long enough to change the states of gates 3 and 1.   If the T pulse is, in fact, shorter than the output pulse, the output complement can be obtained from the output of gate 3.

Again, because of the charging time for C, there is a limit to how frequently the T pulse can occur without affecting the output pulse length.

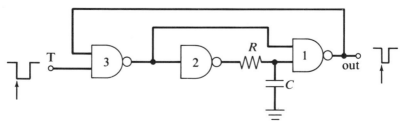

Fig. 5-28   Gated RC delay monostable.   (From P. Sandland, *Electronics*, p. 108, June 26, 1967.)

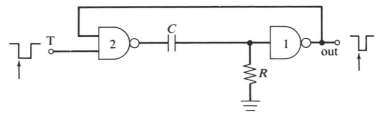

Fig. 5-29   Gated stored-charge monostable.

The duty cycle for these circuits is relatively low since the same RC time constant determines the charging and discharging times. Since the pulse is only about one discharge time constant long, but the charging must go on for about three time constants to be complete, the maximum duty cycle is about $1/(1+3) = 25\%$.

Stored-charge monostable. A circuit almost identical in its action to Fig. 5-28 but which uses only two gates is shown in Fig. 5-29. In this circuit, R is small enough to keep the input to gate 1 at logic **0** (for TTL, 220 Ω or less). The output and T input are normally logic **1** making the gate 2 output **0**. When T becomes **0**, the gate 2 output becomes a logic **1**. Since the potential across C cannot change instantly, the gate 1 input also becomes logic **1** and the output a **0**. The positive charge on C at the gate 1 input discharges through R. When the gate 1 potential falls below the logic threshold, gate 1 reverses state again and the pulse is over. The connection from the output to the input of gate 2 holds the gate 2 output at **1** for the duration of the output pulse.

Experiment 5-11   *Logic Gate Monostable*

The pulse-width, pulse shape, rise and fall time, and maximum duty cycle characteristics of the gated monostable circuit of Fig. 5-28 or 5-29 are experimentally determined.

5-5   The Astable Multivibrator

The astable multivibrator circuit is a logical extension of the monostable multivibrator, that is, neither of the two states is stable. The circuit alternates between two semistable states generating a repetitive signal of alternating **1**'s and **0**'s at the output. This circuit is used as an oscillator or square-wave generator primarily. It is a simple, inexpensive, logic-compatible, and very convenient generator when the frequency stability is not a critical requirement.

The Transistor Astable Circuit

The astable multivibrator is made by using a capacitor coupling for both halves of the basic bistable circuit as shown in Fig. 5-30. Slight differences in the components in the two halves of the symmetrical circuit cause one transistor to turn ON more quickly than the other when the power is first applied. The falling potential at the collector of the first transistor is capacitor-coupled to the base of the other transistor which causes it to move toward cutoff. The rising potential at the second collector is capacitor-coupled back to the base of the first transistor to reinforce the initial trend. This cycle very quickly puts the first transistor in saturation and the other in cutoff. But this state is not stable.

Assume that Q_1 is saturated and Q_2 is cutoff. There is a negative potential on the base of Q_2 which resulted from the drop in potential at Q when Q_1 turned ON. This negative charge on C_1 will discharge through R_1 until V_{BE} of Q_2 reaches 0.7 V and Q_2 begins to conduct. Now the falling potential at Q is conducted through C_2 to turn Q_1 OFF. The rising potential at Q transmitted through C_1 flips Q_2 quickly ON by the regenerative feedback. The sudden drop in potential at $\overline{Q}$ results in a negative potential at the base of Q_1 which is turned OFF. The charge on C_2 now discharges through R_2 while C_1 charges through R_{L1}. When the base of Q_1 reaches the conducting potential, Q_1 begins to turn ON and the half-cycle is complete. The turning ON of Q_1 turns Q_2 OFF, and the potentials are reversed for this half-cycle. The circuit continues through this cycle indefinitely generating complementary square waves at Q and $\overline{Q}$.

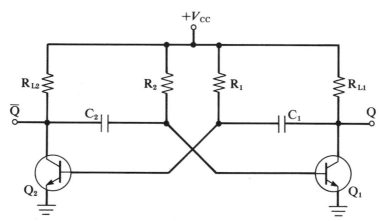

Fig. 5-30   Basic astable multivibrator circuit.

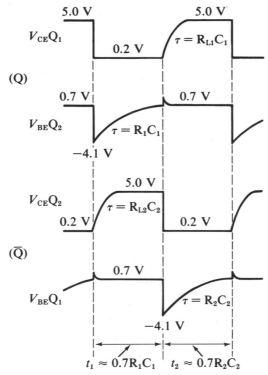

Fig. 5-31   Astable waveforms.

Waveshapes.   The waveshapes at the collectors and bases through one complete cycle are shown in Fig. 5-31.   It is clear from the wave-shapes that both halves of the astable behave just as the Q_2 collector–Q_1 base half of the monostable circuit.   The charging and discharging sequence and the time constant expressions are the same as for the mono-stable.   From Fig. 5-31 it is seen that the total time of one cycle is $t_1 + t_2$.   Therefore, the frequency f is

$$f \approx \frac{1}{0.7} \left(\frac{1}{R_1 C_1} + \frac{1}{R_2 C_2} \right)$$

and if $R_1 = R_2 = R$ and $C_1 = C_2 = C$, to make a symmetrical square wave,

$$f \approx \frac{1.4}{RC}$$

It is possible to make an asymmetric wave by using unequal values for C or R, but the same duty cycle restrictions apply as for the monostable circuit. For this circuit, the short part of the cycle should be at least 10 or 20% of the total cycle time.

Improving rise time. One of the limitations of the basic astable circuit is that the output waveform leading edge is rounded at both outputs. Where this is undesirable, the rise time can be improved in a number of ways. An inverting amplifier switching circuit such as that of Fig. 5-1 can be used at either or both outputs as desired. The basic logic gate memory circuit of Fig. 5-8 can be driven from both outputs. This may have an advantage in ensuring gate-compatible outputs.

Another technique for improving the rise time is to modify the circuit as shown in Fig. 5-32. Charging resistors R_{C1} and R_{C2} and diodes have been added to the basic circuit on each side. Charge and discharge paths are shown assuming that Q_2 is ON and Q_1 is OFF. The negative charge on the base side of C_2 is discharging through Q_2, D_2, and R_{D2}. The discharge time constant remains essentially unchanged by the inclusion of the diode in this path. When Q_1 turns OFF, Q can rise to V_{CC} without charging C_1 because as the potential at Q increases, diode D_1 becomes reverse biased disconnecting Q from C_1. R_{C1} now provides the charging current for C_1, but this time constant does not limit the rise time of Q. For the ON transistor Q_2, R_{C2} is essentially in parallel with R_{L2}. Therefore, to modify an astable circuit for faster rise time, make the

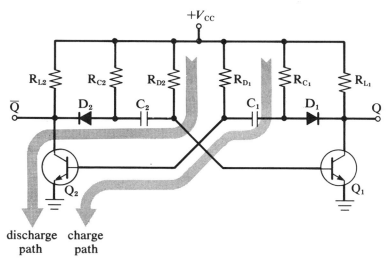

Fig. 5-32 Astable multivibrator with improved rise time.

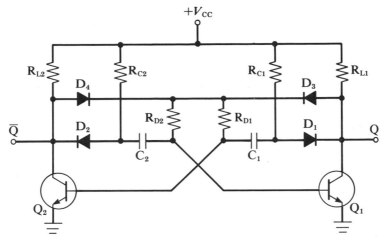

Fig. 5-33   Self-starting, astable multivibrator.

parallel combination of the new R_L and R_C equal to the original R_L value. For faster rise time, keep R_L as small as possible since wiring, transistor, and load capacitance still limit the rise time. However, R_C must be small enough to fully charge C during the transistor OFF time.

Self-starting astable. The circuits of Fig. 5-30 and 5-32 may not begin to oscillate every time they are turned ON. There is a chance that both transistors could saturate simultaneously which is a stable state. The circuit of Fig. 5-33 has no stable state and, thus, can be guaranteed to start oscillating every time. The discharge resistors R_{D1} and R_{D2} are connected by diodes D_3 and D_4 to the transistor collectors instead of to V_{CC}. If both transistors were saturated, neither collector would be positive enough to keep the other base ON. Therefore, that possible stable state is eliminated. Nor can both transistors be OFF for then both would be supplying ON bias to each other. The only possible states are the semistable one ON, the other OFF and vice versa. The circuit of Fig. 5-33 has both options — self-starting and fast rise time.

Experiment 5-12   *Transistor Astable Circuit*

The transistor monostable circuit of Experiment 5-10 is modified to form the astable circuit of Fig. 5-30. The waveforms and time constants of Fig. 5-31 are confirmed and the useful frequency range is determined.

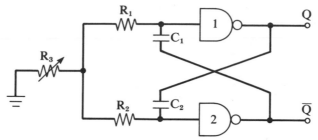

Fig. 5-34   Current-sinking, logic gate, astable multivibrator.

Logic Gate Astable

An astable multivibrator can be made of logic gates, as shown in Fig. 5-34. This circuit appears to be analogous to the stored charge monostable of Fig. 5-29, but if the connections to ground from the gate inputs will sink the **0** level input current, both gates could be stable in the logic **1** output state. Actually the gate input-sinking resistors are chosen to put the gate input level near the logic threshold so the charging and discharging of C_1 and C_2 will put the input voltages above or below the threshold level. Assume that the gate 1 input has just gone below the threshold level making Q become logic **1**. This transition is coupled through C_2 to make the gate 2 input a **1**. The resulting logic **0** at $\bar{Q}$ is applied to the input of gate 1 through C_1. Now the gate input side of C_1 becomes more positive as it is charged by the gate-sinking current, and C_2 becomes less positive as it discharges through R_2. Whichever gate input first crosses the threshold will change the output state and trigger the transition. The frequency is determined primarily by the values of C_1 and C_2. The values of R_1, R_2, and R_3 are chosen to optimize the oscillation, although fine frequency adjustment by R_3 is possible. The frequency is approximately $1/2RC$, where $R = R_1 + R_3$ and $R_1 = R_2$ and $C = C_1 = C_2$.

5-6   The Schmitt Trigger

The Schmitt trigger is a bistable device which is another variation on the coupling techniques used when two transistor switches are connected regeneratively. One collector is coupled to the other base in the familiar way, but the second interconnection is through the transistor emitters. This leaves one base connection free for an input signal. The action of this circuit is such that the state of the output depends on whether the input voltage is above or below a particular value. This circuit has had great application in pulse detecting, signal squaring, and level discrimination circuits. It completes the discussion of basic multivibrator circuits.

The Basic Schmitt Circuit

The basic Schmitt trigger circuit is shown in Fig. 5-35. Assume first that E_{in} is low enough so that Q_2 is cut off. The collector voltage at Q_2 is coupled to the base of Q_1 by the voltage divider R_C and R_{B1} in the familiar way. When Q_2 is cut off and the collector voltage is high, Q_1 will be saturated. The current through Q_1 causes an IR drop across R_E and thus determines the emitter potential of Q_2. The output potential will be $I_1 R_E + V_{CE(sat)}$ in this state. If the input voltage E_{in} begins to rise, a potential will be reached which causes Q_2 to begin to conduct. The decreased potential at the Q_2 collector causes Q_1 to conduct less. This decreases the current through R_E, lowering the Q_2 emitter potential, and causing Q_2 to conduct still more. This regenerative cycle continues until Q_2 is saturated and Q_1 is cut off. The output voltage in this state is $+V_{CC}$. The potential at E_{in} required for this transition is E_{ON}.

For regenerative coupling through the emitters, it is essential that the emitter potential decrease when Q_2 turns ON and Q_1 turns OFF. To ensure this R_{L2} is generally made larger than R_{L1} so that the current through Q_2 is less than $I_{Q1(sat)}$.

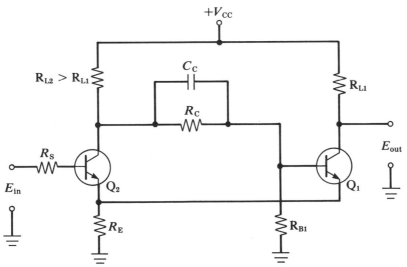

Fig. 5-35 Basic Schmitt trigger.

Now suppose that E_{in} begins to decrease. When E_{in} becomes low enough to bring Q_2 out of saturation, the Q_2 collector voltage increases causing Q_1 to conduct increasing the emitter potential. This regeneration acts quickly to turn Q_2 OFF and Q_1 ON. The E_{in} required for this transition is E_{OFF}. In summary, to turn Q_2 ON, E_{in} must be positive with respect to the emitter potential $I_{Q1(sat)}R_E$ by about $V_{BE(ON)}$ for Q_2 or

$$E_{ON} \approx I_{Q1(sat)}R_E + V_{BE(ON)}$$

To turn Q_2 OFF, E_{in} must be slightly less than $V_{BE(ON)}$ plus the new emitter potential $I_{Q2(sat)}R_E V_{BE(ON)}$. Since $I_{Q2(sat)}$ must be less than $I_{Q1(sat)}$ for the necessary regeneration, it follows that

$$E_{OFF} < E_{ON}$$

Schmitt Trigger Response

The action of the circuit to a varying input voltage is shown in Fig. 5-36. The first section demonstrates how an irregular amplitude waveshape with slow rise and fall times is converted to sharp logic level

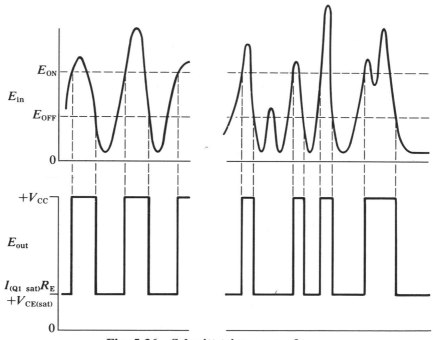

Fig. 5-36 Schmitt trigger waveforms.

transitions as E_{ON} and E_{OFF} levels are crossed. The second section shows pulses of varying amplitude which are converted into clean pulses. Note also that pulses below E_{ON} in amplitude are discriminated against. By setting E_{ON} at an appropriate level, noise or other undesired signal components can be rejected. Signals exceeding E_{ON} become sharp, unambiguous logic transitions. Note, however, that the double peak is not resolved because E_{ON} and E_{OFF} must both be traversed to make each output pulse.

Hysteresis. The difference between E_{ON} and E_{OFF} is called the "hysteresis" voltage or "dead zone." We have seen that some hysteresis is essential to the operation of the Schmitt trigger circuit. Where it is desired the hysteresis can be reduced by reducing the difference between R_{L1} and R_{L2} or by adding another resistor between the emitters of Q_1 and Q_2. Reducing the hysteresis increases the rise and fall time of the output and makes the triggering more sensitive to small noise fluctuations in the signal level. If the hysteresis is reduced to zero or less, the circuit will function simply as a linear amplifier.

Experiment 5-13   *The Schmitt Trigger*

The characteristics of the Schmitt trigger circuit of Fig. 5-35 are determined experimentally. The measured values of F_{ON} and E_{OFF} are compared with the expected values. The amount of hysteresis and the trigger potential are varied while the response of the circuit to a sine-wave input signal is observed.

Applications

The basic Schmitt trigger circuit provides a number of essential functions in wave-shaping, logic-level restoring, squaring, discriminating, and sensing. It has been a virtually indispensable basic circuit. However it has some fairly serious drawbacks. These are primarily that (1) the hysteresis adjustment which may influence the measurement is also critical to the circuit operation, (2) the absolute level of E_{ON} and E_{OFF} is not easy to adjust, and (3) the output signal in the OFF state is $I_{Q1(sat)} R_E + V_{CE(sat)}$ volts above ground potential, making this circuit incompatible with many common logic gate inputs. Circuits have been devised to overcome one or more of these problems. The most versatile and effective circuit is the recently developed high-speed, high gain, integrated circuit amplifier called a "comparator" or "sense amplifier." It performs the function of the Schmitt trigger circuit with adjustable hysteresis and trigger level and easy logic-level interfacing. But it is not an emitter-coupled bistable multivibrator. For description and applications see Chapter 7.

5-7 Wave-Shaping and Generating with Multivibrators

All the basic multivibrator circuits have been described in the preceding sections of this chapter. Their applications are easy to perceive in most cases; counting and storage for the bistable, pulse generator and delay element for the monostable, oscillator for the astable, and squaring and detecting for the Schmitt. However, some more complex problems in generating, wave-shaping, and discrimination have been ingeniously solved with various combinations of the above circuits. A few examples are presented in this section to give an idea of the range of possibilities.

Pulse Train Generator

A series of pulses at a constant repetition rate is required. The ability to vary the pulse width may be desired. An astable multivibrator (MV) triggering a monostable MV, as shown in Fig. 5-37 can be used. This is better than an asymmetric astable MV because the pulse width can be made as small as desired without duty cycle limitations, and the repetition rate is not dependent on the pulse width. The repetition rate could be very precise if a crystal oscillator followed by a Schmitt trigger is used in place of the astable MV.

A Dual Monostable Astable

It is sometimes necessary to make a square-wave generator when working in a logic system where separate astable MV packages are not provided. An astable MV can be made with two monostable MV's connected to trigger each other alternately, as in Fig. 5-38. The trailing

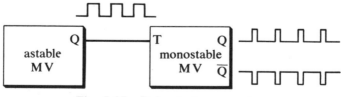

Fig. 5-37   Pulse train generator.

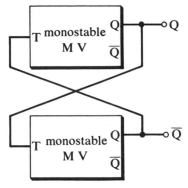

Fig. 5-38   Dual monostable astable.

edge of the signal at $\overline{Q}$ triggers the upper monostable MV to generate a pulse.   At the end of that pulse, the Q signal triggers the lower monostable MV, and so on.   Duty cycle restrictions on the monostable must be observed.   With integrated circuits, this technique may be especially convenient since integrated astable circuits are rare, whereas the monostable is quite common.

Generating a "True" Square Wave

Occasionally it is necessary to generate a square wave which is truly symmetrical, that is, having an exactly equal time in the **0** and **1** state.   Adjusting a monostable for equal ON and OFF times, or adjusting the trigger level on a Schmitt trigger which is squaring a sine wave may only be good to a few percent.   A precise and adjustment-free technique is shown in Fig. 5-39.   The toggle flip-flop alternates on every input cycle producing a square wave with half-cycles as reproducible as the input frequency.   Of course the output frequency is half that of the oscillator or astable MV.

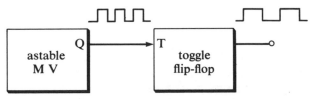

Fig. 5-39   Symmetrical square-wave generator.

Gating an Oscillator

An oscillator generating a continuous train of pulses is to be gated ON and OFF. An ordinary gate cannot be used, because if the gate ON or OFF signal occurs during the pulse, a shortened pulse is not to be allowed. A circuit to perform this function is shown in Fig. 5-40. If S is **0,** the output is **1,** gate 4 is OPEN and cross-coupled gates 2 and 3 have G and $\overline{G}$, respectively at their output. If G = **1** the output of gate 2 is **1,** gate 1 is OPEN and provides the inverted signal S at the output. However if G goes to **0** before S, and gate 3 changes state, gate 2 cannot change since the **0** output holds gate 2 in the **1** state until S returns to **0.** Thus the gate cannot cut short an S pulse. If G now stays at **0,** all inputs to gate 2 are **1** and the **0** output of gate 2 holds the output at logic **1.** The gate is CLOSED.

If the G input goes to **1** while S is **0,** gate 2 goes to a **1** opening gate 1 for the next S pulse. If G goes to **1** while S is **1,** gate 4 is CLOSED and the gate 2-3 flip-flop cannot respond to the change in gate level until S is **0** again. Thus a partial S pulse at the beginning of the gating signal is also blocked.

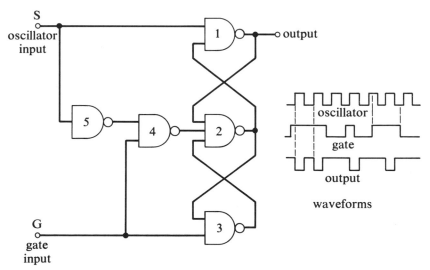

Fig. 5-40 Gating an oscillator. (From G. Maley and J. Earle, *The Logic Design of Transistor Digital Computers*, p. 259, Prentice-Hall, Englewood Cliffs, New Jersey, 1963.)

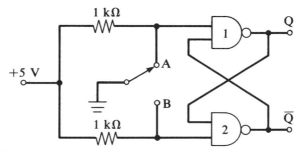

Fig. 5-41   Contact bounce eliminator.

Contact Bounce Eliminator

All solid metal switch contacts are prone to "bouncing" on contact. This can cause difficulty in fast digital circuits if, for example, the switch circuit output is connected to a counting circuit, pulse generator, or register advance.   A single-pole, double-throw switch can be connected to the basic NAND memory circuit, as shown in Fig. 5-41.   In the position shown, $Q = 1$ and $\overline{Q} = 0$.   Erratic contact with A as the switch moves toward B has no effect since the gates cannot change state until gate 2 input is **0**.   This occurs on the first contact with B. $\overline{Q}$ is now **1** and this is not affected by bounce at the B contact.   This circuit assumes the bounce is not so bad that the switch bounces between A and B.

Single Pulse Generator

It is sometimes very convenient to be able to pass a single pulse from a pulse train on command from a gating signal.   A partial pulse should not be allowed.   With such a circuit one could observe a counter or other system responding to a high-frequency signal, but advancing only one step at a time.   The JK flip-flop forms the heart of this circuit as shown in Fig. 5-42.   A **0** level pulse at G clears the two flip-flops. The first trailing edge at S after the clear pulse triggers Q of A to logic **1**. Since $\overline{Q}$ of B is also **1**, a **0** appears at the output.   On the next leading edge, $\overline{Q}$ of B is triggered to a **0** terminating the output pulse.   The flip-flops cannot be triggered to the $Q = 0$ state by subsequent pulses because the **0** at $\overline{Q}$ is connected to K in each case.   Thus, the output pulse is exactly the same length as the signal pulse and only one can occur between each clear pulse.   If a clear pulse occurs while S is at logic **0**, the next leading edge should not trigger flip-flop B.   The **0** at Q of A con-

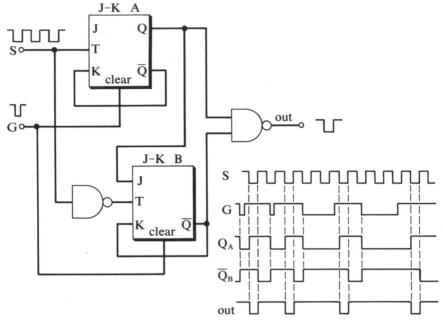

Fig. 5-42   Single pulse generator.

nected to J of B prevents the flip-flop from triggering out of order.   Nor does it matter how long the clear pulse is applied.   The first complete signal pulse after the clear pulse is over will appear at the output.

A Delayed Pulse

A pulse can be generated a fixed length of time after an initial pulse using two monostable MV's.   This is used where two or more phenomena should occur in a fixed time sequence used as pulsing a test circuit after the oscilloscope sweep has been triggered or sampling a signal voltage after a transient has been applied to the system.   The circuit which simply uses one monostable to delay the triggering of a second monostable, which is the pulse generator, is given in Fig. 5-43.

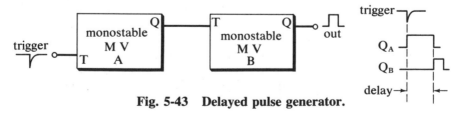

Fig. 5-43   Delayed pulse generator.

Pulse-Width Discriminator

Just as the Schmitt trigger can discriminate against signals which are less than a specific amplitude, it is also useful to be able to sort input pulses by their duration. Figure 5-44 shows a circuit that provides a pulse at one output or another depending on whether the input pulse is wider or narrower than a reference pulse from a monostable MV. The input pulse is shaped by a Schmitt trigger and inverted, so that the leading edge can trigger the monostable. The input and monostable pulses are thus inverted with respect to each other and compared by a NAND gate. If the **0**-going pulse is the wider, it will hold the gate output at logic **1** through the comparison; if not, a pulse results.

Experiment 5-14 *Wave-shaping*

Any of the circuits in this section may be wired and tested. Recommended for particular interest are the oscillator gate, Fig. 5-40; the single pulse generator, Fig. 5-42; and the pulse-width discriminator, Fig. 5-44.

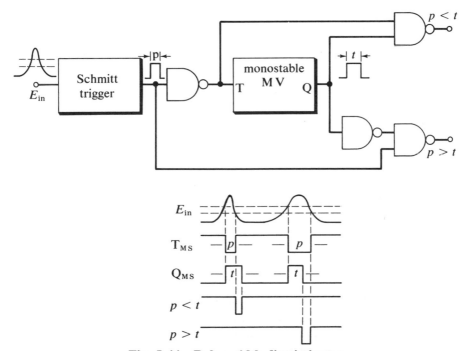

Fig. 5-44 Pulse-width discriminator.

chapter six

Counters, Registers, and Readout

Each flip-flop can store one piece of binary information or "bit." For storing more complex information, groups of flip-flops must be used together. A group of flip-flops arranged to store information is called a *register.* There are many kinds of registers which differ in the way the information is put in, taken out, and read. The logic-level information that is to be stored can be set into each flip-flop of the register simultaneously. A register arranged in this way is a parallel input or parallel entry register. With a serial entry register, the bits of information are set one at a time into one end of a series of flip-flops. The information shifts from one flip-flop to the next as each new bit enters the register until the register is full. A third kind of register advances through a prescribed sequence of different states each time a pulse appears at the input. By observing the state of the register flip-flops at any time, the number of input pulses received can be determined. This is called a *counting register* or simply a *counter.*

This chapter describes these basic kinds of flip-flop registers, their primary applications, and their associated readout and control systems.

6-1 Binary Counters

The register of a binary counter stores information about the accumulated number of input pulses as a binary number. That is, if the state of each flip-flop in the register is sequentially written as a **1** or a **0,** the resulting binary number is equal to the number of input pulses

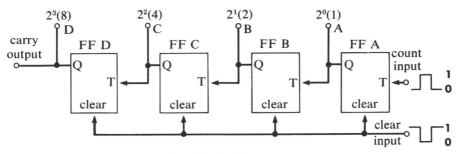

Fig. 6-1 Basic binary counter.

counted. It is a popular counter because it is very simple, stores the maximum possible amount of information per flip-flop, and is related to a basic number system.

A basic binary counter is shown in Fig. 6-1. Toggle or complementing flip-flops are used. The outputs of each flip-flop change state when the T input signal goes from **1** to **0**. Assume that the four flip-flops have been cleared ($Q = 0$) by an appropriate pulse at the clear input. When the first pulse appears at the input, flip-flop A toggles and A becomes logic **1** ($2^0 = 1$). On the next pulse, A returns to logic **0** which toggles flip-flop B. Now B is **1** and A is **0** ($2^1 = 2$). After the next pulse, A is **1** again and B is unchanged ($2^1 + 2^0 = 3$), and so on. The table of waveforms in Fig. 6-2 shows how the A flip-flop is SET after every odd input pulse to represent the number 1. The B flip-flop output represents the number 2, and the C and D outputs represent 4 and 8, respectively. The sum of the values of the SET flip-flops represents the cumulative count at any instant. The outputs could be connected to indicators, and the instantaneous count read out in binary form. The counter of Fig. 6-1 can be extended to any desired number of flip-flops by adding flip-flops E, F, G, and so on,

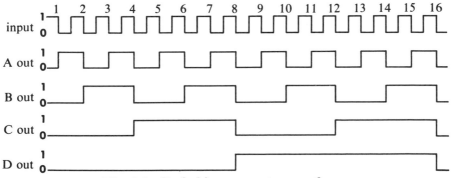

Fig. 6-2 Basic binary counter waveforms.

Table 6-1   Powers of Two

n	2^n	n	2^n
0	1	11	2,048
1	2	12	4,096
2	4	13	8,192
3	8	14	16,384
4	16	15	32,768
5	32	16	65,536
6	64	17	131,072
7	128	18	262,144
8	256	19	524,288
9	512	20	1,048,576
10	1,024		

in like manner. Since each successive flip-flop output represents another binary digit (power of 2), n flip-flops will have 2^n states and can count from zero to $2^n - 1$. In Fig. 6-2 the waveforms of the four-flip-flop (4-bit) counter show the progression from 0 to 15. Then, on the sixteenth pulse the register returns to the zero state. Thus the number of flip-flops required to count or store a particular number in binary form can be determined from a table of the powers of 2 as given in Table 6-1. (A longer table of this type is given in the Appendix.) For instance, to accumulate

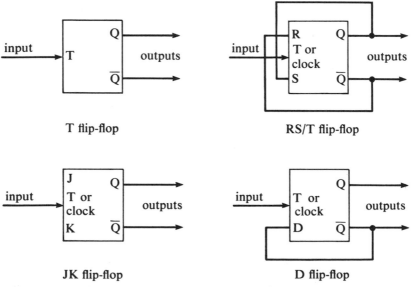

Fig. 6-3   Complementing connections for various flip-flop types.

1000 counts (or store a number to 0.1% accuracy), at least a ten-flip-flop register will be required. Any flip-flop capable of being toggled can be used in the circuit of Fig. 6-1. The methods of connecting the various flip-flop types for complementing operation are reviewed in Fig. 6-3.

Experiment 6-1 *Binary Counting*

The four-bit binary counter of Fig. 6-1 is wired using JK flip-flops. Binary counting is observed by connecting the flip-flop outputs to indicator light drivers and pulsing the input with a push-button pulser. The 16 successive states shown in Fig. 6-2 are verified. Clear connections are wired to a push button for instant clearing of the counting register.

Down-counting. The counter of Fig. 6-1 is called an *up-counter* because the register counts up to the next higher number for each input pulse. A counter can also be made which counts down or subtracts one from the count for each input pulse. Such a counter and its output waveforms are shown in Fig. 6-4. The circuit is identical to the up-counter except that $\overline{Q}$ instead of Q is connected to the following T. When the

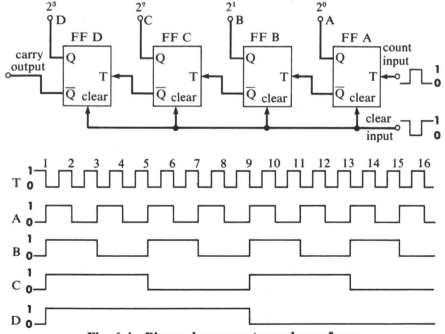

Fig. 6-4   Binary down-counter and waveforms.

first **1–0** transition appears at the input, flip-flop A toggles changing $\overline{Q}$ of A from **1** to **0**. The signal toggles flip-flop B, and so on, resulting in a full count of 15 in the register. The next pulse brings flip-flop A back to **0** for a 14, and so on, until the register is "empty."

Experiment 6-2 *A Binary Down-Counter*

The circuit of Fig. 6-1 is modified to make a down-counter in two ways. First, by connecting the indicator lights to the $\overline{Q}$ outputs of the flip-flops, then by connecting the $\overline{Q}$ output of the first flip-flop to the T input of the next, and so on, as shown in Fig. 6-4. The response to counting and clearing pulses is analyzed and observed.

A counter that can be changed from an up-counter to a down-counter by electrical command is called an *up-down counter*. The circuit simply requires a selector-switch type of gating circuit at each T input to determine whether the toggling signal will come from Q or $\overline{Q}$ of the previous flip-flop. (Selector-switch gates are described in Section 4-8 and shown in Figs. 4-32 and 4-33.) The resulting circuit is shown in Fig. 6-5. When a logic **1** is applied to the up-down (U/D) control, the gates connected to the Q outputs are OPEN and the $\overline{Q}$ gates are CLOSED. With a **0** at the U/D control, the situation is reversed. However, when the U/D control input

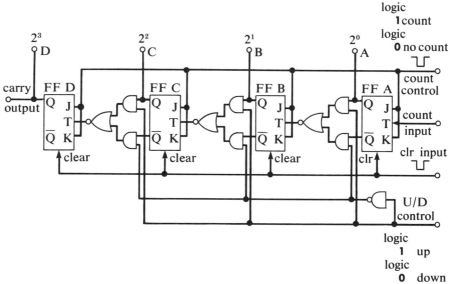

Fig. 6-5   Binary up-down counter.

changes state, a false trigger transition could be generated at the NOR gate output depending on the state of Q. One of the easiest ways to avoid false triggering is to prevent any triggering during the changes in U/D control level. This can be done with JK flip-flops as shown. When **0** is applied to the count control input, all flip-flops are inactivated, the U/D level is changed, and the count control is then returned to a **1** level for counting in the new direction. An up-down counter that does not require a count control signal to prevent false triggering is described later in this section.

Experiment 6-3 *A Binary Up-down Counter*

The binary up-down counter of Fig. 6-5 is wired using AND-OR-INVERT gates or the equivalent NAND gate circuit. Counting is observed in the up and down counting modes. The necessity of applying a **0** to the count control input when the U/D control input is changed is observed.

Readout from a Binary Counter

The simplest form of readout from a binary counter is a row of lights with the light drivers connected to the flip-flop Q outputs, the right-most light being turned on by flip-flop A. Thus each light corresponds to a binary digit and the binary number is read by interpreting an ON light as **1** and an OFF light as **0** in each place. In the early days of digital counters, this readout was common, and it is still used to display the output of some computer registers where the binary presentation has more significance than its decimal equivalent.

Conversion of the binary number to its decimal equivalent for more convenient readout requires a decoding circuit. As we shall see below, the complexity of circuits for decimal decoding from a binary register increases rapidly with increasing bit size of the register. It is not currently practical to decode more than about 4 or 5 bits. Where a decimal readout is required, a binary-coded decimal register, as described in the next section, is generally used.

Octal and hexadecimal. If a decimal readout is not needed, the only objection to the binary readout is the length of the numbers. We see from Table 6-1 that 19 binary digits are required to represent a number needing only 6 decimal digits. More efficient number systems which are very readily decoded from the binary are the octal (base 8) and the hexadecimal (base 16). In these systems the binary number is actually divided

Table 6-2 Comparison of Binary, Decimal, Octal, and Hexadecimal Number Bases

Decimal	Binary	Octal	Hexadecimal
0	0	0	0
1	1	1	1
2	10	2	2
3	11	3	3
4	100	4	4
5	101	5	5
6	110	6	6
7	111	7	7
8	1000	10	8
9	1001	11	9
10	1010	12	A
11	1011	13	B
12	1100	14	C
13	1101	15	D
14	1110	16	E
15	1111	17	F
16	10000	20	10
17	10001	21	11
18	10010	22	12
19	10011	23	13
20	10100	24	14
32	100000	40	20
50	110010	62	32
60	111100	74	3C
64	1000000	100	40
100	1100100	144	64
255	11111111	377	FF

into groups of three (or four) digits and each group is decoded into the 8 (or 16) states that it can represent. The octal, decimal, and hexadecimal equivalents of some binary numbers are given in Table 6-2. As expected, the decimal system is more compact than the octal and less than the hexadecimal. The right 3 digits of the binary number convert to the units place in the octal system according to the relationship established for 0 to 7. The next 3 binary digits convert to the second column in the octal number (the eights place) by the same code. Thus $(110\ 101\ 011)_2$ is $(653)_8$ (the subscript refers to the number base). Using the binary-hexadecimal conversion given in the numbers 0 to $(F)_{16}$, that same binary number is equal to $(1AB)_{16}$. Thus decoding a large binary number to octal or hexadecimal is a simple repetition of the decoding of each group of 3 or 4 binary digits.

A binary-to-octal decoder using NAND gates is shown in Fig. 6-6. This decoder uses a 3-input NAND gate and inverter to obtain the AND function. Each pair of gates decodes one line in the first eight lines of Table 6-2. That is, if $\overline{C}$ and $\overline{B}$ and $\overline{A}$ are 1, then a 1 level appears at the zero output ($0 \cdot 0 \cdot 0 = 0$). The 5 output is 1 when $C \overline{B} A = 1$ [$(101)_2 = (5)_8$]. The decoded octal output can be used to activate the appropriate elements of a neon, numeral indicator lamp, a printer, or other readout device. Generally a decimal readout is employed with the 8 and 9 numerals unused.

Experiment 6-4 *A Binary-to-Octal Decoder*

The binary-to-octal decoder of Fig. 6-6 is wired (or at least four elements of the decoder). The $\overline{A}$, A, B, $\overline{B}$, C, and $\overline{C}$ inputs are connected to the Q and $\overline{Q}$ outputs of three flip-flops arranged as a binary counter. The decoder outputs are connected to indicator lights. The 0, 1, 2, 3, etc., lights are observed to come ON sequentially as the counter is advanced through its 8 states.

The diode matrix shown in Fig. 6-7 is a binary-to-hexadecimal decoder. Each hexadecimal output is the output of a diode AND gate with

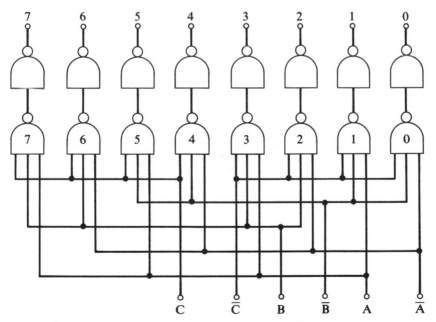

Fig. 6-6 Binary-to-octal decoder.

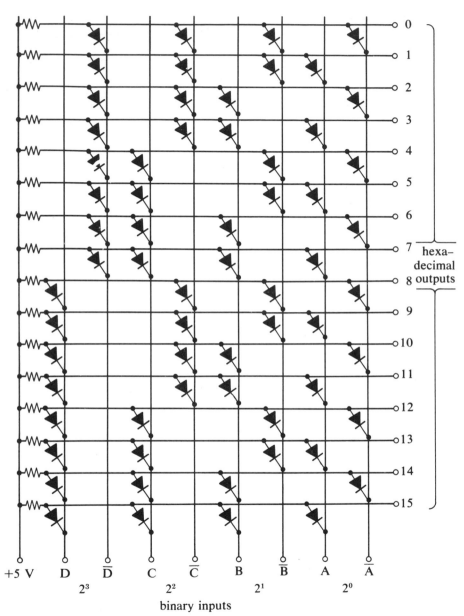

Fig. 6-7 Diode matrix binary-to-hexadecimal decoder.

4 inputs. Only when all 4 binary inputs to a row of diodes are logic **1**, will that output be logic **1**. The diode matrix decoder is completely general. By putting diodes at the desired intersections any required decoding pattern can be achieved.

Synchronous Binary Counting

A synchronous binary counter is shown in Fig. 6-8. The advantage of synchronous operation is the reduction in propagation delays in long binary counter chains. On the eighth and sixteenth pulses in Fig. 6-2,

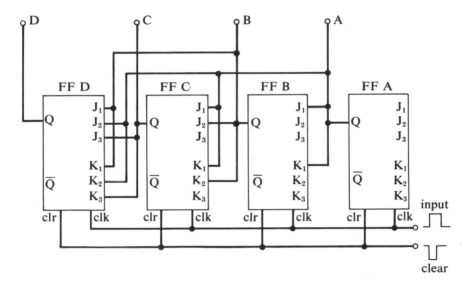

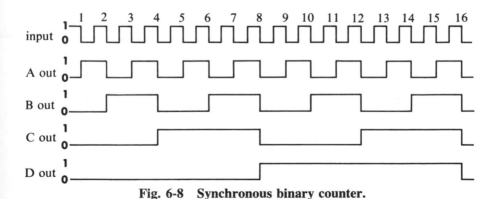

Fig. 6-8   Synchronous binary counter.

all four flip-flop outputs change state. Flip-flop B begins its transition when FF A has completed its change of state, and so on. If the propagation delay is 50 nsec through each flip-flop the transition at the D output will occur 200 nsec after the input signal transition. If the input pulse rate is 10 MHz, the transition at D does not occur until 1 or 2 more input pulses have been counted by A and B. Thus the outputs would not represent the exact count at high count rates. This problem is overcome by triggering all four flip-flops simultaneously and using the J and K inputs to determine which transitions will be allowed.

Note that the waveform chart in Fig. 6-8 is identical to that of Fig. 6-2. Flip-flop A changes state on every other count, and flip-flop B alternates only when A is 1 after the previous count. Therefore, the output Q of A is connected to J_1 and K_1 of B to prevent B from changing state when A is 0. Similarly C should change only when A AND B are 1. Connecting A and B outputs to J and K inputs of C allows C to alternate only on every fourth count. Connections to D are obtained in the same way. Note the convenience of multiple J and K inputs in this application.

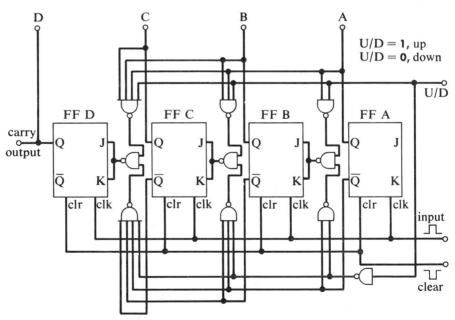

Fig. 6-9   Synchronous up-down counter.

Experiment 6-5 *Synchronous Binary Counters*

The synchronous binary counter of Fig. 6-8 is wired and tested. Then the circuit is modified to make a synchronous binary down-counter and tested. A high-frequency input signal is used and the frequency at each output is measured and related to the input frequency.

Down and up-down counter. In the synchronous binary up-counter, each flip-flop is allowed to toggle or complement if all the less significant digits are **1**'s. In the synchronous down-counter, the J and K inputs of following successive flip-flops are connected to the $\bar{Q}$ output so that each flip-flop complements only if all less significant digits are **0**'s. To make an up-down counter, selector gates are connected at the JK inputs so as to determine whether the Q's or $\bar{Q}$'s of the previous flip-flops affect the JK level. Figure 6-9 gives the circuit diagram for the synchronous binary up-down counter. A **1** at the U/D input commands an up count and a **0** commands a down count. The selector gates in this illustration are made with NAND gates though the AND-OR-INVERT (AOI) gates used in Fig. 6-5 would serve the same purpose.

Experiment 6-6 *Synchronous Up-down Counter*

The synchronous binary up-down counter of Fig. 6-9 is wired and shown to count up and down. It is demonstrated that the count control input required by the asynchronous up-down counter when changing count direction is not required for this circuit.

6-2 Binary-Coded Decimal Counters

The problem of decoding binary information into decimal form was discussed in the first chapter and in the previous section. It was also shown that for binary-to-octal or -hexadecimal decoding, a relatively simple decoding circuit could be used for each 3- or 4-digit group in the binary number. In instruments or systems in which the readout convenience of decimal numbers outweighs the ease of computation and economy of binary numbers, the registers are arranged to store a decimal number in a group of four flip-flops. Up to 16 combinations of output states are available from four binary circuits, but, when they are used to store a decimal number, only 10 of these states are used. Each combination is assigned to one of the decimal numerals 0 to 9. Thus each group of four binaries represents one decimal digit in the stored number.

For instance, to store numbers up to 9999, four groups of four flip-flops each are required. Since the decoding for each group of four flip-flops is identical, more flip-flops and decoding circuits can readily be added to store and readout as large a decimal number as necessary.

The most common storage code for decimal numbers follows the first 10 states of the binary counter as shown in Table 6-3. It is called the 1 2 4 8 or *natural code* because the values of 1, 2, 4, and 8 can be assigned to binaries A, B, C, and D, respectively, and the stored decimal number can be obtained by adding the values of the SET (1 level) flip-flops. There are thousands of other possible BCD codes, some of which have advantages in various applications. Because it has become the industry standard for counting circuits, this section will primarily illustrate 1 2 4 8 code. Some other codes and their advantages are given at the end of this section.

Synchronous BCD Counter

The synchronous decade counting unit of Fig. 6-10 is similar to the binary counter of Fig. 6-8 except that it has a 10-count cycle and a specially coded output. The desired output levels for each input pulse in the cycle are shown for the 1 2 4 8 (natural) code. The J and K inputs are used to make the counter conform to this specific pattern. Flip-flop A alternates with every input pulse so no J or K connections are required. Flip-flop B alternates whenever A is HIGH, but should not go to 1 on the tenth pulse when D is HIGH. Connecting J_1 and K_1 of B to Q of A satisfies the first condition and connecting J_1 of B to $\overline{Q}$ of D prevents

Table 6-3   Binary-Coded Decimal Storage

Decimal No.	Binary Stages			
	D	C	B	A
0	0	0	0	0
1	0	0	0	1
2	0	0	1	0
3	0	0	1	1
4	0	1	0	0
5	0	1	0	1
6	0	1	1	0
7	0	1	1	1
8	1	0	0	0
9	1	0	0	1

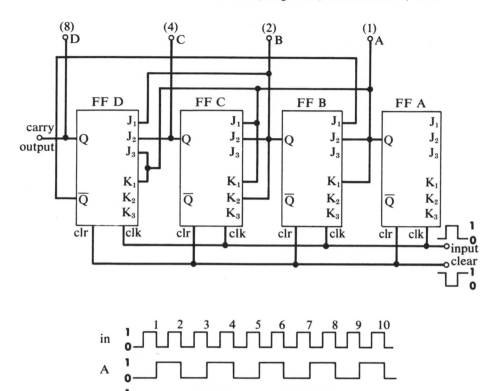

Fig. 6-10   Synchronous BCD counter.

Q of B from becoming **1** when $\overline{Q}$ of D is **0** to satisfy the second condition. The transitions of C are affected only by A and B outputs as in the binary counter. The conditions for the **0–1** transition of D require A, B, AND C to be **1**. However, D should return to **0** the next time A is **1**, so only the A $=$ **1** condition is put on the K input of flip-flop D. This method, i.e., plotting the desired output waveforms and using the J and K inputs to allow transitions only at the appropriate counts, allows counters with any special output sequence to be easily designed and constructed. If multiple J and K inputs are not available or are insufficient, other gates can be used at the J and K inputs.

Consider, for instance, the design of a BCD down-counter. The output waveforms are shown in Fig. 6-11. Satisfy yourself that the waveforms show the count down according to the 1 2 4 8 BCD code. Flip-flop A alternates on every clock pulse so no J or K connections to A are required. On the first pulse, D becomes 1. By looking at the waveforms, it is seen that this transition should occur only when A, B, AND C are 0. Therefore the $\overline{Q}$ outputs of A, B, and C are connected to J_1, J_2, and J_3 of D. On the third pulse, B AND C become 1 and D becomes 0. It is necessary to prevent D from clearing on the second pulse by connecting $\overline{Q}$ from A to K of D. Flip-flop C is prevented from becoming 1 except when A AND B are 0 by connecting $\overline{Q}$ outputs from A and B to J inputs of C. The condition that A is 0 is not sufficient for B to become 1 because B should not rise on the first pulse. It is necessary to

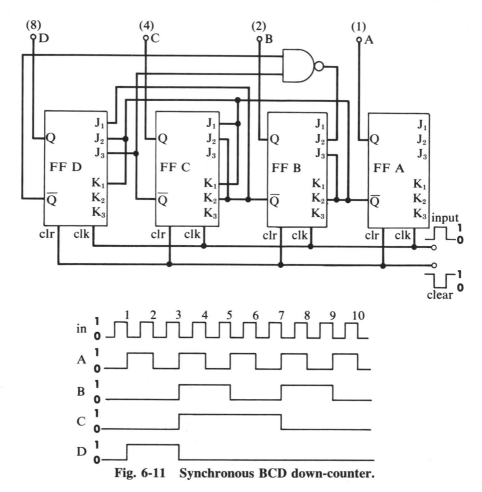

Fig. 6-11   Synchronous BCD down-counter.

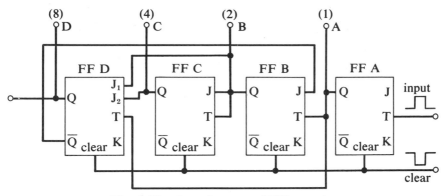

Fig. 6-12 Asynchronous BCD counter.

add the condition that D OR C be **1**. The A = **0** condition is met by connecting $\overline{Q}$ of A to J of B. The C + D = **1** condition is met by the NAND gate connected to J_2 of B. If $\overline{Q}$ of C or $\overline{Q}$ of D is **0**, the **1** at J_2 of B will allow B to SET. To complete the design, the clearing of B can occur whenever A = **0** ($\overline{Q}$ of A to K of B) and the clearing of C should occur when A and B are **0** ($\overline{Q}$ of A to K_1 of C and $\overline{Q}$ of B to K_2 of C).

A BCD up-down counter is made by using gates to switch the J and K input signals from the outputs shown in Fig. 6-10 to the ones shown in Fig. 6-11. The resulting circuit requires 14 NAND gates or 4 rather complex AND-NOR gates.[1]

Experiment 6-7 *Synchronous BCD Counter*

Either the synchronous BCD up-counter of Fig. 6-10 or the BCD down-counter of Fig. 6-11 is wired. The output waveforms of each flip-flop are observed on an oscilloscope and compared to the appropriate figure. The D flip-flop output frequency is measured and compared to the frequency of the input signal.

Asynchronous BCD Counter

The natural code (1 2 4 8) BCD counter is considerably simplified if asynchronous counting is permitted as shown in Fig. 6-12. The output waveforms of this counter are identical to those shown in Fig. 6-10. Flip-flop A is toggled by the input signal. The output of A toggles B on input pulses 2, 4, 6, and 8, but the connection of $\overline{Q}$ of D to J of B keeps

[1] *TI Series 54/74 Integrated Circuits*, Texas Instruments, Inc., 1966.

B from setting on the tenth pulse. Flip-flop C is toggled by the B output. The output of C cannot be used to toggle D because D must clear on the tenth pulse when only the A output changes. Thus D is toggled by the A output but prevented from setting until B AND C are 1.

The problem of delay in the flip-flop affects the maximum count rate in the coded asynchronous counter. Consider a possible delay of 1 μsec per flip-flop. On the eighth pulse, B clears and D sets, and, on the tenth pulse, B must not set because of the **0** from $\overline{Q}$ of D at its J input. The change in level at J must, therefore, precede the tenth pulse. In other words, the 3-μsec delay through flip-flops B, C, and D must be less than the time of two input cycles. Thus the maximum count rate would be $2/(3 \times 10^{-6}) = 670$ kHz, whereas the synchronous counter with the same flip-flops could have counted at rates up to 1 MHz.

The somewhat slower count rate of the asynchronous BCD counter can be more than offset by the inherent speed of integrated circuit techniques. Thus the simpler asynchronous circuit of Fig. 6-12 has been used to make the most popular, integrated circuit BCD counter. In the integrated circuit form, this counter can be obtained with maximum count rates of 10 to 50 MHz. Generally the connection between the Q output of A and the T input of B (and D) is not completed inside the integrated circuit package. Bringing these connections out instead allows the A circuit with 2 stable states to be used independently of the BCD circuit with 5 stable states. Because this BCD counter is made up of a two-counter followed by a five-counter it is sometimes referred to as a *biquinary counting circuit.*

An asynchronous down-counter similar to the up-counter of Fig. 6-12 can be made. The design of such a circuit is suggested as an exercise for the reader.

Experiment 6-8 *Asynchronous BCD Counter*

The asynchronous BCD counter of Fig. 6-12 is wired. The outputs of the flip-flop are connected to indicator light drivers so that the count sequence can be observed to follow the binary pattern.

Memory Register

As we have seen, the counting register advances through a prescribed sequence of steps in response to a series of input pulses. When the input pulses stop, the output levels of the counter flip-flops can be interpreted (decoded) to indicate the number of input pulses that occurred. This information is stored in the counting register until the counter receives additional input pulses or a clear pulse. Presumably, something is to be done with the count information such as decoding and visual readout, paper tape printout, or transmission of data to the control station. Sometimes, it is not possible to make the necessary use of the count information before it is desired to resume the count or clear the counting register to begin again. In such cases, a memory register containing a number of flip-flops equal to the counting register is used. Upon command, the state of each memory flip-flop becomes the same state as the corresponding counter flip-flop, thus storing the count information in the memory register and freeing the counting register to proceed without losing the desired information.

The memory register is now a common feature of most counter-based digital instrumentation. It allows continuous display of the count measurement until the next measurement is completed and a new count is transferred into the memory register, decoding, and readout circuits. A 4-bit memory circuit is shown in Fig. 6-13 in conjunction with one decade of a BCD counter. When the count has been completed, the memory transfer input is pulsed setting the memory flip-flops according to the logic level at the data inputs. The count information now appears at the memory outputs A_n, B_n, C_n, and D_n. When the memory transfer pulse is over, the data inputs are inactive and the counter output can now change without affecting the memory register outputs.

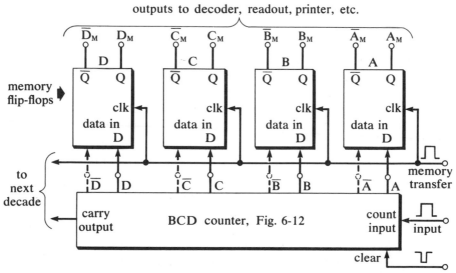

Fig. 6-13 Four-bit memory register with BCD counter.

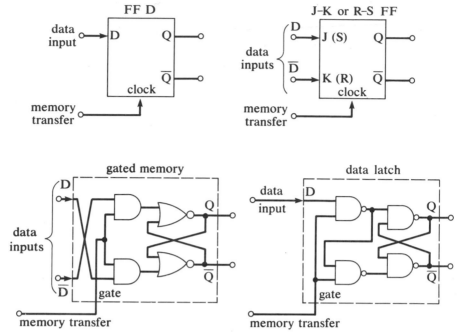

Fig. 6-14 Memory register connections for various flip-flop types.

Virtually any gated or clocked flip-flop can be used in the memory register. The memory register connections for the D, JK, and RS/T clocked flip-flops and the gated memory and data latch are shown in Fig. 6-14. The gated and clocked devices are basically different in their operation. With the clocked flip-flop, data level is transferred to the output flip-flop on one edge of the clock pulse, but with the gated device, the output follows the input as long as the gate is open. Thus the gated memory and data latch can be pulsed for data transfer and storage or held open for continuous transfer from input to output (no storage). Because of their simplicity and the ease of turning the memory function

Experiment 6-9 *Memory Register*

A memory register of one or more bits is wired using the AND-OR-INVERT gate or NAND gate data latch circuit shown in Fig. 6-14. The register inputs are connected to the BCD counter which was wired in Experiment 6-8, and the outputs are connected to indicator lamps. Data hold and transfer operations for the memory register are observed.

on and off, the gated devices are generally used in digital instruments. However, in applications where there is a possibility of the memory output affecting the counter output, the clocked flip-flops must be used to break the feedback cycle.

BCD Decoding and Readout

As mentioned earlier in this section, BCD counters are readily decoded for decimal readout by decoding each 4-bit BCD group at a time. This is readily accomplished by a 10-gate version of the octal decoder circuit shown in Fig. 6-6 or by a 10-output version of the diode matrix decoder shown in Fig. 6-7. The 1 2 4 8 BCD decoding logic is given in Table 6-4. The logic statements are those for the first 10 terms in the hexadecimal decoder. However, the elimination of the other 6 terms allows some simplification of the logic statement as shown in the right-hand column. The state of D only needs to be specified in order to distinguish between 0 and 8 and between 1 and 9. If D is 1, the number must be 8 or 9 and the state of A is enough to distinguish between these. The minimized statement can only be used when the other 6 states (10 to 15) are rigorously excluded. If an undesired state, such as $\bar{A}BCD$, should appear (at power turn on, for instance), the 4 and 8 outputs would both be ON. Using the diode matrix decoder, the complete statement implementation requires forty diodes and the minimized statement, thirty.

Table 6-4 Binary-to-Decimal Decoding Table, Natural Code

Decimal No.	Flip-Flop Outputs				Logic Statement	Minimized Statement
	D	C	B	A		
0	0	0	0	0	$\bar{A}\ \bar{B}\ \bar{C}\ \bar{D}$	$\bar{A}\ \bar{B}\ \bar{C}\ \bar{D}$
1	0	0	0	1	$A\ \bar{B}\ \bar{C}\ \bar{D}$	$A\ \bar{B}\ \bar{C}\ \bar{D}$
2	0	0	1	0	$\bar{A}\ B\ \bar{C}\ \bar{D}$	$\bar{A}\ B\ \bar{C}$
3	0	0	1	1	$A\ B\ \bar{C}\ \bar{D}$	$A\ B\ \bar{C}$
4	0	1	0	0	$\bar{A}\ \bar{B}\ C\ \bar{D}$	$\bar{A}\ \bar{B}\ C$
5	0	1	0	1	$A\ \bar{B}\ C\ \bar{D}$	$A\ \bar{B}\ C$
6	0	1	1	0	$\bar{A}\ B\ C\ \bar{D}$	$\bar{A}\ B\ C$
7	0	1	1	1	$A\ B\ C\ \bar{D}$	$A\ B\ C$
8	1	0	0	0	$\bar{A}\ \bar{B}\ \bar{C}\ D$	$\bar{A}\ \ \ D$
9	1	0	0	1	$A\ \bar{B}\ \bar{C}\ D$	$A\ \ \ D$

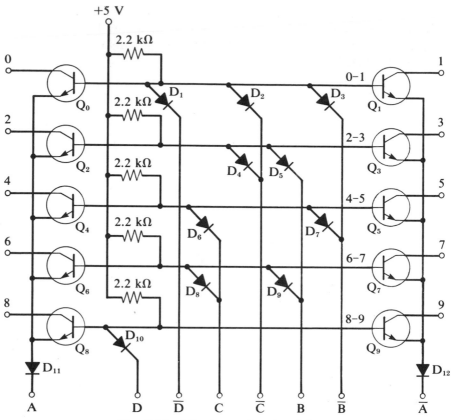

Fig. 6-15   Biquinary decoder-driver.

Decimal Readouts

One of the most common kinds of decimal readouts is the NIXIE[2] tube, which is a neon lamp with one anode and ten separate cathodes. Each cathode is in the form of one of the decimal numerals. The anode is connected to a positive high-voltage (150–200 V) supply through a current-limiting resistor. Whichever cathode is grounded will light. To operate all the elements of the NIXIE tube, a driver (transistor switch) must be connected between each cathode and ground, i.e., ten drivers in all. The driver transistor inputs are each connected to the appropriate decoder output terminal.

Figure 6-15 shows the circuit of a combination decoder-driver frequently used with the NIXIE type of decimal display. The circuit is

[2] Trademark, Burroughs Co.

based on the minimized diode matrix decoder for B, C, and D. If $\overline{B}, \overline{C}$, AND $\overline{D}$ are all at logic **1**, the top horizontal bus marked 0-1 will be logic **1**, but not the others. Referring to the right-hand column in Table 6-4, we see that $\overline{B}\overline{C}\overline{D}$ is TRUE for decimal numbers 0 or 1. Thus the 0-1 bus is connected to the bases of the 0 element driver transistor Q_0 *and* the 1 element driver Q_1. Whether transistor Q_0 or Q_1 will be ON as a result of the logic **1** on the 0-1 bus depends upon the level of A. If A is **0** ($\overline{A} = 1$), the emitter of Q_0 is at about 1 V (the **0** level A output voltage plus V_γ of D_{11}), the base-emitter junction is forward biased, and Q_0 is ON. The potential on the 0-1 bus is thus 1 V plus $V_{BE(ON)} \approx 1.7$ V. Since $\overline{A}$ is at a logic **1** level of +4 to +5 V, the base-emitter junction of Q_1 is reverse biased, and Q_1 is OFF. The remaining nine conditions can be followed in a similar way.

This circuit is popular because of its simplicity. Only twelve diodes are needed in addition to the ten driver transistors. Advantage has been taken of the fact that the BCD condition is duplicated for A and $\overline{A}$. Thus when the signal $\overline{B}\overline{C}\overline{D}$ is generated, it can be used for both $(\overline{B}\overline{C}\overline{D}) \cdot \overline{A} = 1$ and $(\overline{B}\overline{C}\overline{D}) \cdot A = 1$ and does not have to be generated twice. The decoder of Figure 6-15 is called the *biquinary decoder* because it combines a two-state decoder (A and $\overline{A}$) with a five-state decoder ($\overline{B}\overline{C}\overline{D}$, $\overline{B}C$, $B\overline{C}$, BC, and D).

Another frequently used decimal readout device is the seven-segment display shown in Fig. 6-16. Seven independent lamps are arranged in the array shown. Combinations of these lamps are lit to form the shapes of the numerals. Decoding for this readout requires an OR gate for each segment in addition to the BCD decoder of Fig. 6-7. For instance, element A is lit for 0, 1, 4, 5, 6, 8, and 9. Therefore the driver for element A is connected to the output of an OR gate which provides the function $A = 0 + 1 + 4 + 5 + 6 + 8 + 9$ or $A = P_0 + P_1 + P_4 + P_5 + P_6 + P_8 + P_9$ using the *P* notation introduced in Section 4-8. Some elements are used in so many of the numerals (element E appears in eight of them)

Fig. 6-16   Seven-segment decimal display.

it is simpler to list the numerals that they don't appear in; for instance, element C is OFF for numerals 1, 5, and 6. Therefore $\overline{C} = P_1 + P_5 + P_6$ can be used instead of $C = P_0 + P_2 + P_3 + P_4 + P_7 + P_5 + P_9$.

Integrated circuit (IC) decoder-driver packages are now available for either the ten-element cold-cathode displays or the seven-segment displays. The BCD information is connected to 4 input terminals, and the indicator lamp and its supply are connected to the output terminals of the IC decoder-driver. This electronic convenience reduces the choice of readout devices to mechanical, aesthetic, and economical considerations.

Experiment 6-10   *A Complete Decade Counting Unit*

A complete decade counting circuit including the asynchronous BCD counter of Fig. 6-12, the memory register of Fig. 6-13, the decoder-driver circuit of Fig. 6-15, and a neon decimal display tube is studied and operated. Several such circuits are used together to make a 3- or 4-digit decimal counter.

Decade Scaling

Decimal counters are frequently used for decade scaling, that is, to divide an input frequency or pulse rate by an exact factor of 10. As the waveform of Fig. 6-10 shows, the D output of the 1 2 4 8 BCD counter goes to 1 on the eighth pulse and back to 0 on the tenth pulse. Thus 1 output pulse is produced for every 10 input pulses. If the input frequency is exactly 1 MHz, the frequency at the D output will be exactly 100 kHz. The 100-kHz signal can be further divided by additional decade counting units to obtain precise 10-kHz, 1-kHz, 100-Hz, etc., signals. With six decades of division a precise 1.000000-Hz signal can be obtained from a 1.000000-MHz crystal oscillator. This application was described in Section 1-4.

For scaling applications the actual count is never read. Therefore, the count sequence or output code does not matter so long as there is

---

Experiment 6-11   *Decade Scaling*

A circuit that contains a series of several integrated BCD circuits is studied and used for scaling. A standard frequency source is connected to the input, and precise output frequencies of 1/10, 1/100, etc., of the input frequency are observed.

---

one flip-flop which has only 1 output pulse per cycle of 10 input pulses. At one time odd-coded decade counters were used for scaling where a few components could be saved, but now the same integrated circuits are used for both counting and scaling applications.

Scaling circuits are sometimes used at the inputs to counters to divide the input events down to a number which will not overrange the counter. Extremely high-speed scaling circuits are also used at frequency meter inputs to reduce the signal frequency to a value that is within the measurement capability of the basic frequency meter.

Other BCD Codes

The 1 2 4 8 BCD code has been used exclusively thus far in the discussion of binary-coded decimal counters. Though it has become an industry standard for most instrumentation, other codes have been used and have advantages in certain applications. Table 6-5 shows six useful BCD codes. The 8 4 2 1 (or 1 2 4 8), the 4 2 2 1, and the 5 1 1 1 1 codes are *weighted* codes which means that a numerical weight can be assigned to each bit and the decimal number is equal to the sum of the weights of the 1 level bits.

In computation, the process of subtraction is usually accomplished by adding the complement of the subtrahend. The 4 2 2 1 and excess-3 codes are examples of codes for which the complement can be obtained simply by inverting each bit. For instance, if 0010 (decimal 2) in the 4 2 2 1 code is inverted, the result 1101 is decimal 7 in that code. (The number plus its "nines" complement always equals 9.) Note that

Table 6-5   BCD Codes

Decimal No.	Natural 8 4 2 1	4 2 2 1	Excess 3	Gray	2 out of 5	Biquinary 5 1 1 1 1
0	0000	0000	0011	0000	00011	00000
1	0001	0001	0100	0001	00101	00001
2	0010	0010	0101	0011	00110	00011
3	0011	0011	0110	0010	01001	00111
4	0100	1000	0111	0110	01010	01111
5	0101	0111	1000	1110	01100	10000
6	0110	1100	1001	1010	10001	10001
7	0111	1101	1010	1011	10010	10011
8	1000	1110	1011	1001	10100	10111
9	1001	1111	1100	1000	11000	11111

the same is true for the excess-3 code, but that the inverse of 0010 in the 8 4 2 1 is 13, an unallowed BCD number. The excess-3 code is most commonly used in BCD computation. For some reason, the 4 2 2 1 code which is both weighted and self-complementing has not evolved as a favorite in either the instrumentation or computation fields.

Gray codes have the characteristic that only 1 bit changes state from one numeral to the next. These are used extensively in position encoders because the maximum reading error during the transition from one incremental position to the next is the adjacent numeral. If the 8 4 2 1 were used for position encoding, all bits to be changed must change at precisely the same position to avoid large error outputs during such transitions as 3 to 4 and 7 to 8. This precision usually imposes unreasonable mechanical tolerances. The Gray codes, however, are neither weighted nor self-complementing and, thus, generally require conversion to one of the other system codes.

Many "error-detecting" codes have been devised. These codes include some means of checking to see if the number has been transmitted or processed correctly. Additional bits are added to the number to impose some condition that can be readily detected. For instance, a bit can be added which makes an even number of 1's in the number. If an error is made in 1 bit, the test for an even number of 1's will fail, and the error is detected. This is the simple parity check. This is often sufficient, though it clearly does not guarantee the detection of two errors. The two-out-of-five code is another example of an error-detecting code. The probability of undetected errors for this code is low. However, like the Gray codes, it must be translated into a more useful machine code once the signal is past the error-inducing hazards.

There are literally thousands of useful codes for binary, octal, decimal, and hexadecimal numbers. The techniques of converting the various binary-coded numbers to decimal form or to some other binary-coded form generally follows the same approach as illustrated earlier in this section.

6-3 Modulo-M Counters

The modulus of a counter is the number of input pulses for a complete count cycle. For instance, the four flip-flop binary counter of Fig. 6-1 has a modulus of 16. If it starts at 0000, in 16 pulses it will be at 0000 again. Similarly, a three flip-flop binary counter is a modulo-8 counter and the 8 4 2 1 BCD counter of Fig. 6-10 is a modulo-10 counter. Many counting and scaling applications require counters with a modulus other than 2^n or 10^n. Sometimes it is desirable to be able to select or vary the modulus of a counter as needed.

A counter can be made to have a particular modulus in several ways—the flip-flops can be wired to repeat the output cycle every M counts (fixed-modulus counter), or a decoding circuit can be used to detect the $M - 1$ count and stop or reset the counter, or the counter may be preset to a value from which it will proceed until it is full or clear. Examples of each of the above techniques will be illustrated in this section.

Fixed-Modulo Counters

The modulus of a binary counter with n flip-flops is 2^n. Therefore, simple binary counting circuits provide modulo-2, -4, -8, -16, -32, etc., counters. In each case the binary counter provides the largest modulus attainable with a given number of flip-flops. The number of flip-flops required to count to any given modulus will be at least equal to the number required to count to the next larger power of 2. For instance, a modulo-5 counter will require three flip-flops ($2^2 < M < 2^3$), a modulo-28 counter, five flip-flops, and so on.

A modulo-4 counter followed by a modulo-5 counter would result in a modulo-20 counter, since 1 cycle of the 4-counter advances the 5-counter one step, resulting in $4 \times 5 = 20$ counts for the combination. In general, the modulus of a sequence of counters is the product of the moduli of each counter in the sequence. Conversely a counter of modulo M can be made up of a sequence of counters having moduli m_1, m_2, m_3, . . . , where m_1, m_2, m_3, etc., are integer factors of M. For example, a modulo-30 counter could be made of modulo-2, -3, and -5 counters in sequence. The availability of a few counting circuits having moduli of the smaller prime numbers will allow counters of a great number of moduli to be constructed.

Modulo-3 counter. Two flip-flops are required for a modulo-3 counter. One possible circuit is shown in Fig. 6-17. As shown by the count sequence table, the counter outputs follow the natural binary sequence. When both flip-flops are CLEAR, there is a **1** at J of flip-flop A (J_A) and a **0** at J_B. On the first input pulse, FF A can set, but FF B cannot. Now there is a **1** at J_B. On the second pulse, FF B sets and FF A clears. The resulting **0** at J_A prevents FF A from setting on the third pulse, but FF B clears leaving both flip-flops cleared to begin again.

The modulo-3 counter can be combined with additional flip-flops (modulo-2 counters) to make modulo-($3 \times 2^n = 6$, 12, 24, etc.) counters.

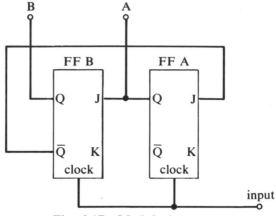

Fig. 6-17 Modulo-3 counter.

If the binary counter(s) precede the modulo-3 counter, the natural binary count sequence will be preserved. However, if a binary counter follows the modulo-3 counter, the count will go from 010 to 100 on the third pulse causing the number of pulses and the binary number output to disagree. In scaling applications, this will not matter since the modulus is the same in either case. Binary stages may be added to make a modulo-6 or -12 counter in either the synchronous or asynchronous mode. In the synchronous mode, Q of each binary flip-flop is connected to J and K of all

---

Experiment 6-12 *Modulo-3 and -6 Counters*

The modulo-3 counter of Fig. 6-17 is wired and the count sequence observed. A modulo-6 counter is made by connecting a T binary before the modulo-3 circuit and then by connecting the T binary after the modulo-3 circuit. The count sequences of the two modulo-6 counters are compared.

Count Sequence Table for Fig. 6-17

Count n	Outputs	
	B	*A*
0	0	0
1	0	1
2	1	0
3, 0	0	0

Count Sequence Table for Fig. 6-18

	Outputs		
Count n	C	B	A
0	0	0	0
1	0	0	1
2	0	1	0
3	0	1	1
4	1	0	0
5, 0	0	0	0

succeeding stages. The asynchronous counting mode is achieved by connecting Q of the binary output to the input connection of the modulo-3 counter.

Modulo-5 counter. The synchronous and asynchronous BCD counters of Figs. 6-10 and 6-12 were made up of a 2-counter followed by a 5-counter. The modulo-5 part of the asynchronous circuit is shown in Fig. 6-18. Flip-flops A and B are wired as an ordinary modulo-4 binary counter. Flip-flop C is clocked by the input pulses but cannot set until FF A and FF B are 1. When FF C is set (after the fourth pulse) the $\overline{Q}$ connection from FF C to J of FF A prevents FF A from setting. Thus, on the fifth pulse FF C clears and flip-flops A and B stay clear. Preceding binary stages can be added to make modulo-10, -20, -40, etc., counters with natural binary-coded output.

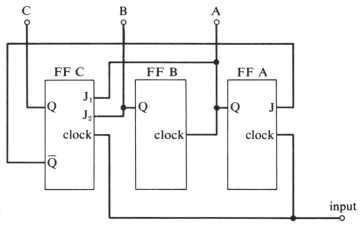

Fig. 6-18   Modulo-5 counter.

Count Sequence Table for Fig. 6-19

	Outputs		
Count n	C	B	A
0	0	0	0
1	0	0	1
2	0	1	0
3	0	1	1
4	1	0	0
5	1	0	1
6	1	1	0
7, 0	0	0	0

Modulo-7 counter. Another example is used to illustrate fixed-modulus counter design. The modulo-7 counter shown in Fig. 6-19 requires some gating in order to achieve the desired count sequence and modulus. The desired count sequence table shows that flip-flop A is to alternate state except on the sixth pulse when it is to stay clear. Thus J_A should be **0** when B AND C are **1**. This can be written as $\overline{J}_A = B \cdot C$ or $J_A = \overline{B \cdot C}$ and implemented by gate 1 as shown. Flip-flop C changes state on each **1–0** transition of FF B which requires only a simple Q-to-

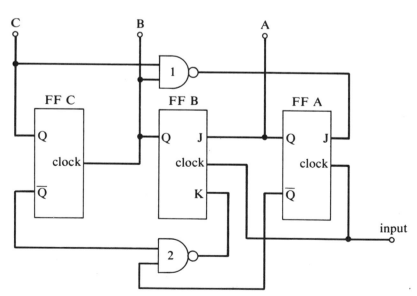

Fig. 6-19   Modulo-7 counter.

clock connection between FF B and FF C. Flip-flop B sets on the first pulse after FF A is 1 which is done by clocking FF B with the signal and connecting Q of FF A to J of FF B. Flip-flop B is to clear on the fourth pulse (after A is 1 and on the sixth pulse (after C is 1). Therefore, K_B should be 1 when A OR C are 1. This is written $K_B = A + C = \overline{\overline{A} \cdot \overline{C}}$ and implemented by the gate 2 circuit. Thus gates can be used with the JK flip-flop to obtain the desired count sequence. Of course, this circuit can be used with binary circuits to provide modulo-7, -14, -28, etc., counters and with modulo-3 counters to provide modulo-21, -42, etc., counters and so on.

Experiment 6-13 *The Use of Gates in Fixed-Modulus Counter Design*

The modulo-7 counter of Fig. 6-19 is wired and tested. The frequency division (scaling) of this circuit is checked with a digital frequency meter, and, if possible, the maximum input frequency is determined. The design and testing of a modulo-11 or -13 counter is suggested as an exercise.

Modulus control by direct clearing. In all previous examples of fixed-modulus counters, the counter circuit uses logic gates and/or J and K flip-flop inputs to control the counting sequence and bring the counter to zero after the maximum count has been reached. Another approach is to apply a pulse to the direct clear inputs of the flip-flops when the desired maximum count is reached. A counter of this type with a modulus of 11 is shown in Fig. 6-20. On the count of 11 the output will be 1011 and all inputs to the NAND gate will be 1. The resulting 0 output from the gate clears all of the flip-flops. The clear pulse is very short since the clearing of the flip-flops returns the gate output to a 1. The counter is then free to count again to 11. Note that reset occurs on the eleventh count so 10 is the maximum count and 11 is the modulus of this counter.

A counter of any modulus can be quickly and easily made in this way. The flip-flop outputs which are 1 when the desired modulus is reached are connected to the clear control gate. It is not necessary to insure that the flip-flops which are to have a 0 output actually have a 0. For instance, the gate output in Fig. 6-20 will be 0 for a count of 1011 or 1111 since the level of flip-flop C is not detected. However, since the counter resets at 1011, the count of 1111 can never be attained. In general, the smallest count that satisfies the 1 output requirement must have 0's at the outputs of the other flip-flops.

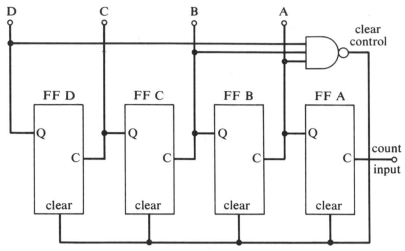

Fig. 6-20   Modulo-11 counter with direct clearing.

The disadvantages of this simple direct-clearing approach to modulus control are loss of the direct clear inputs of the flip-flops for other purposes and some loss of counting rate. The minimum time between input pulses must be the sum of the counter propagation time, the gate delay time, the reset time, another gate delay time, and the minimum time between clear and trigger pulses for the flip-flop.

Experiment 6-14   *Modulus Control by Direct Clearing*

The direct-clearing modulo-11 counter of Fig. 6-20 is wired and the count sequence is observed. If possible, the maximum input frequency for reliable counting is determined. Counters with several moduli between 1 and 16 are wired by connecting other combinations of flip-flop outputs to the gate input.

Variable Modulus Counters

For some counting and scaling applications, it is desirable to be able to change the modulus of the counter quickly and easily. This is generally done by allowing an ordinary binary or BCD counter to advance until the desired maximum count is detected and then stopping or clearing the counter. A general block diagram of a variable modulus counter is shown in Fig. 6-21. A circuit is used to compare the outputs of the counting register with inputs representing the desired maximum count. When the desired count is reached, the digital comparison circuit output changes logic level. The comparator output can be used to close the counting gate thus stopping the counter, or to reset the counter, or to cause the counter register to clear on the next input pulse.

An asynchronous binary counter which can be programmed by external inputs to stop at any count from 1 to 15 is shown in Fig. 6-22. The output of gate 1 controls the counter. As long as the count control is **1**, the counter will advance. This will be the case until all inputs to gate 1 are **1**. A **1** output results from gates 2 through 5 if either the control input or $\overline{Q}$ of the corresponding flip-flop is **0**. Thus, if the 2^1 control input is **1**, the output of gate 3 will be **1** only when B is **1**. The counter advances then until all outputs are **1** for which **1** levels have been applied to the control inputs. At this count, the outputs A, B, C, and D match the control inputs 2^0, 2^1, 2^2, and 2^3 exactly, the count control becomes **0**, and flip-flop A is prevented from changing due to the **0** at its J and K inputs. A complete digital comparison has not been made by gates 1 through 5 since a **0** at a control input produces a **1** at its gate output regardless of the state of the corresponding flip-flop. However, as in the case of Fig. 6-20, the lowest count for which the **1** states match must have **0**'s at the remaining flip-flop outputs. When this self-stopping counter is cleared or reset by a pulse to the clear input, it will advance again toward the programmed maximum count as pulses appear at the count input.

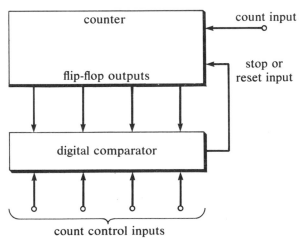

Fig. 6-21   Variable modulus counter.

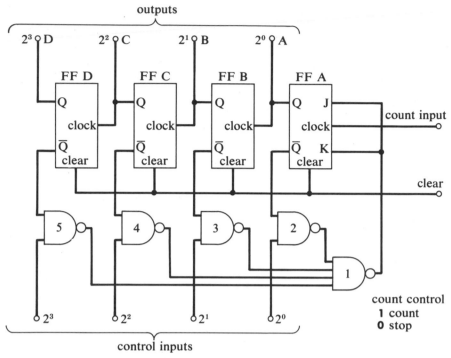

Fig. 6-22 **Counter that stops at preset count.**

Experiment 6-15 *Self-stopping, Variable Modulus Counter*
The counter of Fig. 6-22 is wired. It is observed that the modulus
of the counter is equal to the binary number applied to the control
inputs. The level of the count control signal is observed to change
when the preset count is reached.

Automatic Recycling

An adjustable modulus counter which recycles automatically is
shown in Fig. 6-23. This counter senses the correspondence between
control inputs and flip-flop outputs with the same circuit as the previous
example. When the inputs and outputs match, a **0** is applied to the J
input and a **1** to the K input of each flip-flop so that the entire register
will clear on the next count pulse. Providing for a possible transition of

all flip-flops on any input count requires a synchronous counting circuit. Gates 6 through 12 are used to mix the usual synchronous counting J and K input connections with the count control signal. OR gates are used at the K inputs since a **1** from the count control must take priority over other K inputs. AND gates suffice at the J inputs since a **0** at any J input prevents the flip-flop from setting. Note that NAND gates 6, 9, and 12 are drawn using the OR-equivalent symbol.

Direct clearing of variable modulus counters is done by connecting the count control output to the direct clear inputs of the flip-flops. The comparison circuit of Fig. 6-23 (gates 1–5) could be connected to the direct clear inputs in a synchronous counting circuit such as Fig. 6-8 which would eliminate the need for gates 6 through 12 in Fig. 6-23.

Many other circuits for variable modulus counters are used, but all follow the basic approach shown in Fig. 6-21. The flip-flop outputs can also be decoded and then compared with an octal, decimal, or other input signal. The stop or reset functions can be derived in many ways depending on the gates and flip-flops to be used and the requirements for the circuit.

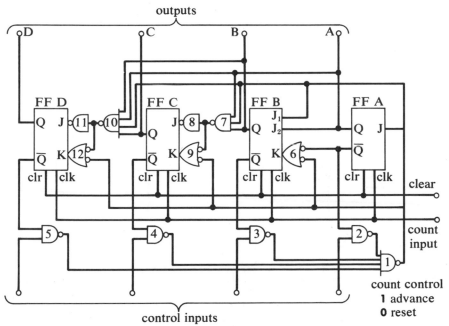

Fig. 6-23 Automatic cycling variable modulus counter.

Experiment 6-16 *Automatic Recycling, Variable Modulus Counter*

The circuit of Fig. 6-22 wired in the previous experiment is modified by connecting the count control signal to the clear input instead of to J and K of FF A. A direct-clearing, automatic recycling, variable modulus counter results. The relationship between the binary number set into the control inputs and the counter modulus is noted. The circuit of Fig. 6-23 may also be wired and tested if desired.

Preset Counters

A *preset counter* is a counter that can be "preset" to perform a particular count. That is, to count to a preset number and stop, or to start at a preset number and count down to zero, or to start at one preset num-

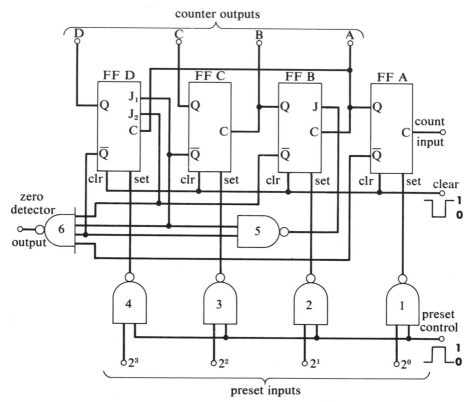

Fig. 6-24 Preset BCD down-counter.

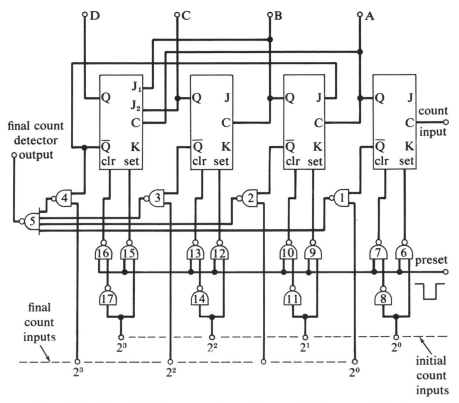

Fig. 6-25 Preset BCD up-counter with presettable count detector.

ber and count to a second preset number. Preset counters are forms of the variable modulus counter. The technique of stopping or clearing a counter when a particular count has been reached has been discussed previously in this section. Examples of counters that are preset to a particular count before beginning the count are presented below.

A BCD down-counter which can be preset to any count and which indicates when it has counted down to zero is shown in Fig. 6-24. The counter is first cleared by applying a **0** to the clear input. Then a **1** is applied to the preset control input causing those flip-flops to set for which a **1** level exists at the preset inputs. Thus the preset number is transferred into the counting register. Now the counter is allowed to count. When the counter has counted down to zero the output of gate 6 becomes **0**. The zero detector output signal can be used to stop further counting by a connection to J and K of flip-flop A or to indicate to another circuit that the countdown has been completed. This kind of preset counter is used, for instance, to measure how long it takes to accumulate the preset number of counts. Additional decades can be ganged together to achieve any required count capacity. Ten-position BCD-coded panel switches are frequently used to provide the logic levels to the preset inputs for each decade.

The counter of Fig. 6-24 must be cleared before the preset control pulse is applied to insure that only the desired flip-flops will be set. A positive preset circuit which uses both direct set and clear inputs is illustrated in Fig. 6-25. When the preset input goes to **1**, the logic levels at the initial count inputs are transferred to the corresponding flip-flops by gates 6 through 17. (Remember that a **0** at the set or clear input performs the set or clear function.) This circuit also includes a presettable count detector like the one first shown in Fig. 6-22. The output from gate 5 will become **0** when the count matches the levels at the final count inputs. This signal can be used to stop, reset, or preset the counter according to the application requirements.

Experiment 6-17 *A Preset BCD Down-Counter*

The preset BCD down-counter of Fig. 6-24 is wired. The zero detector output is used to stop automatically the counter on the count of zero by connection to the J and K inputs of FF A or to the clear bus. The circuit is tested for a number of preset input values.

6-4 Shift Registers

As stated at the beginning of this chapter, a register is a group of flip-flops that is used to remember binary or binary-coded information. The flip-flop inputs are connected to the data sources, and the outputs to the read circuits. If there are no interconnections between the flip-flops, it is called a parallel input, parallel output register, i.e., all bits of information are entered at once and read out at once as with the memory register of Fig. 6-13. It is frequently convenient to be able to shift the information over from one flip-flop to the next in the register. Registers where this is possible are called *shift registers*. Shift registers are used for accepting binary data in serial form, for aligning decimal places for binary addition, and for shifting the multiplicand in binary multiplication.

Serial Entry

A basic shift register circuit is shown in Fig. 6-26. Data is entered in the form of a logic level at input I. If the complement of I is avail-

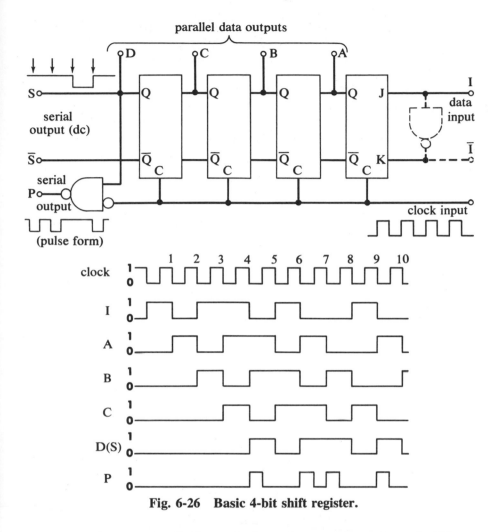

Fig. 6-26 Basic 4-bit shift register.

able, it is connected to $\bar{I}$; if not, the inverting gate shown is used. When the clock input goes to **1**, the level at I is entered into the master section of FF A to appear at Q of FF A when the clock signal returns to **0**. Thus, the information at I appears at A one clock pulse later. On the next clock pulse, the information at A will appear at B while the new information at I will be transferred to A. The waveform chart of Fig. 6-26 shows the response of the shift register to a hypothetical signal at I. Note how the information I is shifted to the next flip-flop output on each clock pulse, appearing at A one pulse later, at B two pulses later, and so on. The output at S has exactly the same form and sequence as that at I except it appears four clock pulses later. A logic-level train (serial-form binary signal), such as I, can be delayed by any desired number of clock pulses by using that number of flip-flops in the shift register circuit. If a pulse-form output is desired, it may be obtained by gating S with a signal derived from the clock source as shown.

Another application of the shift register is the conversion of data from serial to parallel form. Note that the first four levels at I in Fig. 6-26 are 1011 and that after four clock pulses the outputs DCBA are 1011 in parallel form (simultaneous appearance on separate lines). If eight flip-flops were used, an 8-bit word could be read into the register in eight clock pulses and appear at the eight flip-flop outputs simultaneously. This application is most useful for converting serial digital information telemetered from a remote source to a parallel output suitable for printing or computer input.

The basic shift register can be wired with all three types of clocked flip-flops. A comparison of their interconnections is shown in Fig. 6-27. The JK and RS flip-flop connections are identical since their truth tables are identical for the states $R = \bar{S}$ ($J = \bar{K}$). The D flip-flop, somewhat limited for general-purpose applications, is seen to provide the minimum interconnections for the shift register with no input-inverting gate required.

Experiment 6-18   *A Serial Input Shift Register*

The basic shift register shown in Fig. 6-26 is wired. Data and clock pulses are applied from manually operated switches to observe the shifting action of the circuit.

Parallel Entry

As in the case of the counting register, the shift register can be modified for parallel data entry (preset). A 4-bit shift register with parallel data entry is shown in Fig. 6-28. When the enter input is pulsed with a

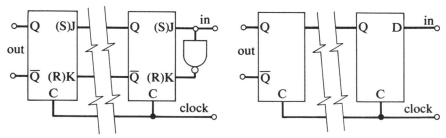

Fig. 6-27   Shift register circuits for JK, RS, and D flip-flops.

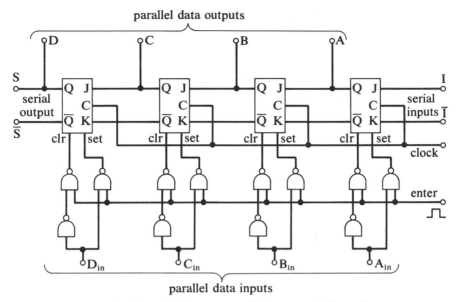

Fig. 6-28   Parallel input, parallel output shift register.

1 pulse, the logic levels at the data inputs are set into the corresponding flip-flops by the direct set and clear inputs. If the clock input is now pulsed, the data will be shifted one flip-flop to the left for each clock pulse. This circuit is used to align the decimal places of two binary numbers prior to addition or subtraction and to shift the multiplicand for each step in a binary multiplication.

Another obvious application of Fig. 6-28 is as a parallel-to-serial information converter. Data entered at the parallel inputs will appear at S in serial form as the clock input is pulsed.

Experiment 6-19   *A Parallel Input Shift Register*

Parallel data inputs are added to the circuit of the previous experiment to make the circuit of Fig. 6-28. Data is entered into the register via the parallel input and then observed in serial form at S as the circuit is clocked.

Shift Right/Left Register

The shift registers in Figs. 6-26 and 6-28 shift the data to the left on each clock pulse. The flip-flops could be wired to shift the data to the right by connecting Q and $\bar{Q}$ of B to J and K of A, Q and $\bar{Q}$ of C to J and K of B, and so forth. The serial data input would then be at J and K of FF D, and the output at Q and $\bar{Q}$ of FF A. The circuit would, in fact, be identical to Fig. 6-26 if the output labels were changed from DCBA to ABCD, left to right, respectively. Therefore, there is no electrical difference between a shift right and a shift left register (unlike the up- and down-counters where there is a difference).

In some cases, it is desirable to be able to shift the data in a register to either the right or left as required. Such a register is a *shift right/left register* and is obtained by gating the J and K inputs of each flip-flop from the Q and $\bar{Q}$ outputs of the flip-flops on either side of it. A shift right/left register is shown in Fig. 6-29. Type-D flip-flops are used because only one input needs to be gated. Each D input is connected to the output of an AND-OR-INVERT gate. The left AND gate is connected to $\bar{Q}$ of the flip-flop to the left, whereas the right AND gate is connected to the flip-flop output to the right. The level at the shift control input determines whether the left or right AND gate in each AND-OR-INVERT gate will be active. When the shift control input is **1**, a **0** is applied to the left AND gates blocking them. The D inputs then obtain their information from flip-flops to the right and thus shift left when clocked. When the shift control input is **0**, the left AND gate is active and infor-

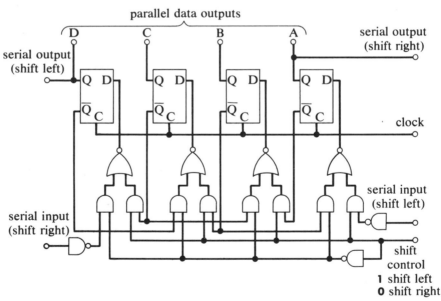

Fig. 6-29   Shift right/left register.

mation is shifted to the right. Following the pattern of Fig. 6-29 a shift right/left register of any required length may be made. Frequently the direct parallel data entry of Fig. 6-28 is combined with the shift right/left logic of Fig. 6-29 to provide a "full-treatment" serial/parallel input, serial/parallel output, shift right/left register.

Circulating Register

Like other registers, the shift register can be used to store information. If an 8-bit word is clocked serially into an 8-bit (eight flip-flop) shift register, it is stored in the register. The word can then be read out in parallel immediately or in serial form during the next eight clock pulses. It is sometimes desirable to be able to read the word out in serial form without losing it, that is, without destroying its storage. This is accomplished by entering each bit back into the data input I as it appears at the output S. If the clock input were pulsed continuously, the word would then circulate through the register continuously and appear at the serial output S every n clock pulses, where n is the number of flip-flops in the register.

Parallel data outputs are often omitted from the shift register designed for circulating applications. This means that only five or six external connections to the register are required (input, output, clock, and power) *no matter how many flip-flops are used.* For this reason, serial input/output shift registers for circulating storage applications were one of the first circuits developed for large-scale integration (LSI). LSI shift registers of 100-bit lengths are made by several manufacturers and are becoming an increasingly popular form of data storage.

A circuit for controlling a circulating register is shown in Fig. 6-30. When the circulation control is **1**, the register output is connected to the register input and the data input gate is OFF. When the circulation control goes to **0**, the data at the data input will be entered into the register.

Ring counters. Another application of the feed-around connection is made by entering a single **0** or **1** in a parallel output shift register with all other flip-flops in the opposite state. The position of the **1** or **0** will move through the flip-flop outputs on successive clock pulses to give n distinguishable states. In the case of the 4-bit shift register of Fig. 6-26, these 4 states would be 0001, 0010, 0100, and 1000. The result is an automatic-cycling modulo-n counter called a *ring counter.* The ring counter is actually a counter and decoder in one circuit as can be seen from the output states. This advantage is offset somewhat by the inefficiency of requiring one flip-flop per bit and having to provide a circuit to introduce the required starting state.

The circuit in Fig. 6-31 is a ring counter that is self-starting and -correcting. Any starting or counting error will be corrected within one cycle. The Q and $\bar{Q}$ outputs of flip-flop A are connected to the J and K inputs of FF B to pass the state of FF A on to FF B with each clock pulse. Likewise, B is connected to FF C, and so on. However, in this case E is not connected back to FF A. Instead, FF A is controlled at the J inputs and allowed to go to **1** only when B, C, and D are **0**. If this is not the case, FF A will feed **0**'s to B on each clock pulse until B, C, and D are all finally **0**. Then, on the next clock pulse, FF A will go to **1**. With no inhibition at the K input, FF A will return to **0** on the next clock pulse and the **1** transfers to FF B. FF A must remain **0** until the **1** has passed to FF E. On the next pulse, FF A will go to **1** and the cycle begins again.

Experiment 6-20 *Circulating Registers and Ring Counters*

The parallel entry shift register of the previous experiment and Fig. 6-28 is made into a circulating register by connecting the S and $\bar{S}$ outputs to the I and $\bar{I}$ inputs. A 4-bit word is entered by the parallel data input and then circulated repeatedly through the register as the circuit is clocked. The word in dc or pulse form (cf. Fig. 6-26) may be observed on an oscilloscope.

The number 0001 is entered into the above circuit to make a ring counter. The sequence of the circuit is observed as pulses are "counted" at the clock input. Wiring and testing the 5-bit ring counter of Fig. 6-31 is optional.

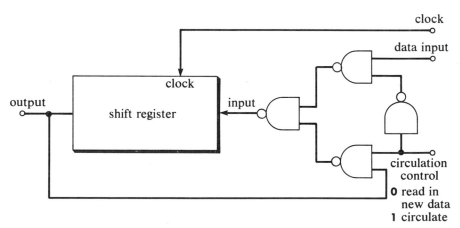

Fig. 6-30   Circulating register.

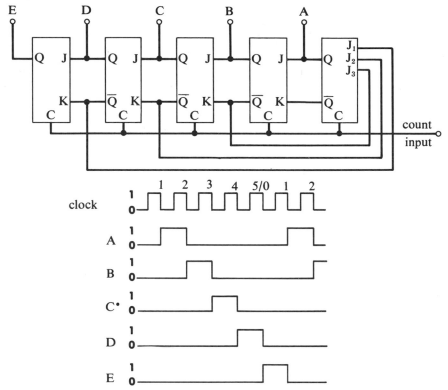

Fig. 6-31 Self-starting and -correcting 5-bit ring counter.

6-5 Counting Measurements

Earlier sections of this chapter have shown how flip-flops are used to count and store information. Such circuits have virtually endless applications in digital instrumentation and computation. However, in every case, some auxiliary circuits will have to be used to control what is counted or stored and when. This section uses typical counting measurements to illustrate the counter and register control for particular applications. First the problem of gating the count signal is discussed, then an automatic measurement cycling circuit is developed.

Gating the Counter Input

The simplest counter input gate is shown in Fig. 6-32. When the gate time input is **0**, the AND gate is CLOSED. Pulses at the count input

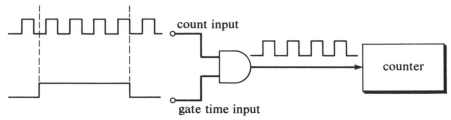

Fig. 6-32   Counter input gating.

will be counted only when the gate time input is **1**.   This simple gate would be satisfactory where the gate time signal is only **1** when the count measurement is to be made and will not become **1** again until the count reading is taken, the counter reset, and the next count measurement is desired.   To insure that this is the case usually requires that the gate time signal itself be gated or controlled.   Control is certainly required when the gate time signal is a continuous square wave from a standard clock oscillator.   In other words, the problem is not that of gating, but that of gating just once per count measurement.

There are numerous circuits for solving this problem.   One of the most direct is to use two flip-flops — one to start the count and the other to stop it.   Each flip-flop is triggered by the appropriate part of the gate time input signal.   Such a circuit is shown in Fig. 6-33.   The start and stop flip-flops are normally cleared so that Q_{FFA} is **0** and $\overline{Q}_{FFB}$ is **1**.   Gate 2 is closed by the **0** input.   On the first **1–0** transition of the gate time signal, FF A triggers ON opening gate 2 and the count begins.   The inverting gate 1 at the C input of FF B causes FF B to trigger on the **0–1** transition of the gate time signal.   The resulting **0** at $\overline{Q}_{FFB}$ closes gate 2, ending the count.   The counting interval in this case is exactly the duration of the **0** level at the gate time input.   Neither FF A nor FF B can be triggered to clear by the gate time signal because the connections from $\overline{Q}$ to K of each flip-flop prevent them from clearing.   Thus gate 2 cannot open again until a **0** pulse is applied to the reset input.   Note also that the start and stop flip-flops cannot be triggered in the wrong order because of the connection from Q_{FFA} to J_{FFB}.   This circuit may be recognized as Fig. 5-42, the single-pulse generator.   The letters in parentheses in Fig. 6-33 refer to the corresponding inputs of the count of gate control of the general-purpose counter timer of Figs. 1-12 to 1-18.

The signal that controls the gate time interval may take a variety of forms.   Frequently, the period of the gate time signal is the desired timing interval.   The T flip-flop converts each input cycle into successive half-cycles of a square wave.   The resulting gate control circuit for counts per period is shown in Fig. 6-34.   If the gate time signal is a 1-Hz signal

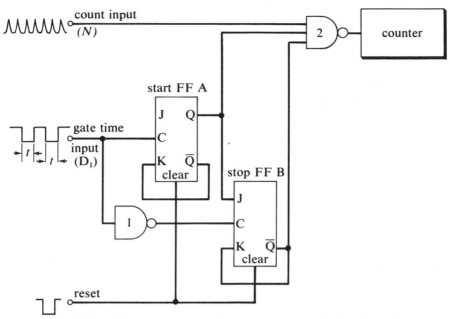

Fig. 6-33 Count gate control, counts per gate interval mode.

from a standard time oscillator, the counter will indicate counts per second or frequency. Conversely, if a 1-MHz signal is applied to the count input, the counter will indicate the number of microseconds in one period of the gate time input signal. This, then, is the mode of gate control used for period and frequency measurements.

A third kind of gate time signal is shown in the count gate control

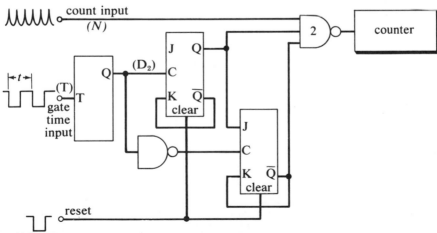

Fig. 6-34 Count gate control, counts per gate time period mode.

circuit of Fig. 6-35. An event results in a pulse at the start input and another event causes a pulse at the stop input. The gate time interval is the time between the two events. If a standard time source is connected to the count input, the time between the two events will be measured and stored by the counting register.

The circuits for the three described modes of count gate control differ only by the connections to the C inputs of the two flip-flops. For use in a general-purpose instrument, it would be desirable to use selector gates at those inputs so that a single gate control circuit could be used in several modes.

Automatic Measurement Cycling

Most modern digital instruments provide for automatic repetitive measurements. This involves controlling a sequence of events within the instrument generally as follows: (a) the count gate control is allowed to be triggered; (b) the count gate is opened and closed for the measurement interval; (c) the counter information is transferred to the memory; (d) an adjustable display time (or repetition delay time) passes; (e) the counter is cleared; (f) the count gate control is allowed to be triggered again; and so on. A typical circuit that provides such a sequence for frequency or period measurement is developed below.

The circuit of Fig. 6-36 contains a series of BCD counters such as in Fig. 6-12, memory registers such as Fig. 6-13, and decoder-driver circuits such as Fig. 6-15. The counter is gated by the count control cir-

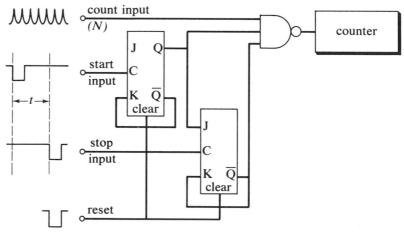

Fig. 6-35   Count gate control, start-stop mode.

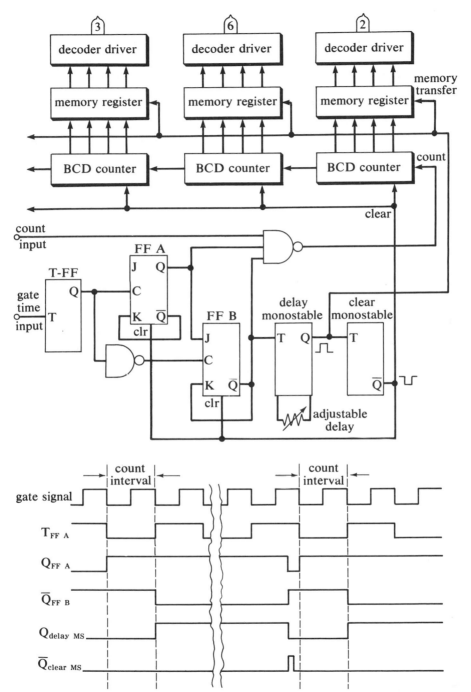

Fig. 6-36 Automatic cycling frequency or period meter.

cuit of Fig. 6-34. When the count interval is over, $\overline{Q}_{FFB}$ goes from **1** to **0** triggering the delay monostable. The output of the delay monostable is used to activate the transfer of information from the counting to memory registers. The duration of the delay monostable output pulse is adjusted for the desired display time or measurement repetition rate. At the end of the display time, the display monostable triggers the clear monostable. This monostable generates a short **0** pulse to clear the counting registers and the flip-flops in the gate control circuit.

In this example, monostable multivibrators connected in sequence were used to provide the timing of the automatic memory transfer, display time delay, and register-clearing cycle as well as to provide the signals to activate the memory transfer and register-clearing functions. More complex sequences for other applications can easily be devised using combinations of flip-flops and monostable circuits in this way.

Experiment 6-21 *Count Gate Control and Automatic Cycling*

A complete period or frequency meter is wired as shown in Fig. 6-36. A 1.000-Hz signal is connected to the gate time input and a variable frequency source to the count input. The frequency of the variable source is measured, and the measurement sequence is observed. The counting of the counter can be watched if the memory is defeated by connecting the memory transfer input to a logic **1**.

By connecting other signals to the count and gate time inputs and by modifying the count gate control circuit, as suggested in Figs. 6-33 and 6-35, the period, duration, or time relationship of a number of signals can be measured. With the addition of a decade scaler this circuit can be used to measure multiple period average and frequency ratio as well.

---

chapter seven

Digital and Analog–Digital Instruments and Systems

Many basic phenomena and techniques are utilized in scientific instrument systems to obtain information about material composition, temperature, pressure, position, and other physical and chemical parameters. The desired information is encoded in the form of signals (electrical, optical, mechanical) that can be readily sorted, converted, decoded, and finally displayed on a readout device in terms of numbers and desired units.

For example, information about the elemental composition of an aluminum alloy can be encoded by a spark source as characteristic electromagnetic radiation (optical emission signals). These optical signals can be received and sorted by a spectrometer into wavelengths characteristic of each element in the aluminum alloy. The optical signals can then be converted to electrical signals by photodetectors, and the electrical signals can be manipulated and utilized to provide a numerical readout in units of concentration for each element in the sample. Several conversions of data from one form to another are present in this example. In scientific instruments it is common practice to convert (transduce) the form of the data several times in order to display the readout information in the desired form and units. The conversion techniques and devices are carefully chosen to provide the required selectivity, sensitivity, and precision.

The output data are sometimes used for automatic control of the input parameter, and, in this case, interconversions of the output signal information are often required to provide the necessary feedback signals to control the input parameter. It is the purpose of the first section of this chapter to consider some general concepts of converting data from one form to another, especially with respect to electrical interconversions.

Next, analog operations are described with emphasis on operational amplifier circuits, specifications, and techniques. The timing and sequencing of events and signals that are so important in measurement and control systems are described and several useful circuits are presented. Digital and analog circuits are then combined to illustrate important types of digital-to-analog and analog-to-digital converters for scientific instrumentation. This is followed by a brief discussion of digital computation.

7-1   Data Domains and Interconversions

When a physical quantity such as temperature, pressure, or pH is measured with the aid of an electronic instrument, the quantity is converted to an electrical signal. The electrical signal is then processed and displayed. For a time, the desired information is contained in the form of an electrical signal, a particular characteristic of which is related in a known way to the measured quantity. The many characteristics of *electrical signals* that are used to convey information can be categorized in three groups: amplitude (A), time interval (Δt), and digital (D). The amplitude category includes all electrical quantities such as voltage, current, resistance, capacitance, wherein the magnitude of the quantity is related by a *continuous* function to the desired information. The time interval category includes those signals for which the time relationship between parts of a waveform or between different waveforms is a *continuous* function of the desired information. Signals for which the frequency, pulse width, or phase angle are related to the desired quantity are examples of time interval signals. A digital signal represents a specific number that is related *discontinuously* in integer steps to the desired information. The digital signal contains the number as logic levels at a group of terminals simultaneously (parallel form), as a succession of logic levels at a single terminal (serial form), or as a number of pulses in a pulse train (count serial form).

Domains

The categories analog, time interval, and digital can be considered *domains* in which the quantity can exist in electrical form. For this discussion let us relegate all other (nonelectrical) forms of informations to the "physical" domain P, but with the realization that various intra-conversions may occur within this domain. In the physical domain, the desired quantity could be position (position of balance pointer, meter needle, recorder pen, etc.), strain, temperature, light intensity, thermal conductance, etc. The three electrical and one nonelectrical data domains are shown in Fig. 7-1. The measured or controlled quantity must exist in one or more of these domains at any instant.

Interdomain Converters

If the data domains are represented as shown in Fig. 7-1, then scientific instruments can be classified according to the interdomain conversions that are performed in them. Interdomain converters can be classified as physical quantity-to-digital (P-D), digital-to-analog (D-A), analog-to-time interval (A-Δt), etc. Numerous examples exist for each of the twelve possible interdomain conversions. A few instruments that illustrate the data domain concepts are described in this section. Many more examples will be described in greater detail in the remainder of this chapter.

Devices that convert a physical property into some type of electrical signal, (P-A, P-Δt, and P-D converters) are usually referred to as *input transducers*; similarly, converters to a physical property (A-P, Δt-P, and D-P) are called *output transducers*. A few examples of input and output transducers will be described for illustrative purposes, but generally these devices fall outside the scope of this text.

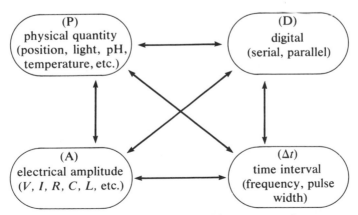

Fig. 7-1   Data domains and interconversions.

Temperature Measurement

Using the data domain concept, an instrument or measurement system can be designed by identifying the domain of the quantity to be measured and the domain of the desired output form, and then using the appropriate interdomain converters to convert from the initial to final domains. The example of temperature measurement is used here to illustrate this procedure. Temperature is in the P domain. Since this book is primarily about digital electronics, let us presume that the desired output form is a number for readout, printout, or computer input. Thus the desired instrument will convert temperature in the P domain to an electrical signal in the D domain.

P-D converter. One of the few direct P-D converters available at this time is the encoded wheel as shown in Fig. 7-2. A separate light-photodetector pair is used to monitor information for each concentric ring with a light and detector on opposite sides of the wheel. The photo-detectors provide logic level outputs that depend on whether the wheel is opaque or transparent at that position. For the example in Fig. 7-2 the position of the wheel is known within 1 part in 16 by decoding the binary outputs from the four detectors. For greater accuracy additional rings and photodetectors are required. Temperature could be converted directly to mechanical position of the shaft of the encoded wheel (an intra-P domain conversion) by means of a bimetallic coil as illustrated in Fig. 7-3. Thus a thermometer with a digital output could be designed using a direct P-D conversion.

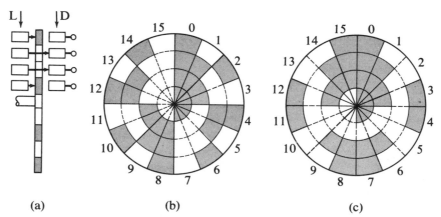

(a) (b) (c)

Fig. 7-2 Coded wheels: (a) end view showing light(L)–detector(D) pairs for each concentric ring; (b) opaque and transparent regions for "natural" code; (c) for Gray code.

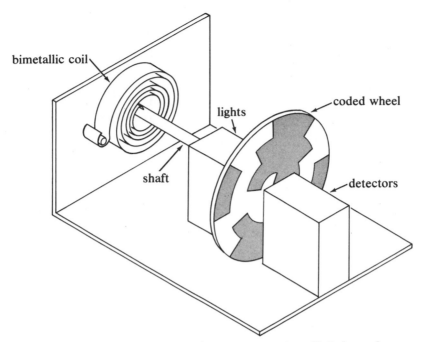

Fig. 7-3 **Direct conversion of temperature to a digital number.**

P-A-Δt-D converter. Interdomain converters do not always follow the most direct route in providing the desired interconversion. For instance, the direct P-D converter described above might not be sufficiently accurate, fast, sensitive, or economical for the desired application. If there is not a good system available for direct P-D conversion, a temperature transducer to another domain can be used, followed by additional interdomain conversions. For instance, the resistance of a thermistor or fine platinum wire is related to the temperature (P-A conversion). When the thermistor is used in a Wheatstone bridge circuit, as shown in Fig. 7-4, a voltage is produced at the bridge output that is related to the temperature. A digital voltmeter can then be used to convert the bridge voltage to a digital form.

Recall the ramp and voltage-to-frequency converter types of DVM's described in the first chapter, Figs. 1-19 and 1-20. Both of these instruments perform a conversion of voltage (A domain) to pulse width or frequency (Δt domain) followed by a second conversion by counter and standard time oscillator to the digital form (Δt-D). Thus, from the measured temperature to the DVM readout, as shown in Fig. 7-4, there are several interconversions, i.e., P-A-Δt-D. Not the most direct route, indeed! It is possible to make an A-D conversion without traversing

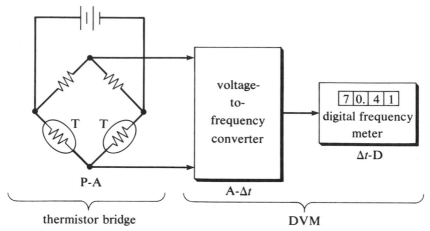

Fig. 7-4 Digital temperature measurement using a thermistor transducer.

the Δt domain. Several A-D conversion techniques are described in Section 7-5.

   P-Δt-D converter.   A transducer that converts temperature to frequency (P-Δt) is the basis of an accurate instrument for temperature measurement. The input transducer is a quartz crystal which has a resonant frequency that is linearly dependent on temperature. The quartz crystal is used in a crystal oscillator, the frequency of which is measured with a digital frequency meter (Δt-D converter). The measurement system, a P-Δt-D instrument, is shown in Fig. 7-5. Because the frequency of the quartz crystal changes about 1 kHz/°C out of 28 MHz, the frequency f_t is mixed with a standard frequency f_s to provide a difference frequency $f_t - f_s$ that can be measured and calibrated more easily.
   Even though the signals f_t or $(f_t - f_s)$ may be logic level signals, they are *not* in the digital domain. The information is contained in the frequency and the frequency is a continuous function of the temperature, not a specific number represented by the succession of logic levels. However, the standard clock opens the gate to the counter for a precise time interval, and it is this operation that provides an absolute number of pulses at the counter input. The signal at this point, being a specific number of logic level changes, is in a *count serial digital* form. The counter converts the count serial digital signal to a *parallel digital signal* at the flip-flop outputs [binary-coded decimal (BCD) or binary]. Thus the clock and gate perform the Δt-D conversion. A clear understanding of the difference between the Δt and D domains is important to an under-

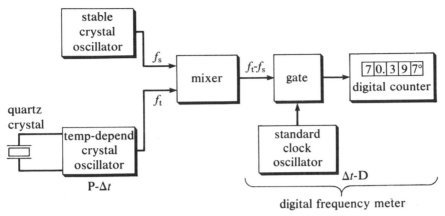

Fig. 7-5 **Digital temperature measurement using a quartz crystal transducer. (Courtesy of Hewlett-Packard Co., Inc.)**

standing of the various forms, applications, and limitations of digital instrumentation. In addition, a careful analysis of the required initial and final domains and the available techniques of domain interconversion could reveal shortcuts in many cases.

Light Measurements

One further example of P-D interdomain conversion, before proceeding to specific circuit techniques, is for the quantitative measurement of light. In the ultraviolet and visible spectral regions the photomultiplier (PM) tube, illustrated in Fig. 7-6, is a sensitive and precise input transducer. It is the purpose here to illustrate that it is possible to arrive at a digital readout (in D domain) that is related to the incident light (from the P domain) by several routes of interdomain conversion.

The photons of light strike the photocathode of the PM tube, and photoelectrons can be ejected from the cathode with an efficiency dependent on the energy ($h\nu$) of the photon and the type of cathode surface. Each photoelectron that is ejected by a photon is attracted to the first dynode, because the dynode is maintained at a positive voltage relative to the photocathode. Depending on the voltage and dynode characteristics, a certain probable number of secondary electrons (e.g., 4 or 5) will be ejected from the dynode for each photoelectron striking it. The secondary electrons from the first dynode are attracted to the second dynode where each electron ejects several more secondary electrons— and likewise on around the successive dynodes until the electrons are

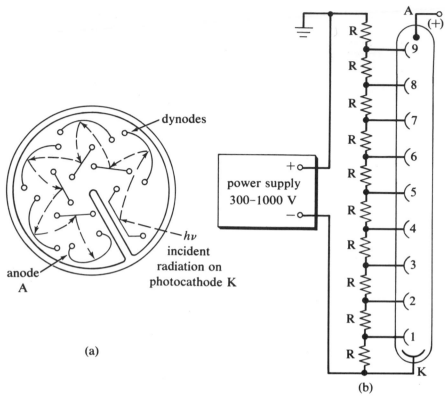

Fig. 7-6 **Photomultiplier tube with nine dynodes: (a) cross-sectional view; (b) schematic.**

collected at the anode. Thus, each photoelectron (which was the result of one photon of light) can be internally amplified within the PM tube typically by about 10^4 to 10^8 times by the process of secondary electron emission. Consequently a single photoelectron ejected at the photocathode results in a relatively large packet of electrons arriving at the anode, usually with a transit time spread of only a few nanoseconds. Because the charge of the electron is 1.6×10^{-19} coulombs/electron, and assuming an internal electron gain of 10^6 for each photoelectron, and a transit time interval of 5×10^{-9} sec, the average current during the pulse interval is $1.6 \times 10^{-19} \times 10^6/(t \times 10^{-9}) = 3.2 \times 10^{-5}$ A.

If the rate of photoelectron ejection and the frequency response of the measurement system, illustrated in Fig. 7-7, are such that individual current pulses can be resolved, the number of pulses can be counted. Because the count of current pulses is related to the number of photons

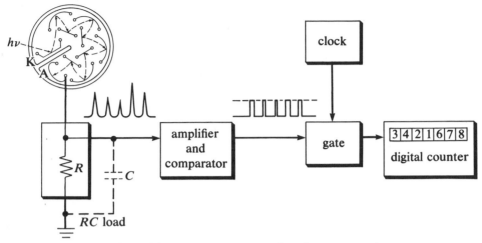

Fig. 7-7   Measurement system for photon counting.

that strike the photocathode, the measurement is called *photon counting*. If the counts are accumulated during the entire period of some light-emitting event, the conversion is from an inherently discrete physical quantity (photons of light) to a digital count of current pulses that represents the magnitude of the event. In this case the instrument system is essentially a one-step P-D converter, as illustrated in Fig. 7-8a.

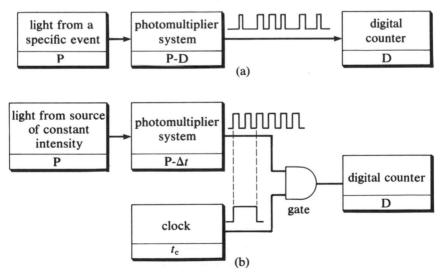

Fig. 7-8   Digital readout systems for photon counter: (a) P-D; (b) P-Δt-D.

Many applications of the PM tube are for various types of spectro-photometers where they are used for the quantitative measurement of light intensity or radiant power, i.e., the number of photons per unit time per unit solid angle.   In this case, as illustrated in Fig. 7-8b, the photo-multiplier system provides a pulse train with an average frequency that is proportional to light intensity (P) so that the domain conversion is from P to Δt.   To obtain a number for the D domain, it is necessary to measure the number of discrete photon pulse intervals during a unit time interval t_c from an accurate clock.   The overall conversion system is thus P-Δt-D. It is also interesting to note that the generation of the accurate clock time interval t_c is generally the result of a P-Δt conversion, such as con-verting the physical properties of a quartz crystal with an electronic oscillator and scalers to obtain a voltage signal of accurate pulse duration.

Although it is possible to make quantitative light measurements by using the photon-counting techniques, most measurements with the PM tube utilize a more indirect interdomain route to obtain a digital readout. That is, the PM tube is operated at light levels and with measurement circuits that essentially integrate the current pulses so that an average dc current (P-A conversion) is obtained that is directly proportional to light intensity.   Now to obtain a digital readout the current is typically converted to a frequency (A-Δt conversion) proportional to the current, and a selected unit time interval t_c is provided by the clock to make a digital frequency measurement.   The net result is a P-A-Δt-D conver-sion, as illustrated by the block diagram in Fig. 7-9.

The diagrams in Fig. 7-8 and 7-9 indicate that the direct route would seem the most reasonable, and studies have indicated significant inherent advantages of photon counting for many types of spectrophotometery. The recent advances in electronic instrumentation should certainly make

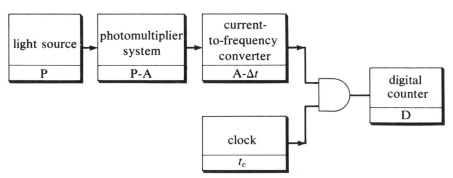

Fig. 7-9   Conversion of light intensity to proportional current and subse-quent conversion to digital readout.

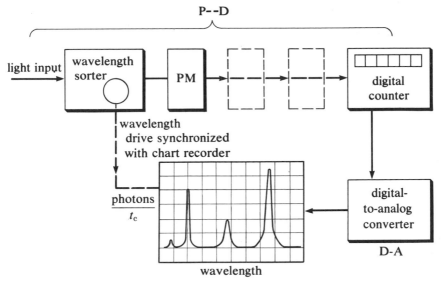

Fig. 7-10 **Digital-to-analog converter for recorder readout of photons/t_c vs wavelength.**

more common the direct conversion routes with instruments utilizing photomultipliers. Likewise, it seems probable that many new conversion systems will be developed that can provide more direct routes for the ultimate P-D conversion, rather than through the analog (A) domain. However, as of now most input transducers provide P-A conversion, and it is thus important to consider the types of analog-digital (A-D) converters that provide the digital information required for modern display and processing of information.

When the desired information exists only in the digital domain it is frequently important to display the data on a recorder or oscilloscope as a continuous function of another quantity. To provide such an analog readout, it is important to have digital-to-analog (D-A) converters, as illustrated by the example in Fig. 7-10. Specific circuits of D-A converters are described in Section 7-4.

7-2 Analog Concepts and Operations

For scientific measurement and control systems, information about specific physical quantities is eventually transduced to either electrical signals in the analog form (where the magnitude of voltage, current, resistance, time interval, etc., are continuous functions of the physical quantities) or into the digital form (where discrete bits of information, with each bit represented by a **1** or **0** logic voltage level, form a digital word that expresses the value of the physical quantity).   In the previous section the discussion emphasized that most scientific measurement and control systems, at present, involve analog operations, even though the desired form for data processing and readout is digital.   Therefore, in working with scientific digital instrumentation, it is necessary to consider operations on the electrical analog information because it will frequently enter the system at some point.   Also hybrid analog–digital systems will often be the most efficient in providing scientific information in the desired form.

The basic concepts of operational feedback that are important for an understanding of analog circuits are presented first.   We have found that an analysis of an electromechanical operational feedback system (servo system) is especially instructive in gaining an understanding and intuitive "feel" for these concepts.   Subsequent to this discussion the basic analog building block, the all-electronic *operational amplifier* (OA) is introduced.   Both linear and nonlinear operations are then presented. The OA's and the electromechanical system are utilized for specific practical circuits and instruments.   The section is concluded with a discussion of techniques for analog gating.

Electromechanical Operational Feedback System

The accurate measurement of an unknown dc voltage e_u is generally based on a comparison procedure as illustrated in Fig. 7-11.   A continuously variable reference voltage e_r is adjusted in value until it is equal to the unknown voltage e_u.   The representation in Fig. 7-11a indicates that e_r is manually adjusted while visually observing a voltage null detector.   In Fig. 7-11b a chopper alternately connects e_u and e_r at the amplifier input and any difference voltage ($\Delta e = e_u - e_r$) is amplified by the servo amplifier to provide a drive signal for a servo motor (M).   The shaft of the motor is mechanically coupled to vary e_r and also to move an indicator (chart pen, scale pointer, mechanical counter) of which the positional value is a known function of e_r.

If the servo system were ideal, even the slightest difference voltage between e_u and e_r would provide a sufficient motor signal to correct the difference.   However, there is always a small uncertainty or error e_s, so that $e_u = e_r \pm e_s$.   Many servo systems in potentiometric chart recorders maintain the error to within a few microvolts for full scale spans of 1 mV or more.   For certain specialized instruments the error is reduced to as

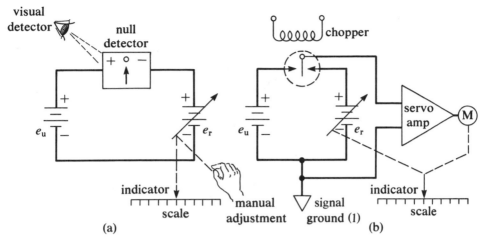

Fig. 7-11   **Comparison measurements of unknown voltage: (a) manual potentiometric system; (b) automatic self-balancing (servo) system. Dashed lines indicate mechanical linkages. See Footnote 1.**

low as a few nanovolts.   For practical purposes the error signal is often negligible, and e_r effectively follows e_u.

After changing the input voltage e_u, there is a certain time lag during which the motor adjusts e_r to the new value.   For servo recorders the response time will typically allow full scale travel of the pen in about 1 sec.   The frequency response of the system is thus quite limited.   This is one of the major differences of the servo capability as compared to the relatively wide-band operation of the OA discussed below.

When the servo is at balance the accuracy of the measurement of e_u depends on the accuracy of the reference voltage e_r and the magnitude of e_s.   The gain A of the servo amplifier should, of course, be made very high so that e_s is negligible.   Then the amplifier gain could vary over rather wide limits because of component aging, temperature, etc., and it would not affect the measurement accuracy.

In the automatic self-balancing system of Fig. 7-11b the voltage sources were both referenced to the signal[1] ground and alternately

[1] Note that a triangular symbol is used for ground in this circuit.   In an analog circuit with low signal levels it is advantageous to separate the common point of the signal circuit (triangular symbol) from the power supply common (standard ground symbol).   This prevents IR drops in the power supply common leads and connectors from affecting the voltage levels in the signal circuit.   The two commons are usually connected together at a single point near the lowest-level signal input.

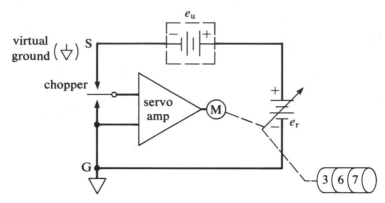

Fig. 7-12 Automatic comparison measurement of unknown voltage.

switched by the chopper across the amplifier input. An alternative method is shown in Fig. 7-12. The voltages e_u and e_r are connected with their polarities in opposition, and one end of the combination is connected to system ground and the other end to contact S. The other input contact G is connected to system ground. Therefore, the system will come to balance (e_s negligible) only when points S and G are practically at the same potential. Because contact G is connected to ground, the contact S must be virtually at ground potential. The servo system essentially maintains the input point S at *virtual ground* by automatically varying e_r to maintain it equal to e_u.

When at balance in the potentiometric mode (Fig. 7-11b and 7-12) the input impedance is said to be "infinite" because if $e_r = e_u$ no current will flow from the input voltage source e_u. Note also that the servo amplifier can be ac coupled (RC and/or transformer coupling) because the input signal is chopped. Therefore dc drifts in the servo amplifier need not affect the measurement accuracy.

Experiment 7-1 *Characteristics of a Servo Potentiometric Recorder*

The damping, dead zone, voltage spans, and other characteristics of a self-balancing potentiometer are investigated. In the potentiometric measurement mode the servo forces a reference voltage to follow the input voltage. The input impedance at balance is extremely high.

Current measurements. The unknown voltage source e_u in Fig. 7-12 can be replaced with a resistor (or other passive element) as shown in Fig. 7-13. Then if an input current (from some current transducer such as the phototube shown) is passed through the resistance R_f, there is a voltage drop $e_u = iR_f$. And, again, if the polarities of e_u and e_r are in opposition the servo will automatically maintain point S at virtual ground so that $-e_r = e_u = iR_f$. Because e_r is thus proportional to the current, the indicator can be made to read directly in units of current.

This method of measuring current has several unique features. Regardless of the transducer current the servo system maintains the points S and G at the same potential so that the supply voltage across the transducer remains constant. The two leads from the transducer are connected to ground and virtual ground which is very favorable for optimum signal-to-noise ratio. Also, currents from several transducers can be connected to the input S without interaction of sources.

In Fig. 7-13b, several current sources are shown connected at the input S so that the total current $i_T = (i_1 + i_2 + \cdots i_n)$. The voltage drop across R_f is $e_u = (i_1 + i_2 + \cdots i_n) R_f = -e_r$. The indicator for e_r is, therefore, proportional to the sum of the input currents. Point S is often called the *summing point*. Note that because the servo maintains nearly zero potential drop between input points S and G the measurement system presents essentially zero resistance to any input current. Thus the system is ideal for making current measurements.

Experiment 7-2 *Current Measurements Using Electrome-chanical Operational Feedback*

A servo recorder is modified to the operational feedback mode so as to make a sensitive current recorder with full-scale current spans of 10^{-8} and 10^{-7} A.

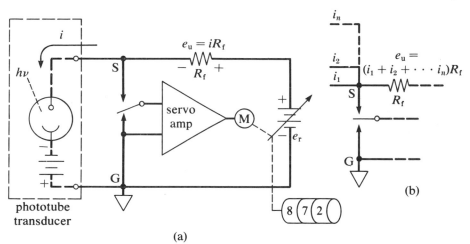

Fig. 7-13 Current measurements: (a) from a photodetector; (b) summation of input currents.

Voltage measurements. In Fig. 7-14 an input voltage source e_{in} is shown connected through an input resistance R_{in} to the input of the servo feedback system. In this case e_{in} and R_{in} are a current source, and because points S and G are held at the same potential, the input source sees essentially zero resistance between S and G so that the input current is determined only by R_{in}. Therefore $i_{in} = e_{in}/R_{in}$. This assumes that the internal resistance of e_{in} is negligible compared to R_{in} or has been included in it, and that the small voltage drop that might exist between S and G is also negligible compared to e_{in}.

If the only path for the input current is through the resistance R_f in Fig. 7-14, then at balance the reference voltage $e_r = -e_{in}(R_f/R_{in})$. If the full-scale value of $e_r = 1$ V, $R_f = 10$ kΩ and $R_{in} = 1$ kΩ, then the full-scale value for $e_{in} = 0.1$ V. The input resistance in this example is quite low, and there is the possibility of significantly loading the voltage source.

For a high input voltage the input resistance R_{in} could be very large and the source loading would be insignificant. For example, if $e_r = 1$ V, $R_f = 1$ MΩ and $e_{in} = 1000$ V, then the input resistance R_{in} could be 1000 MΩ.

Two rather subtle assumptions were made in presenting the characteristics of the electromechanical feedback system. One of these is that the leakage current of the servo amplifier is negligible compared to i_{in}. With a field-effect transistor (FET) input this is usually a good assumption because the leakage current is then typically less than 10^{-12} A. Also, the internal resistance of e_r was neglected in the calculations. This is only justified if the internal resistance is negligible compared to R_f.

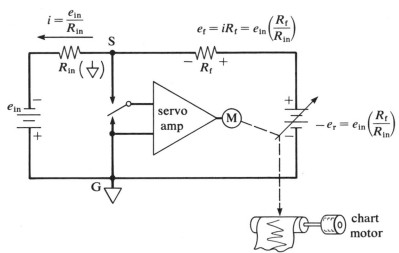

Fig. 7-14   Voltage measurements with servo operational feedback system.

Experiment 7-3	*Voltage Measurements Using Electro-mechanical Operational Feedback*

A modified servo recorder is used in the operational mode for voltage measurements. A comparison is made between the operational and potentiometric modes for voltage measurements.

Feedback and input elements and networks. Instead of using resistors as in the previous example, it is possible to substitute other passive elements (capacitors, diodes, etc.) or combinations of these elements (networks) for R_{in} and R_f. Each type of feedback and input network imparts certain characteristics to the circuit and makes it perform a different measurement, computation, or control function. These functions can be classified as continuous linear and nonlinear operations and as discontinuous nonlinear operations.

These operations and functions will be considered after introducing the all-electronic operational amplifier so that OA as well as electromechanical servo circuits can be illustrated.

The Operational Amplifier

The operational amplifier (OA or op. amp.) is a low-drift, high-gain, dc amplifier that is generally used with negative feedback elements or networks. It basically performs the same service as the electromechanical servo in the previous section, but its compactness and wide-band performance make it more generally applicable. Like the servo it is always working to keep its input signal (error signal) essentially zero. With an ideal OA the error signal is negligible and the function performed is dependent only on the type and values of the external feedback and input elements.

Ideal operational amplifier. Referring to Fig. 7-15, the ideal OA would have infinite input impedance Z_{in} so that amplifier input current $i_s = 0$, an extremely high gain A so that a negligible voltage e_s would be required to produce the output voltage e_o, and it would maintain these conditions from dc to very high frequencies. Unfortunately the gain A of the amplifier is sometimes not sufficiently high and, thus, the error signal e_s is not negligible. This is especially true at higher frequencies where the gain A drops off. Basic relationships between output voltage e_o and the input current or input voltage are developed in the next sections. The stepwise derivation shows the assumptions in arriving at the

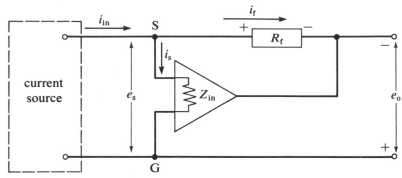

Fig. 7-15   Operational amplifier analysis.

ideal expressions that are dependent only on the feedback and input elements.

Experiment 7-4   *Familiarization with Operational Amplifiers*

The adjustments and connections for a solid-state discrete component OA and an integrated-circuit OA are investigated.

Output voltage–input current relationship.   The functional relationship for the OA circuit of Fig. 7-15 is readily obtained as follows:

Because the amplifier input voltage e_s is amplified by the gain A, and the amplifier is connected in the inverting mode,

$$e_o = -A e_s \qquad (7\text{-}1)$$

The voltage across R_f is $(e_s - e_o)$.   Therefore

$$e_s - e_o = e_s + A e_s = e_s(1 + A) \qquad (7\text{-}2)$$

and the current i_f through R_f is, thus,

$$i_f = \frac{e_s - e_o}{R_f} = \frac{e_s(1 + A)}{R_f} \qquad (7\text{-}3)$$

The input current i_{in} primarily passes through R_f to give the current i_f.

It is assumed here that i_s is negligible in comparison to i_f so that practically

$$i_f = i_{in} \tag{7-4}$$

Therefore the input resistance R_s of the OA system to the input signal current is

$$R_s = \frac{e_s}{i_{in}} = e_s \frac{R_f}{e_s(1+A)} = \frac{R_f}{1+A} \tag{7-5}$$

From Eq. (7-5) it can be seen that if $R_f = 100$ kΩ and $A = 50{,}000$, the input resistance $R_s = 10^5/(5 \times 10^4) = 2\ \Omega$. Such a low resistance is ideal for a current-sensing device. If point G is at ground potential then point S is virtually at ground potential, as for the servo system. Note, however, that, if R_f is made very large and A drops off, then R_s can become significant and introduce errors.

Equation (7-3) can be rearranged and i_{in} can be substituted for i_f [Eq. (7-4)] so that

$$e_s = \frac{i_{in}R_f}{1+A} \tag{7-6}$$

and because $e_o = -Ae_s$, it follows that

$$e_o = -Ae_s = -\frac{A}{1+A}R_f i_{in} \tag{7-7}$$

and, if $A \gg 1$,

$$e_o = -i_{in}R_f \tag{7-8}$$

Equation (7-7) also indicates the need to maintain a very high A so that the simple basic expression of Eq. (7-8) will be true in practice.

Experiment 7-5 *OA Current-to-Voltage Amplifier*

The basic $e_o = -i_{in}R_f$ relationship is verified for a range of currents and resistances. Limitations of this expression under certain conditions are observed.

Output voltage–input voltage relationship. As shown by Eq. (7-5) the input resistance of the OA system at the summing point in the inverting mode is very low so that in connecting a voltage source it is important to add an input resistance R_{in} as shown in Fig. 7-16. The input current i_{in} from the signal source e_{in} is

$$i_{in} = \frac{e_{in} - e_s}{R_{in}} \tag{7-9}$$

As before, it is assumed that i_s is negligible so that $i_{in} = i_f$, and substituting from Eq. (7-9) for i_{in} and from Eq. (7-3) for i_f,

$$\frac{e_{in} - e_s}{R_{in}} = \frac{e_s - e_o}{R_f} \tag{7-10}$$

When e_s approaches zero (because of high gain A), Eq. (7-10) reduces to

$$\frac{e_{in}}{R_{in}} = -\frac{e_o}{R_f} \tag{7-11}$$

or

$$e_o = -e_{in}\left(\frac{R_f}{R_{in}}\right) \tag{7-12}$$

This relationship assumes that the amplifier is capable of producing the necessary output current and that e_o is within the maximum voltage limits.

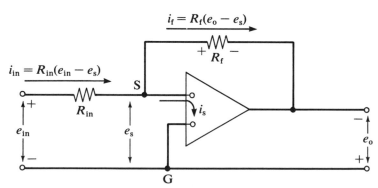

Fig. 7-16   Operational voltage amplifier.

If e_s is negligible, it can be seen from Eq. (7-9) that $i_{in} = e_{in}/R_{in}$, and thus input current depends only on input resistance and voltage.

Experiment 7-6 *OA Inverting Voltage Amplifier*

The basic $e_o = -e_{in}(R_f/R_{in})$ relationship is verified for several ratios of resistances and for a range of input voltages. Voltage and current limits are noted.

Differential input. Most OA's have balanced differential inputs so that signals can be applied to both the inverting (−) and noninverting (+) input terminals. Ideally, the output voltage of a difference amplifier depends only on the potential difference between the − (inverting) and + (noninverting) inputs and $e_o = -Ae_s$ where $e_s = e_+ - e_-$. In practice, the actual magnitude of e_+ will affect the output voltage, even if e_s is constant. Therefore, it is more accurate to write $e_o = A_d(e_- - e_+) + A_c e_+$ where A_d is the amplifier gain for the *difference mode signal* and A_c is the gain for the *common mode signal*. The measure of merit for a difference amplifier is the ratio A_d/A_c, the *common mode rejection ratio* CMRR. Values for the CMRR are typically between 1000 and 20,000 although higher values are available in specially designed amplifiers.

Experiment 7-7 *Differential OA Inputs*

The use of the OA as a differential amplifier is investigated and certain limitations are noted.

Limitations. The problems of low gain and amplifier input current are apparent from the previous discussions. Also, there are voltage and current offsets, drifts, noise pickup, common mode error, frequency response limitations, and other factors to consider when selecting an OA. Several references are given in the Bibliography in which the many types and characteristics of operational amplifiers are described in detail.

Linear Operations

Voltage follower. The servo voltage comparison system illustrated in Fig. 7-11 varied a reference voltage e_r so that it followed the input voltage e_u. A voltage follower can also be made with an OA as illus-

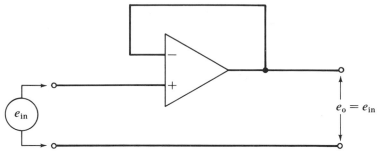

Fig. 7-17   Voltage follower.

trated in Fig. 7-17. The output voltage is fed directly back to the inverting input and the input signal is connected to the noninverting input. If
A is very high the error signal e_s will be negligible. Therefore, $e_o = e_{in}$
over a wide frequency range if the OA also has good CMRR. Note that
the output polarity is the same as the input polarity, and the input impedance is very high.

Experiment 7-8   *The OA Voltage Follower*

The OA voltage follower is shown to have high input and low output
impedance. Its use as an impedance transformer over a wide frequency range is tested. Comparisons are made with the servo
voltage follower.

Adder.   The addition of two voltage signals is illustrated in Fig. 7-18
utilizing the servo feedback system (Fig. 7-18a) and using the OA (Fig.
7-18b). The expressions relating input and output signals are also
given in Fig. 7-18. If $R_f = R_1 = R_2$, then $e_o = e_1 + e_2$. Many additional
input voltages can be summed in the same way.

The servo system can provide its own indication of the output voltage by mechanically linking an indicator to e_r. A separate readout system
would be required for the OA system. The OA can operate much faster
than the servo which is a big advantage for some applications. But
both systems can provide the same basic addition function.

Experiment 7-9   *The Adder*

Several voltages are added and subtracted with the OA adder and the
servo adder. Comparisons are made between the two systems.

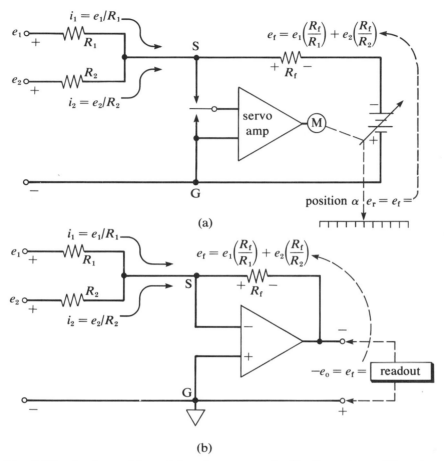

(a)

(b)

Fig. 7-18 **Analog adder: (a) using servo feedback system; (b) using operational amplifier system.**

Inverting amplifier. The OA can be used as a fixed-gain inverting amplifier. For example, with only one voltage input e_1 in Fig. 7-18, $e_o = -e_1(R_f/R_1)$. If $R_f/R_1 = 10 \text{ k}\Omega/1 \text{ k}\Omega = 10$, then $e_o = -10e_1$. Note that if $R_f = R_1$ the amplifier is a unity-gain *inverter* so that $e_o = -e_{in}$.

The unity-gain inverter is a useful circuit function not only to change signs of a signal but also to lower the impedance. Typically the output impedance is less than 1 Ω; the input impedance is R_1 and depends on the value selected.

Integrator. It is possible to integrate input currents or voltages by substituting a capacitor for the feedback resistor in either the servo or

OA systems. The integration of an input current is illustrated in Fig. 7–19 using the servo system.

The servo always works to maintain the potential difference between S and G very small. As before, the reference voltage e_r must be varied by the servo so that it equals the voltage across the feedback element (e_C in this case), i.e., $e_r = -e_C$. The voltage across a capacitor $e_C = Q/C$, where Q is the charge stored on a capacitor of capacitance C. The charge on the capacitor is equal to the integral of the current during the charging period t, so $Q = \int_0^t i_f\, dt$. Therefore,

$$e_r = -e_C = -\frac{Q}{C} = -\frac{1}{C} \int_0^t i_f\, dt \qquad (7\text{-}13)$$

The OA voltage integrator illustrated in Fig. 7-20 operates in principle the same as the servo current integrator. The input current $i_{in} = i_f = e_{in}/R$, and substituting for i_f in Eq. (7-13)

$$-e_o = e_C = \frac{1}{C} \int_0^t \frac{e_{in}}{R}\, dt = \frac{1}{RC} \int_0^t e_{in}\, dt \qquad (7\text{-}14)$$

The integrator is important in analog computation, in instrument readouts for displaying the area under a peak or averaging random noise, in A-D converters (Section 7-5), and in information storage (Section 7-3).

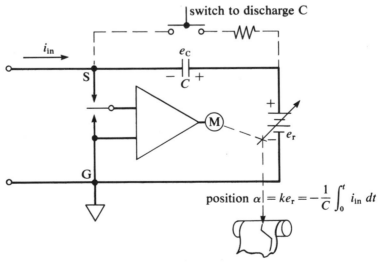

position $\alpha| = ke_r = -\dfrac{1}{C} \int_0^t i_{in}\, dt$

Fig. 7-19   Servo current integrator.

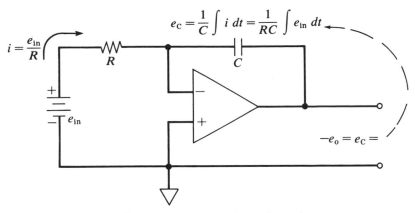

Fig. 7-20   Operational amplifier voltage integrator.

Experiment 7-10   *The Integrator*

Servo and OA integrators are connected and their characteristics are observed.   Several applications of the integrator are illustrated.

Differentiator.   The OA circuit in Fig. 7-21 is a differentiator.   The input current through C is $i_{in} = C\, de_{in}/dt$.   Because $i_f = i_{in}$ and $i_f R = e_f$,

$$e_o = -e_f = -i_f R = -RC\frac{de_{in}}{dt} \tag{7-15}$$

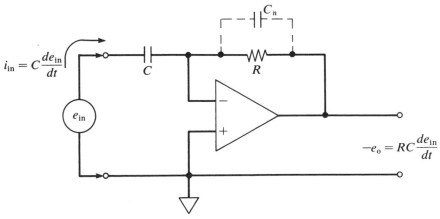

Fig. 7-21   Operational amplifier differentiator.

Noise pulses with a larger de/dt than the signal waveform will be enhanced more than the signal by the differentiator. To minimize the problems of noise enhancement, the upper frequency response of the differentiator is generally limited by putting a small capacitor C across the feedback resistor.

When a problem such as a rate measurement or the solution to a differential equation can be solved by either integration or differentiation, the much quieter integration technique is always preferred. References to the application of integration techniques are given in the Bibliography.

Experiment 7-11   *The OA Differentiator*

The output of an integrator circuit is differentiated with an OA differentiator to restore the original input to the integrator and thereby test the circuit under certain conditions.

Nonlinear Operations

Logarithmic amplifier. The log amplifier is one of the most important nonlinear continuous function circuits. It can be used in multipliers and dividers as well as being applicable for log conversions and compression of data.

A servo feedback log current recorder is shown in Fig. 7-22. This circuit is especially useful for recording the logarithm of the input current from photomultipliers and ionization detectors. The current–voltage relationship of diodes and other semiconductor junctions is logarithmic (Chapter 2) over several decades. The relationship between the feedback voltage e_f across the diode and the current through it can be expressed as

$$e_f = k_1(\log i_{in} + k_2) = k_1 k_2 + k_1 \log i_{in} \qquad (7\text{-}16)$$

where k_1 is a temperature-dependent constant (about 60 mV) and k_2 is another constant (e.g., 12) for a specific selected diode. Therefore, with an input current $i_{in} = 10^{-8}$ A, $-e_r = e_f = 60(-8 + 12) = 240$ mV.

The reference voltage e_r would generally consist of two parts, as shown in Fig. 7-22: a fixed part $(k_1 k_2)$ and a continuously variable part $(k_1 \log i_{in})$. The full-scale travel of the pen can be set to an integral number of decades (n) of input current by making the full-scale variable portion of e_r equal to nk_1 millivolts.

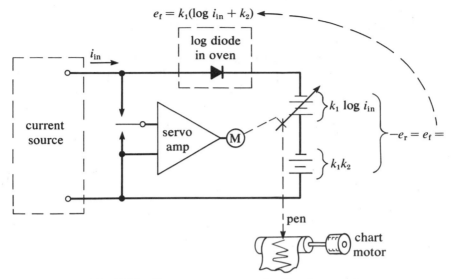

$$e_f = k_1(\log i_{in} + k_2)$$

log diode in oven

i_{in}

current source

servo amp

M

$k_1 \log i_{in}$

$k_1 k_2$

$-e_r = e_f =$

pen

chart motor

Fig. 7-22   Log current servo feedback recorder.

An OA logarithmic amplifier that utilizes the semiconductor junctions of a transistor for the feedback element is shown in Fig. 7-23a. The transistor junctions provide a logarithmic response to the input current. Two log elements are used in the OA circuit of Fig. 7-23b to obtain an output equal to the log of the ratio of two input currents.

Many combinations of log, log ratio, and antilog circuits can be used to provide multipliers, dividers, and other useful circuits for measurement, control, and computation. The use of sample-and-hold circuits (Section 7-3) in conjunction with OA log circuits provides accurate readouts for spectrophotometry and other scientific instruments.

Experiment 7-12   *Logarithmic Amplifier*

A semiconductor diode or transistor is used as a nonlinear resistance in the feedback path of either the servo or OA system so as to provide logarithmic readout of input currents or voltages.

Comparator. The comparator indicates whether a voltage or current is greater or less than a selected reference voltage or current. Operational amplifiers make excellent comparators as illustrated by the circuits in Fig. 7-24. In Fig. 7-24a the reference or comparison voltage e_r is applied to the noninverting input and the input signal is applied to the inverting input. The OA is operated "open-loop," with no negative

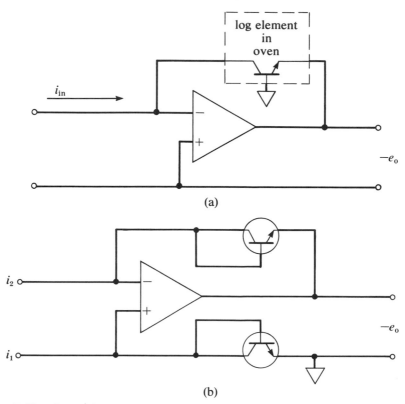

(a)

(b)

Fig. 7-23 Operational amplifier log circuits: (a) log of current; (b) log of ratio.

feedback to reduce its gain. Thus, when the inverting input is negative with respect to the reference voltage at the noninverting input by even a few millivolts the output voltage will be at the maximum positive voltage output of the OA (e.g., +10 V). Similarly, if the inverting input is positive by a few millivolts with respect to the comparison voltage e_r the output will be at the maximum negative voltage of the OA (e.g., −10 V).

With an open-loop gain of 10,000 and an output maximum of 10 V, an input difference signal of only $e_o/A = 10/10^4 = 10^{-3}$ V is required to drive the amplifier to limit. Therefore if the input is more than ±1 mV of the reference voltage the output will be at plus or minus limits (±10 V).

With only a 2 mV change in input signal required at the threshold level to achieve a 20-V swing in amplifier output, a small amount of high-frequency noise on the input signal could result in erratic output behavior in the transition region. For this reason, a portion of the output signal is often fed back to the noninverting input to move the threshold voltage away from its initial value as soon as an output transition has occurred.

Assume that the input signal is more negative than e_r so that the output $e_o = +10$ V. The $+10$ V is divided by the feedback network to provide 20 mV that is summed with the reference voltage (through the two 100-kΩ resistors). Thus the threshold voltage is increased by $+10$ mV because of feedback. If the input signal rises above this level the output will suddenly shift to -10 V. The feedback signal now lowers the original threshold by -10 mV. This means that the input signal must become 20 mV more negative than the previous trigger level to cause the reverse transition. The feedback has, thus, introduced a hysteresis

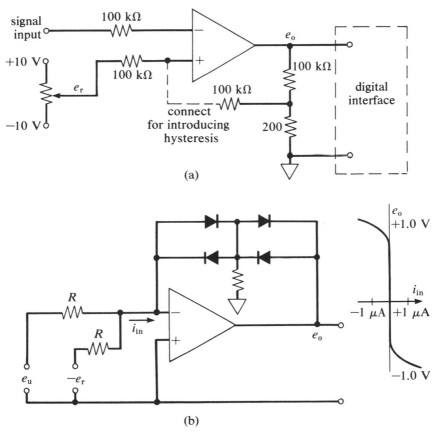

(a)

(b)

Fig. 7-24 Operational amplifier voltage comparator: (a) system operates at amplifier voltage limits and with hysteresis; (b) diodes in OA feedback limit output and provide exponential roll-off near null.

effect that decreases sensitivity but improves noise immunity. The feedback can be adjusted to provide optimum performance for a specific application.

Another comparator is shown in Fig. 7-24b. Both signal and reference voltages are summed at the inverting input through input resistors R. If the reference voltage is of opposite polarity to the input signal, there will be zero input current when $e_u = -e_r$. The output is limited to about ± 1 V by the diodes in the feedback. These diodes also provide a logarithmic roll-off in the region of the null point. Hysteresis can again be provided with a small amount of feedback to the noninverting input.

Voltage comparators are used as voltage level crossover detectors, input interfaces to digital circuits for counting, threshold detectors for control applications, and for many other functions. The overall accuracy, resolution, common mode rejection, hysteresis, input and output characteristics, turn ON and OFF times, operating temperature range, etc., should all be considered for the specific application.

Experiment 7-13 *The Comparator*

The integrated OA comparator on the comparator/V-F card is investigated and the sensitivity, hysteresis, and other performance characteristics are determined. The use of the servo as a comparator is also noted.

Analog Gating

The transmission of an input signal as an exact reproduction to the output of a circuit only during selected time intervals and with zero output at other times is an important function that is performed by analog sampling gates. An example of a precision analog gate is shown in Fig. 7-25. A pair of metal oxide semiconductor field-effect transistor (MOSFET) switches are operated by drive signals so that one is ON when the other is OFF and vice versa. When Q_1 is ON and Q_2 OFF, the OA operates as a regular unity-gain inverting amplifier. The gate is open.

By switching Q_1 OFF and Q_2 ON the gate is closed. The input signal is effectively disconnected because the output is essentially connected to the summing junction when Q_2 is ON.

Experiment 7-14 *Analog Gating*

An OA analog gate is constructed and used to gate analog voltage signals to an integrator.

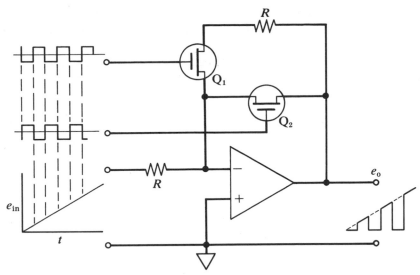

Fig. 7-25 Precision analog gate.

7-3 System Timing, Sequencing, and Sampling

Scientific experiments and instrumental ·measurements frequently require a number of operations to be carried out in a specified sequence. In many cases, the precise timing of some of the operations may be very important. The first part of this section describes several techniques which can be used to control the sequence and relative time of a number of operations.

If the measured quantity is changing with time, it may be important to know or control the exact time and duration of the measurement. Techniques for the "sampling" of a varying signal at specific times are reviewed and the general considerations for interdomain conversion of slowly and rapidly changing signals are discussed.

Sequencing

Automatic sequencing of the operations required for measurement or control is a necessity for total automation but may also provide increased convenience and speed and less chance of errors for many additional applications. Consider the photographic observation of the impact of a bullet hitting a glass bottle. The sequence of operations is (a) turn out room light, (b) open the camera shutter, (c) fire the rifle, (d) trigger the strobe light flash to illuminate the bullet and target at the desired instant, (e) close the camera shutter, and (f) turn on the room lights. If any step occurs out of sequence, the experiment will be spoiled. In addition there is one critical time in the sequence – the time between the firing of the rifle (c) and the triggering of the strobe light (d) since this determines what part of the impact phenomenon will be observed. An automatic precision electronic sequencer is clearly called for in this experiment.

In this section it is assumed that all required operations are ON-OFF operations (not continuous adjustments) and that each operation to be performed can be controlled by a logic level signal. Therefore, a sequencer is a circuit or device that provides a logic level signal output for each operation to be controlled and that determines the time relationship of the logic level changes appearing at the outputs. An electromechanical sequencer can be made by operating switch contacts by cams attached to a motor shaft. One cam and switch is required for each operation. Commercial electromechanical sequencers are available with adjustable cams for varying the sequence as required.

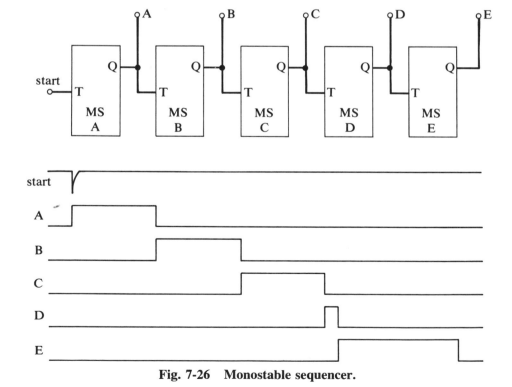

Fig. 7-26   Monostable sequencer.

Monostable multivibrator sequencers. An electronic sequencer can be made with monostable (MS) multivibrators as shown in Fig. 7-26. The output pulse of each MS triggers the input of the next. Thus pulses appear at the A, B, C, D, and E outputs in succession which can be used to actuate the required operations. Note that the time at the start of each pulse is the sum of the durations of all previous pulses. If the pulse width of MS A is increased, the time of the beginning of all successive pulses (relative to the start pulse) is increased. This is a convenient feature when it is important to maintain or control the time *between* each successive operation.

The serial triggered sequencer can be inconvenient if operation C, for instance, is to have a fixed time relationship with the start pulse while the time of operation B is to be varied. In this case, a parallel-triggered circuit such as Fig. 7-27 might be more desirable. The time of the termination of each pulse is related to the start pulse only by its own duration, so the "absolute time" of each operation can be varied independently. Note that the D pulse is to terminate shortly after C terminates. If the

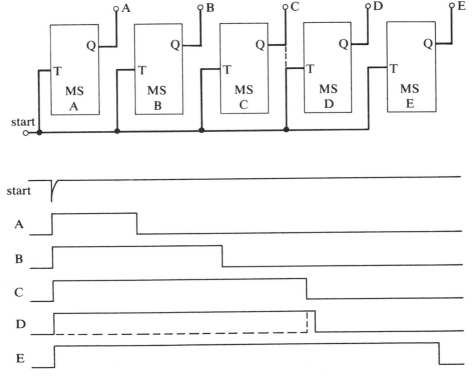

Fig. 7-27 Parallel-triggered monostable sequencer.

time difference between C and D must be known and constant (such as the case when C triggers the oscilloscope sweep and D initiates the event to be observed), extremely stable multivibrators will be required. A better solution would probably be a combination of the serial and parallel trigger approaches as indicated by the dashed lines in Fig. 7-27. If MS D is triggered by MS C, the sequence of C and D and the time interval between C and D can be controlled by conventional MS circuits without affecting the time independence of the other outputs.

When working with MS sequencers, one must keep in mind that the better circuits will have a relative pulse duration stability of about 1% under laboratory conditions. In the serial triggered case, the time errors for the start of the pulse are cumulative but the duration of each pulse is within the single pulse-width error limits. In the parallel-triggered case, the time error for the end of each pulse is the error for a single pulse width, but the error for the time interval between two pulse terminations is as much as 2 times the total pulse duration error or about 2% of the time from the start pulse to the time interval in question. Careful thought must be given to the critical time intervals to evaluate the relative merits of serial and parallel approach for each operation.

Ramp and comparator sequencer. An analog approach to the design of a sequencer is shown in Fig. 7-28. A ramp signal which has a duration longer than the complete sequence is generated by an OA integrating circuit. When the shorting switch across the integrating capacitor is opened, the sequence starts. As the ramp voltage reaches the comparison voltage of each comparator, the comparator output changes logic level as shown in the waveform chart. This circuit is analogous to the parallel-triggered MS sequencer in that each comparison voltage is set independently of the others. The time error for each comparator output transition depends on the stability, linearity, and magnitude of the ramp signal, the stability and sensitivity of the comparators, and the stability and resolution of the comparison sources. A good situation would be a 5-V ramp amplitude with $\pm 0.1\%$ stability and linearity; comparators with ± 1-mV sensitivity and ± 3-mV stability; comparison sources of ± 5-mV accuracy and ± 1-mV stability and noise. In this case the stability of the sweep source (± 5 mV) limits the reproducibility to $\pm 0.1\%$ of the *ramp duration*. The accuracy of comparison sources and linearity of the sweep make the absolute time error as much as $\pm 0.2\%$ of the ramp duration. Note that errors for this kind of sequencer are relative to the duration of the complete sequence.

A kind of "serial" sequencer can be made by putting the comparison voltages E_{RA}, E_{RB}, etc., in series. Although not a true serial sequencer,

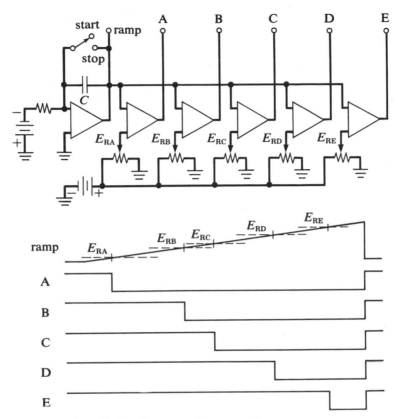

Fig. 7-28   Ramp and comparator sequencer.

the possibility of getting close events out of order is reduced.   A one-step sequencer of the ramp-comparator type is used to trigger the delayed sweep generator in oscilloscopes that have delayed sweep feature.

Experiment 7-15   *Ramp and Comparator Sequencer*

A ramp generator and one or more comparators are wired to make the sequencer of Fig. 7-28.   The adjustability and reproducibility of the sequence are tested.

Counting sequencer.   Each of the two previous sequencers are quite limited in relative and absolute timing accuracy.   When the time relation between operations in a measurement affects the accuracy of the measurement, more accurate time intervals will generally be required.   An extremely accurate approach is to count out precise time increments from a stable crystal oscillator. A sequencing system designed on this principle is shown in Fig. 7-29. When the start signal opens the counting gate, the counter counts up at a constant rate. The output level of each flip-flop is compared with the digital inputs by the equality detector circuits. When the counter output is equal to one of the digital inputs the output of that equality detector will be logic 1 for one period of the crystal oscillator. For instance, if the crystal oscillator has a frequency of 1.000000 MHz and if the digital A input signal has the logic levels corresponding to a counter output of 1467, then exactly 1467 μsec after the opening of the gate, a 1-μsec pulse will appear at output A. If digital B input is set at 1567, a pulse will appear at the B output exactly 100 μsec after the pulse at A.

The counting sequencer combines the advantages of independent adjustment with high accuracy. The accuracy of the timing sequence is dependent on the accuracy of the oscillator. The resolution (finest adjustability of time interval) is equal to one period of the crystal oscillator. The maximum duration of the sequence is the length of time required for the counter register to reach capacity for the oscillator frequency used.

The counting sequencer of Fig. 7-29 is made up of familiar circuits. The counter may be binary or BCD. The gating of counters was discussed in Section 6-5. The equality detector circuit is shown in Fig. 4-35. The digital inputs are generally derived from switches. A separate switch for each flip-flop can be used. Ten-position rotary or thumb-wheel switches with BCD-coded contact arrangements are very convenient

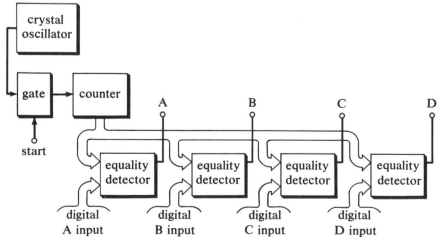

Fig. 7-29 Counting sequencer.

for this application. A separate switch register is required for each input. The counting sequencer is an elegant and rather expensive approach. The equality detectors contain many gates, and the switch registers require considerable panel space. Therefore, the counting sequencer is often used in conjunction with other less expensive and less accurate sequencers to provide the necessary precision time intervals.

A much simplified equality detector was introduced in the variable modulus counter of Fig. 6-22. The count control output in this circuit changes from **1** to **0** when the desired count is reached. Since the decoding is not complete, the count control may make additional **1–0** transitions if the count is allowed to proceed further. Such a detector could be used in Fig. 7-29 if the output were used to SET a flip-flop which would then be SET at the time of equality and remain SET through the rest of the sequence. Later changes in level of the equality detector output would have no effect. The counter could be wired to clear itself and the output flip-flops and close the gate at the end of the sequence.

Ring-counting sequencer. It was mentioned in Section 6-4 that the ring counter (circulating register) is a counter and decoder in one. For short sequences, the ring counter is a very useful sequencing circuit. For instance, if ten operations are to be performed at 0.01-sec intervals and this cycle is to be repeated many times over, a ten-bit ring counter and a 100-Hz oscillator would be a good choice for the sequencer.

A ring counter with 10 stable states can be made with only five flip-flops (FF) by reversing the circulating connections as shown in Fig. 7-30. The consequence of the reversed connection between FF E outputs and the J and K inputs of FF A is that FF E passes the opposite of its state onto FF A on each clock pulse. If the counter begins in the cleared state, FF A will be SET on the first clock pulse. Then A and B will be **1**; then A, B, and C, and so on, as shown in the count sequence table. This kind of counter is called a *switch-tail* or *Johnson* counter. The gate connected to E and $\overline{D}$ provides a self-starting and correcting count sequence. When D is **0** and E is **1** the other three flip-flops are cleared to establish state 9 in the count sequence. No matter what state the counter flip-flops are in, they are sure to satisfy the $E \cdot \overline{D}$ condition within nine counts. At first it might seem as though ten 5-input gates would be required for decoding, thus eliminating any decoding advantage the switch-tail counter may have. However, since only 10 of the possible 32 states are used, only two flip-flop states need to be specified for complete decoding. The decoding gates shown simply test the position of the **1-0** boundary and whether the **0**'s lead or lag the **1**'s.

The switch-tail counter and decoder has recently become quite

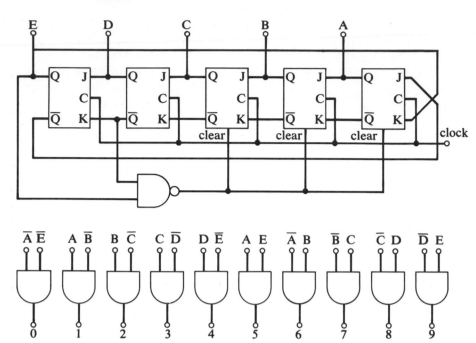

Fig. 7-30   Ten-state switch-tail counter and decoder.

Count Sequence Table for Fig. 7-30

	Outputs				
Count n	E	D	C	B	A
0	0	0	0	0	0
1	0	0	0	0	1
2	0	0	0	1	1
3	0	0	1	1	1
4	0	1	1	1	1
5	1	1	1	1	1
6	1	1	1	1	0
7	1	1	1	0	0
8	1	1	0	0	0
9	1	0	0	0	0

popular for cycle sequencing in small digital computers. Integrated circuit shift register and gate packages make it quite economical. Furthermore, its synchronous design makes it well-suited for high-speed operation.

Experiment 7-16   *Switch-Tail Counter and Decoder*

An 8-state switch-tail counter and decoder is wired following the design shown in Fig. 7-30. The sequence is observed with indicator lights and a dual-trace oscilloscope. The sequence accuracy is measured with a digital timer.

Sampling

Sampling is the technique of controlling the time at which information will be transferred or converted. Both digital and analog signals can be sampled. The sampling of digital signals is accomplished by logic gates. Several examples have been introduced in previous chapters. The first was the multipole switch of Fig. 4-31, that transfers a n-bit parallel digital signal to the outputs *only* when the gate control signal opens the gate. Similarly the gated memory circuit of Fig. 5-10 transfers the input information to the memory flip-flop when the input gates are opened by the clock signal. Digital information is set into the preset counter registers of Figs. 6-24 and 6-25 through gates activated by the preset control signal.

The gating of analog signals requires the faithful transmission of the signal when the gate is OPEN and the isolation of input and output signals when the gate is CLOSED. A circuit that accomplishes this has been described in Section 7-2,

Digital sample-and-hold. When digital information is admitted through logic gates to a binary register as in the case of the gated memory register of Fig. 6-13 or the preset counter of Fig. 6-25, a kind of "sample-and-hold" operation is performed. That is, the digital information at the gate inputs is *sampled* during the memory transfer pulse when the input gates are OPEN and transferred to the flip-flop register where it is *held* or stored until the gates are opened again. The sample-and-hold operation is a very convenient and often used circuit in digital instruments and computers.

Analog sample-and-hold. Sampling of analog signals is done through analog switching devices as described in Section 7-2. As in the case of the digital memory, there is often a need to hold the sampled value of the analog signal after the sampling interval is over. This is generally accomplished by charging a capacitor with the signal value during the sample interval, then measuring the voltage across the capacitor with a high input-impedance amplifier during the hold period. A reliable sample-and-hold circuit that uses the voltage follower amplifier is shown in Figure 7-31. The input signal charges capacitor C through resistance R when the sampling switch is in the "sample" position. The time constant RC must be short compared to the rate of change of the input signal that it is desired to sample, but long enough to average high frequency noise that might be superimposed on the signal. Thus the follower amplifier input and output will follow the input signal variations.

At the desired instant the switch is changed to the "hold" position isolating the input signal and leaving the input potential at that instant across capacitor C at the amplifier input. Ideally, this voltage will be maintained (held) indefinitely at the amplifier noninverting (+) input (and, consequently, at the output). However, finite currents at the amplifier input or through the sampling switch cause the voltage across C to change with time. This drift in the hold voltage can be minimized by using an amplifier with very low input current and a switch with very high OFF resistance. Using a larger capacitance at C will also reduce drift though this might be incompatible with keeping the RC time constant small enough for the + input to follow accurately the input signal. The minimum possible value of R is the ON resistance of the switch plus the output impedance of the signal source. The latter can be made very low by using a voltage follower amplifier between the signal source and the sample-and-hold amplifier.

A variation on the follower sample-and-hold circuit is shown in Fig. 7-32. A capacitance C_2 and a switch S_2 are put in the feedback loop of the follower amplifier. During the sample mode, the circuit behaves exactly as the circuit of Fig. 7-31. During the hold mode, switches S_1 and S_2 are open. If the amplifier + and − input currents are equal and if $C_1 = C_2$ the voltage across the capacitors will drift at the same rate keeping the output voltage constant. Since the differences between the + and − input currents for OA's are generally 10 times less than their

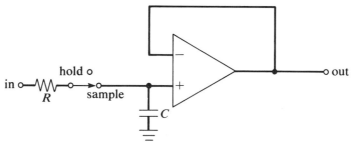

Fig. 7-31   Voltage follower sample-and-hold circuit.

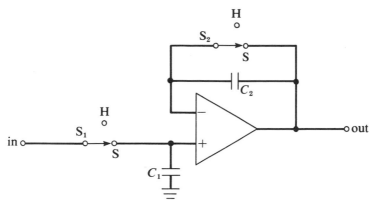

Fig. 7-32 **Input current compensated sample-and-hold circuit.** S = sample; H = hold.

absolute magnitudes, this modification can increase hold times substantially.

Another form of the sample-and-hold circuit is shown in Fig. 7-33. The circuit is basically that of an inverting amplifier. During the sample mode, the switch is closed. The capacitor reduces the impedance of the feedback circuit at high frequencies, but for frequencies substantially below $1/2\pi RC$ the impedance of the feedback circuit is essentially R_f. At these frequencies, the output voltage (and thus the potential across C) is $-e_{in}(R_f/R_{in})$. This circuit can thus provide amplification of the input signal.

When switched to the hold mode, the circuit of Fig. 7-33 becomes an integrator with no signal current to the summing point. The output voltage will thus be held at the voltage across C indefinitely except that amplifier and switch leakage currents cause this voltage to drift with time. In the hold mode, resistors R_{in} and R_f form a voltage divider be-

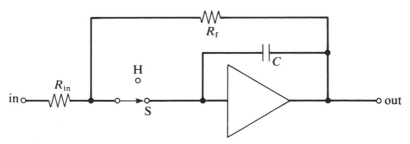

Fig. 7-33 **Inverting sample-and-hold circuit.** S = sample; H = hold.

tween the input and output which should have negligible effect on the output voltage.

The simple circuit of Fig. 7-33 can be improved by adding additional switches as shown in Fig. 7-34. The switch S_2 in Fig. 7-34a is closed during the hold mode. This keeps the junction of R_{in} and R_f at ground potential as it was in the sample mode. Since the summing point of the OA is also very near ground potential, the potential across S_1 is very low and the current leakage through S_1 is thus minimized. Two more switches, S_3 and S_4, are added to the capacitor circuit in Fig. 7-34b. S_4 is closed in the sample mode allowing C to charge through the low output impedance of the OA and the ON resistance of S_4. This can greatly increase the frequency response in the sample mode. In the hold mode, S_4 opens and S_3 closes, putting capacitor C with the last amplifier out-

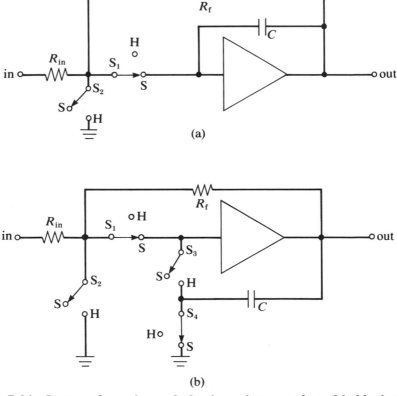

(a)

(b)

Fig. 7-34 **Improved versions of the inverting sample-and-hold circuit. S = sample; H = hold.**

put voltage in the feedback loop as in the other cases. The frequency response characteristic in the sample mode directly affects the minimum amount of time in the sampling mode required to obtain an accurate sample. This is sometimes called the *minimum aperture time*.

The sample-and-hold operation illustrated above can be used with integrator, differentiator, difference, logarithmic, and other OA circuits to combine two operations in a single circuit. One of the major applications of analog sample-and-hold circuits is to sample a varying signal at a particular instant and hold it long enough to make an A-D conversion of the sampled value. A differential sample-and-hold circuit can be used to subtract background from signal plus background. Determination of the ratio or log ratio of two time-separated signals from the same detector (as in double beam spectrophotometry) can be accomplished by using log circuits together with a sample-and-hold circuit.

An ideal sample-and-hold circuit would have an infinitesimal aperture time and an infinite holding time. Even with improved circuits and high-quality switches and OA's, a compromise between aperture and holding times must be made in most cases. The value of the holding capacitor C is adjusted to give the required holding time and the duration of the sampling aperture is chosen accordingly.

If the measurement requires a ratio of aperture-to-hold time that is too small to be achieved in an actual circuit, two or more sample-and-hold circuits can be cascaded. The first is set for the required aperture time and provides an output that holds for the longer aperture time of the second circuit. A sequencing circuit would be used to actuate the switches of the two sample-and-hold circuits in sequence.

Experiment 7-17 *Sample-and-Hold Measurements*

A sample-and-hold circuit is used to measure the amplitude of a periodic signal. Then a differential sample-and-hold circuit is wired and used to measure the difference in potential between two different parts of a signal waveform.

Error-sampled sample-and-hold circuit. All the above examples of sample-and-hold circuits assume that the input RC circuit allows the signal voltage value to appear at the amplifier input during the sampling aperture time. Even without the consideration of hold time, there is a minimum sampling aperture time that can satisfy this condition as determined by the minimum amplifier input capacitance and the minimum sampling switch ON resistance. When a sampling aperture is used which is less than 4–5 times the *RC* time constant of the input circuit, the hold voltage will not be equal to the signal voltage during the sampling aperture.

A circuit that corrects the sampling error and allows extremely brief sampling apertures is shown in Fig. 7-35. Assume that the memory output is equal to the input signal voltage during the previous sampling interval. This voltage is applied to the input amplifier through the feedback resistor R_f. The amplifier input capacitance is held at the previous signal voltage. The input potential change between sampling intervals will be called the *error voltage*. When the sampling gate opens, the amplifier input voltage begins to change by the amount of the error voltage. However, when only a fraction of this change has taken place, say 0.1, the sampling gate closes. The voltage change at the amplifier input, representing only one-tenth the desired change is amplified and capacitor-coupled to the memory gate. The memory gate is open longer than the sampling gate so that there is sufficient time for the amplified fraction of the error voltage to be applied to the memory amplifier. The

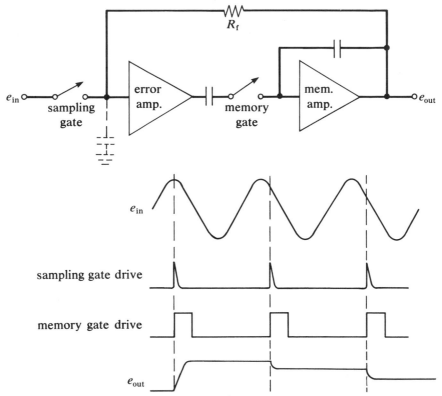

Fig. 7-35   Error-sampled sample-and-hold amplifier.

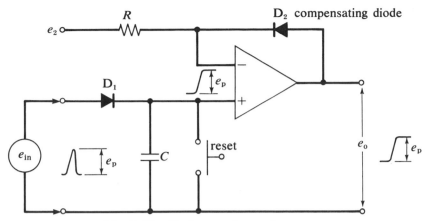

Fig. 7-36 Peak sampler and memory.

overall gain of the amplifier, memory gate, and memory is adjusted to compensate exactly for the fraction of the error voltage change lost due to the short sampling aperture. Thus the memory output voltage correctly represents the input voltage. The output voltage now charges the input capacitor up to the hold voltage through R_f and the circuit is ready for the next sampling.

This circuit is used in "sampling" type of oscilloscopes where apertures of the order of 50 psec have been achieved.

Peak sampler and memory. The peak sampler shown in Fig. 7-36 is similar to the sample-and-hold circuit of Fig. 7-31, except that the diode D_1 enables the peak voltage e_p of the input signal to be held on the storage capacitor C and at the output. After C is reset all positive input signals e_{in} will cause diode D_1 to conduct and charge the memory capacitor. Assuming that the input charge time is small compared to the time the input signal remains at its peak, the capacitor will be charged to e_p-e_{D1}. When the voltage e_{in} goes below its peak value diode D_1 is reverse biased and the peak voltage will be held on C.

To compensate for the voltage drop e_{D1} across the diode so that $e_o = e_p$, a compensating network should be added. This is made of D_2, R, and e_2. Then the inverting input $e_{(-)} = e_o-e_{D2}$ and the noninverting input $e_{(+)} = e_p - e_{D1}$. Therefore, if e_{D1} is about equal to e_{D2}, and because the OA forces $e_{(-)} = e_{(+)}$, it follows that $e_o = e_p$.

The time that the output remains at the peak value depends on the reverse leakage resistance of D_1, the capacitor leakage, and amplifier input leakage.

Experiment 7-18   *Peak Height Measurement*

The peak sampler and memory circuit of Fig. 7-36 is wired and used to sample and store the peak voltage of a sine-wave input signal. The input-output relationship is measured using a dual trace oscilloscope.

Digital Conversion of Varying Quantities

Data exist in the digital domain as electrical signals that represent specific numbers. If a quantity that varies continuously with time is to be converted to the digital domain, the resulting number can only be true for a specific instant in time. It is not possible, therefore, to make a truly continuous digital record of a varying quantity. What can be done is to sample the varying quantity at successive instants in time and convert each sampled quantity to the digital domain. The numerical result of each conversion is then stored in order in memory registers, or recorded on punched cards or paper tape, or by magnetic recording devices. If the sampling rate is high enough for the varying quantity to change only slightly between each sample, the digital record can quite accurately represent the amplitude vs time behavior of the sampled quantity. The maximum sampling rate is limited by the response speed of the sampling system and the time required to convert the sampled quantity to the digital domain. To be complete, the digital record should also contain information on the absolute or relative time of each sampling.

Real time converting. When the time variation of quantities in the various data domains is considered, a third dimension (time) needs to be added to the data domain diagram of Fig. 7-1 as shown in Fig. 7-37. Here each interdomain conversion is shown as a slice across the real time continuum. The time required to perform the conversion is indicated by the "thickness" of the slice. Neither the exact time nor the duration that the data exist in each domain within each conversion slice is indicated in Fig. 7-37 because they depend entirely upon the specific techniques used.

An example of real time, successive digital conversion is the digital recording gas chromatograph shown in Fig. 7-38. Gas chromatography is a means of separating, identifying, and determining the concentration of constituents in a mixture of volatile chemical compounds. The sample in the syringe is vaporized unto a column containing a nonvolatile liquid distributed on an inert support. The sample components are then partitioned between the liquid phase and a flowing carrier gas such as nitrogen. The components that are more soluble in the liquid are slowed in

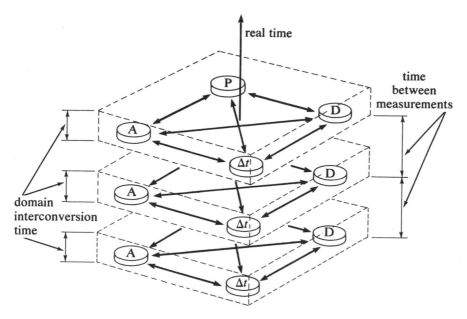

Fig. 7-37   Successive interdomain conversions in real time.

their passage through the column, the less soluble components passing through more quickly. The sample components thus emerge one at a time from the end of the column and are swept by the flow of carrier gas through a detector.

As each compound emerges from the separation column it is detected by a transducer that produces an analog signal related to the concentration of the compound in the carrier gas. One of the most popular transducers is the *flame ionization detector*. The effluent gas is heated in a H_2-air flame causing ionization of the sample compounds. By applying a voltage (about 300 V) between two metal plates on opposite sides of the flame, the ions and electrons are attracted to the plates resulting in a current I. One of the plates is connected to the summing point S of an operational amplifier used as a current-to-voltage (I-V) converter.

For several years, gas chromatographic analysis has been done by graphically recording the detector output vs time to give a plot similar to the waveshape shown on the analog chart recorder readout in Fig. 7-38. Each compound is identified by comparing its retention time on the column (time of the peaks' appearance) with that of known compounds. The amount of each compound in the sample is related to the area under the peak for that compound. Many techniques have been devised for recording peak areas and retention times directly to eliminate the need

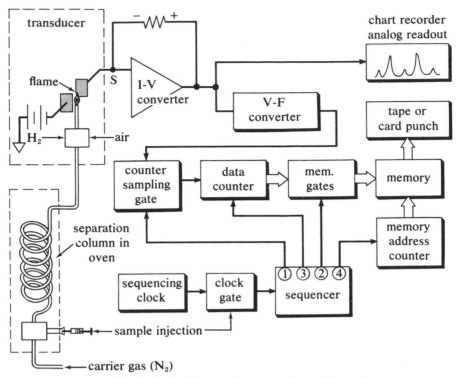

Fig. 7-38   Digital recording gas chromatograph.

for tedious graphical analysis. The most recent and best techniques have involved digital integration and recording. Now many chromatography laboratories are using a small dedicated computer that takes the data in digital form, identifies each peak, calculates peak areas corrected for base-line drift and overlapping peaks as well as the amount of each compound relative to an internal standard or relative to the whole sample, and prints out the completed analysis.

The digital recording of chromatographic curves could be accomplished as shown in Fig. 7-38. The output of the operational amplifier current-to-voltage converter is connected to a voltage-to-frequency converter. Since chromatographic peaks are generally more than 5 sec long, the response speed of the I-V and V-F converters is generally much faster than the rate of change of the input signal. Thus the data in the A and Δt domains follow or "track" the measured quantity continuously in real time. The clock is set to provide pulses at some desired measurement rate, for instance, 5 Hz or once every 0.2 sec.

When the sample is injected, the sequencing clock gate is opened. Each clock pulse triggers the sequencer which provides in sequence: (1) a pulse to open the counter gate for an exact time in order to put a digital measurement of the frequency in the data counter register, (2) a pulse to open the memory gates to transfer the frequency measurement to the memory, (3) a pulse to reset the data counter, and (4) a pulse to advance the memory address counter so that the next measurement will go to the next memory location. The total time for this sequence is the *domain interconversion time* that must, of course, be shorter than the selected 0.2 sec time between the measurements. With a V-F converter which provides 100 kHz output for full-scale input, a 0.1-sec counter gate time will produce 10,000 counts full scale. The count contains more significant digits than is justified by the reproducibility of other variables inherent in the chromatography procedures, so the gate time and measurement interval are reasonable.

In the circuit of Fig. 7-38, the counter gate performs the sampling operation. The sampling gate is open for 0.1 sec giving an average of the input signal over the gating interval. This is a reasonable approach for the chromatography signal because the sample aperture time is short compared to the meaningful variations in the input signal but long enough to give some noise reduction by time integration.

The average detector output for 0.1 sec is recorded in successive locations in the memory at 0.2-sec intervals. When the measurement is completed, the computer is programmed to reset the memory address counter to the location of the first data point and then read data out of successive memory locations, calculating peak areas and retention times as it goes along. No additional time information is required as long as the time between measurements is constant and some time reference point, such as sample injection time or time of first peak, is identified.

It is not necessary to record digital data at regular time intervals. If, for instance, a photometric emission or absorption spectrum were to be digitized, it would be more useful to record the photodetector output at regular wavelength intervals than at regular time intervals. This could be accomplished in several ways. A pulse output from the measurement sequencer could be provided which would advance the monochromator wavelength drive system the desired wavelength increment (using a stepper motor). Another approach would be to obtain pulses from the wavelength drive system for each wavelength increment and use these stepper pulses instead of a clock to trigger the measurement sequencer. The information stored in the memory would thus be the detector output vs wavelength increments, the digital equivalent of *X-Y* plotting.

Experiment 7-19   *Peak Area Measurements*

A voltage-to-frequency converter and a gated counter are used to make a voltage-time integrator. The experimental circuit is part of the digital recorder shown in Fig. 7-38. Integration, or peak-area measurements, are made on several input waveforms.

Data point recording. In each of the above examples, it was possible to devise a scheme that resulted in an incremental digital record providing a good approximation of the continuous signal variation. However, for rapidly changing signals the signal may change substantially over the minimum possible domain interconversion time. In this case, it is not possible to obtain closely spaced digitized data points by successive digital conversions in real time. For some well-characterized curves, only a few data points may be required to make the desired measurement.

The digitization of the response of an RC network to a voltage pulse is shown in Fig. 7-39. The rate of voltage change across R_1C when the

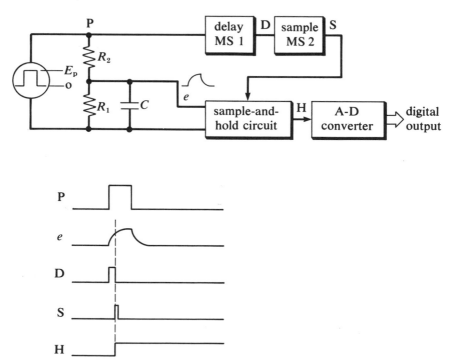

Fig. 7-39   **Digital conversion of a single point on R_1C pulse response curve.**

pulse is applied might be too fast for most A-D conversion techniques so a sample-and-hold amplifier is used. The sampling aperture is kept small compared to the signal rate of change and the hold output remains constant long enough for the A-D converter to convert and store the data. The pulse is applied to the network $R_1 C$ through a known resistor R_2. The same pulse triggers the delay monostable, MS 1. After the delay interval, the output from MS 1 triggers the sampling aperture monostable MS 2 to make the measurement. The curve for the response of $R_1 C$ to a step potential change is

$$e = E_p \left(\frac{R_1}{R_1 + R_2} \right) \left[1 - \exp\left(-\frac{t}{R_1 C} \right) \right]$$

This equation contains two unknowns, R_1 and C. If e is measured for two times, t, simultaneous equations could be solved for R_1 and C. Therefore, R_1 and C can be measured by setting a known delay time, applying a pulse and recording the resulting number, and then setting a different known delay time and pulsing again.

Equivalent time converting. Additional data points can be obtained for confirmation of the equivalent network equation by additional data point measurements at different delay times. To reproduce the original response curve, one would plot the number obtained at the output of the A-D converter vs *the sampling delay time.* Note that for this type of sampling, the sampling delay time is *equivalent* to real time in the reconstruction of the original signal curve. If many data points along the charging curve were desired, the delay monostable could be set a little longer before each pulse is applied, and any number of data points on the curve could be obtained. Note that the time required for the A-D conversion has no effect on the minimum *equivalent time* between successive data points. The requirements for this type of sampling are a sampling gate aperture that is short compared to the rate of change of the signal and a signal that is truly repeatable.

An application of repetitive sampling in measuring diode switching characteristics is shown in Fig. 7-40. The pulse applies a sudden forward bias to the diode D and the resulting current develops a voltage across R. It is this waveform that is to be sampled. The pulse signal triggers a fast sweep generator and advances a counter. The output signal, F, of the sweep generator is connected to a fast comparator, where F is compared with X, the output of the D-A converter. Since the counter has only one count, the D-A output voltage is low and the comparator output

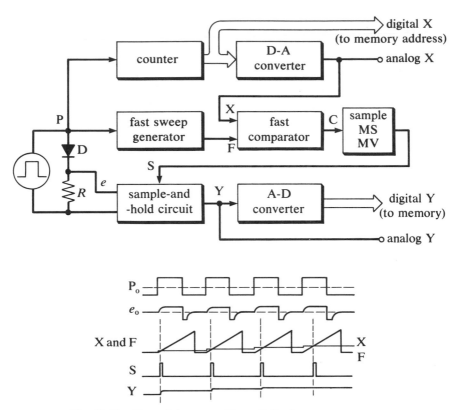

Fig. 7-40   Repetitive sampling of diode response curve.

inverts very near the beginning of the fast sweep.   The comparator inversion causes the sampling gate MS to trigger and take a measurement. When the next pulse is applied, the counter will have a two count, X will be a little larger, and the sweep generator will have to sweep a little longer before the comparator output flips to trigger the measurement.   The voltage at X is linearly related to the sampling delay and is thus an analog output proportional to *equivalent* time.   The counter provides an equivalent time output in parallel digital form.   For digital recording the digital Y output is connected to the memory, and the digital X output to the memory address.   For analog recording, the analog X and Y outputs are connected to the X and Y inputs of an X-Y recorder or oscilloscope. This sampling technique is frequently used when the rate of change of the signal is too fast for the recording instrument to follow in real time.

Experiment 7-20 *Data Point Sampling*

Two MS multivibrators and a sample-and-hold circuit are used to measure a point on a repetitive waveform as shown in Fig. 7-39. The sample-and-hold output voltage is plotted as a function of the delay time to reproduce the original waveform with data points. A suggestion is made for a circuit that will automatically sequence the delay and allow the output waveform to be recorded.

7-4 Digital-to-Analog Converters

In the first three sections of this chapter the concepts of interdomain conversions have been implemented with several illustrations. The significance of D-A and A-D conversions has been emphasized. It is the purpose of this section to describe some specific circuits commonly used to make digital-to-analog conversions.

Weighted-Resistor Network D-A Converter

A circuit that decodes or translates a digital signal to an analog voltage is shown in Fig. 7-41. The value of each summing resistor is inversely proportional to the weighted value of the digital bit that actuates the series switch. For example, if a 1-MΩ summing resistor is to have a weight of 1, then a 500-kΩ resistor is used for the bit with a weight of 2, a 250-kΩ resistor a weight of 4, and a 125-kΩ resistor a weight of 8. The resistor network could be chosen for straight binary, or grouped as in Fig. 7-41 for conversion of the binary-coded decimal information from several decade-counting units.

Each resistor is connected through a switch to a stable reference voltage supply V_R. The other end of each resistor is connected to the summing point of an operational amplifier. If all switches are open the input current is zero and the output voltage $e_o = 0$. If there is a count of 328 in the decade-counting units, then BCD outputs provide the logic level 1's to close the switches in series with the 100- and 50-kΩ resistors (3rd DCU), the 500-kΩ resistor (2nd DCU), and the 1.25-MΩ resistor (1st DCU). For this example the output voltage,

$$-e_o = iR_f = V_R \left(\frac{1}{50} + \frac{1}{100} + \frac{1}{500} + \frac{1}{1250}\right) 10^{-3} R_f$$
$$= 3.28 \ (10^{-5}) V_R R_f$$

and if $V_R = 10$ V and $R_f = 10$ kΩ, then $-e_o = 3.28$ V. The values of V_R and R_f are chosen to provide the most convenient value for e_o in the specific application.

In general terms for a 3-decade D-A decoder,

$$-e_o = (100n_3 + 10n_2 + n_1)V_R R_f 10^{-7} \tag{7-17}$$

where n_3, n_2, and n_1 are the numbers (0 to 9) in the 10^2, 10^1, and 10^0 DCU's, respectively.

The accuracy of the D-A conversion depends on many factors including the stability of V_R, the resistance and voltage drop across the switches, and how accurately the wide range of resistance values are matched.

To decrease the number of values of matched precision resistors, the circuit of Fig. 7-42 could be used. This circuit is basically the same as the weighted resistor technique shown in Fig. 7-41 except that the reference voltage V_R is also weighted.

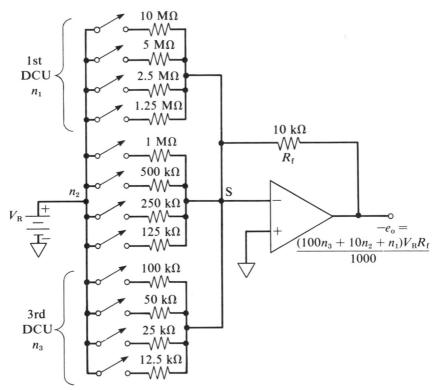

Fig. 7-41 Digital-to-analog converter using weighted resistor network.

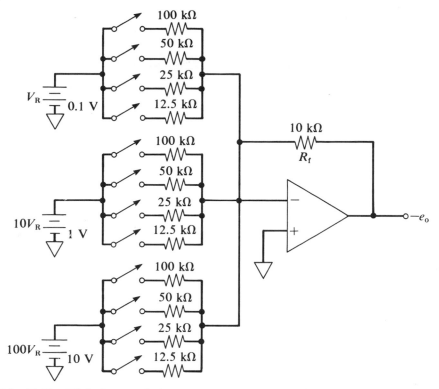

Fig. 7-42 Digital-to-analog converter using weighted resistors and weighted reference voltages.

As before, four values of input resistors are used to translate the number (0 to 9) within any decade to a proportional analog current. However, resistor sets of the same values can be used for each number because the reference voltage for each set is weighted relative to the other sets so as to convert the decades (10^2, 10^1, etc.) to proportional analog currents.

Another possible modification is to use sets of resistors of the same value (as in Fig. 7-42), one reference voltage V_R, a separate OA for each set, but with weighted feedback resistors R_f for each decade, and also a summing amplifier to sum the outputs from the OA's.

Ladder Resistor Network

To decrease the number of values of precision matched resistors needed for the input network, a resistor "ladder" can also be used as

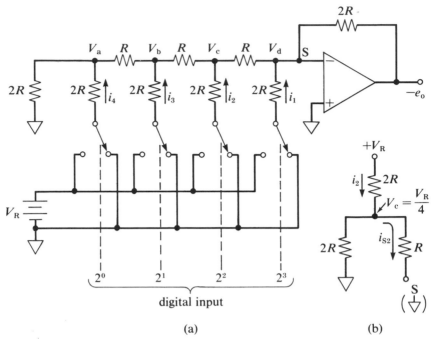

Fig. 7-43 Digital-to-analog converter with ladder resistor network.

shown in Fig. 7-43a. Only two resistor values R and $2R$ are used in the ladder network, but an extra set of switch contacts and more resistors are required to make a converter.

Starting with the most significant bit (2^3) shown in Fig. 7-43a, if this bit is logic **1**, then V_R is connected through $2R$ to the summing point so as to provide a current $i_1 = V_R/2R$. The current to the summing point because of a logic **1** for the 2^2 bit is a fraction of i_2. Current i_2 divides through the resistor R connected to the summing point and through the ladder network to the left of the node as shown in Fig. 7-43b. The voltage source V_R should have negligible internal resistance so that the resistance between each source and the ladder is exactly $2R$ whether the digital inputs are **1**'s or **0**'s.

Kirchoff's laws are used to solve for i_{in} by summing the currents at each node on the ladder. When these equations are written in terms of the iR drops across each ladder resistor, the following four equations result:

$$i_4 = \frac{V_a - V_b}{R} + \frac{V_a}{2R} = \frac{V_R(2^0) - V_a}{2R}$$

$$i_3 = \frac{V_b - V_c}{R} - \frac{V_a - V_b}{R} = \frac{V_R(2^1) - V_b}{2R}$$

$$i_2 = \frac{V_c}{R} - \frac{V_b - V_c}{R} = \frac{V_R(2^2) - V_c}{2R}$$

$$i_1 = i_{in} - \frac{V_c}{R} = \frac{V_R(2^3)}{2R}$$

where $V_R(2^0)$ is V_R or 0 V, depending upon whether the 2^0 digital input is **1** or **0**, respectively. When these equations are solved for i_{in},

$$i_{in} = \frac{V_R(2^3)}{2R} + \frac{V_R(2^2)}{4R} + \frac{V_R(2^1)}{8R} + \frac{V_R(2^0)}{16R}$$

and therefore, if $R_f = 2R$, the output voltage is

$$-e_o = i_{in}R_f = V_R(2^3) + \frac{V_R(2^2)}{2} + \frac{V_c(2^1)}{4} + \frac{V_R(2^0)}{8} \tag{7-18}$$

If $V_R = 8$ V, then the binary number $0101 = (8/2) + (8/8) = 5$ V.

Serial D-A Converter

In the D-A converters already presented the digital information is entered into the circuit simultaneously (parallel entry). If count serial entry of digital information is practical or desired, an integrator can be used as the D-A converter. An OA integrator D-A converter is shown in Fig. 7-44. Each unit in the digital number is represented by a repro-

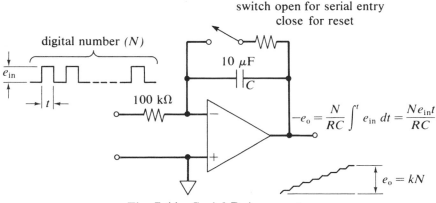

Fig. 7-44   Serial D-A converter.

ducible rectangular voltage pulse of amplitude e_{in} and pulse duration t. Therefore, the voltage across the integrating capacitor is increased by $e_{in}t/RC$ for each pulse. For a serial digital number that provides N pulses to the input of the integrator the analog output voltage is

$$-e_o = Ne_{in}t/RC \tag{7-19}$$

If $e_{in} = 1$ V, $t = 10^{-3}$ sec, $R = 100$ kΩ, and $C = 10$ μF, then the number 3672 would give an output voltage,

$$-e_o = 3672\,(1)\,10^{-3}/10^5 \cdot 10^{-5} = 3.672 \text{ V}$$

The reset switch is closed to discharge the capacitor prior to the next serial entry of a digital number. The values of e_{in}, R, C, and t can all be varied for optimum conversion characteristics for a specific situation.

Count-Rate Meter

The serial D-A integrator of Fig. 7-44 can be readily modified to make a count-rate meter. The input is not a digital number with a finite number of unit pulses as in the previous example but a continuous pulse train of some frequency. In other words, the input information is not in the digital domain but is in the Δt domain. If the reset switch is opened for an exact finite time t_c the OA output voltage just before closing the reset switch is

$$-e_o = (N_{rate}t_c)e_{in}t/RC \tag{7-20}$$

where the count rate $N_{rate} = N/t_c$.

Repeated integration of the input pulses for time intervals of t_c will provide a sawtooth output voltage of amplitude e_o. An RC filter connected to the output of the OA integrator can filter the sawtooth OA output. If the time constant of the RC filter is $RC > 5t_c$, then the output of the filter will be a dc voltage of magnitude e_o.

Given that $t_c = 0.1$ sec, $t = 10^{-6}$ sec, $R = 10$ kΩ, $e_{in} = 1$ V, and $C = 0.1$ μF, then, from Eq. (7-20), $e_o = 1.000$ V for a count rate input of 10,000 counts/sec.

The classical count-rate meter utilizes a so-called *diode pump* shown in Fig. 7-45. The R_1C_1 time constant is chosen so that the capacitor C_1 is completely charged by uniform input pulses of voltage amplitude e_{in}. For one pulse the capacitor charge $Q = C_1e_{in}$. For N pulses in time

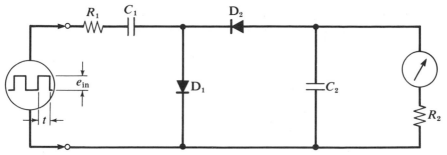

Fig. 7-45   Count-rate meter using a diode pump circuit.

interval t_c, each adding a charge Q, the total change in charge $\Delta Q = NC_1 e_{in}$, and the average rate of change of charge or average current i_{av} is

$$i_{av} = \frac{\Delta Q}{t_c} = \frac{NC_1 e_{in}}{t_c} \qquad (7\text{-}21)$$

It is the function of the circuit in Fig. 7-45 to provide an average current through the meter that is directly proportional to N/t_c as expressed in Eq. (7-21).   When the positive voltage pulse of amplitude e_{in} is applied to the input, the diode D_1 conducts, and C_1 charges to e_{in} if the input pulse width $t > 5R_1C_1$.   This assumes that the diode resistance is negligible compared to R_1 and the voltage drop across the diode is negligible compared to e_{in}.   When the input pulse starts back toward zero the diode D_1 is reverse biased and the cathode of D_2 goes negative so that D_2 conducts and the capacitor C_1 discharges through D_2, the current meter, and the resistances R_2 and R_1.   The capacitor C_2 is used to average the current through the meter, and it should be much larger than C_1.   The circuit values are chosen so that C_1 discharges essentially completely before the next input pulse.   Therefore, the input circuit is in essentially the same condition as for the previous input pulse and another pulse of amplitude e_{in} will again provide a charge $Q = C_1 e_{in}$.   Again this charge will be *pumped* through the meter when the input pulse returns to zero.   The result is an average meter current proportional to input pulse rate.

The sensitivity or range of the meter can be easily changed by varying C_1 or e_{in}.   This circuit is frequently used in nuclear radiation rate meters.   The output pulses from a radioactive particle detector are fed to a monostable multivibrator to produce pulses of constant amplitude e_{in} and selected pulse width t.   The uniform pulses from the monostable MV are fed to the input of the diode pump circuit.

The diode pump network, consisting of C_1, D_1, and D_2, can be connected to the input of an OA integrator to provide a stair-step voltage output.

Experiment 7-21 *A Weighted-Resistor D-A Converter*

A 3-decade weighted-resistor D-A converter is constructed. The digital output from the memory of three decade counting units is converted to an analog voltage and recorded on a strip chart recorder.

Experiment 7-22 *A Serial D-A Converter*

The OA integrator is constructed and used as a serial D-A converter. Also, the circuit is modified and used as a count-rate meter.

7-5 Analog-to-Digital Converters

The outputs of most transducers are analog signals. Therefore analog-to-digital converters are essential if information about physical phenomena is to be processed, stored, and displayed as digital data.

The voltage comparison technique described for servo and OA comparators in Section 7-2 is also the basis for a group of A-D converters. Included in this voltage comparison group are the continuous null balance, the successive approximation, and the counter staircase-ramp methods. These methods all utilize D-A converters to provide a variable reference voltage that follows the input voltage.

Another group of A-D converters uses capacitor-charging techniques that are frequently found in digital voltmeters. The voltage-to-frequency converter and the pulse-width and up-down integrator converters are included in this group. These techniques all involve interdomain conversion from A-Δt-D.

In attempting to arrive at the optimum combination of accuracy, speed, noise immunity, and economy for a specific application, each technique has certain attributes and qualifications. These are indicated as each technique is presented.

Continuous Balance A-D Converter

The direct continuous comparison of an unknown voltage e_u and a variable reference voltage e_r can be performed as illustrated in Fig. 7-46. The voltage e_r is generated by a D-A converter that receives information

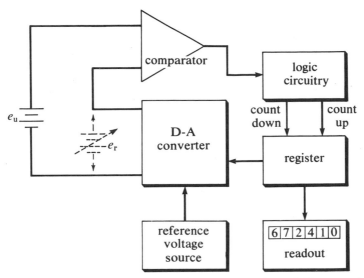

Fig. 7-46 Continuous balance voltage comparison A-D converter.

from a digital register. When $e_u > e_r$ the comparator and logic circuitry enable the register to count UP. As the digital number increases, e_r increases (through the D-A converter) until it is equal to e_u.

Similarly if $e_u < e_r$ the register counts DOWN so as to decrease e_r until equal to e_u.

The conversion of the analog e_u to a digital output is, thus, performed by using the digital circuitry to produce an analog reference voltage that is continuously nulled with the unknown voltage. The circuit is essentially a "digital potentiometer," and the digital output *tracks* a slowly varying input. In the presence of noise on e_u the instrument will *hunt* (attempt to follow the noise). It is analogous to the electromechanical servo system of Section 7-2. The feedback loop is always closed, always keeping the digital output equal to the analog input.

Staircase-Ramp A-D Converter

A variable reference voltage e_r can be generated with a serial entry digital-to-analog converter (Section 7-4). The output voltage of a staircase-ramp D-A converter increases by a small increment for each unit count fed to the input. As shown in Fig. 7-47, the control unit resets the digital counter and the D-A converter to zero. The start pulse opens the gates so as to enter successive pulses from the clock generator into

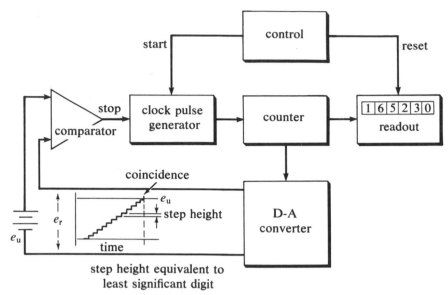

Fig. 7-47 Staircase-ramp voltage comparison A-D converter.

the counter and to the D-A converter. When the staircase output voltage e_r from the D-A converter increases to a value that coincides with e_u, the comparator output reverses polarity and sends a stop pulse to the clock gate. This fixes a number in the counter that is proportional to e_u at the moment of coincidence with e_r. The feedback loop for the staircase A-D converter is open except at the instant when the generated ramp exceeds e_u.

Successive Approximation A-D Converter

One of the fastest and most commonly used A-D converters is illustrated in Fig. 7-48. The conversion process consists of starting with the most significant bit and successively trying a logic **1** in each bit of a D-A converter. The output of the D-A decoder is compared to the input voltage e_u as each bit is tried. If the D-A output is larger than e_u the **1** is removed from that bit and the next most significant bit is tried. When e_u is larger, the **1** remains in that bit. After the least significant bit has been tried the digital word in the D-A converter is equivalent to the analog input voltage.

The digital output word can be taken serially or in parallel. If taken serially the output of the comparator must be suitably gated as each bit is tried. If taken in parallel the flip-flops that drive the D-A converter

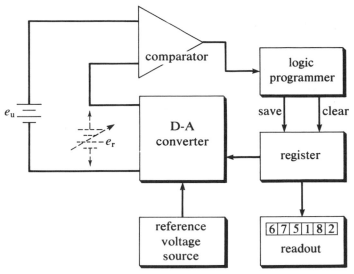

Fig. 7-48   Successive approximation voltage comparison A-D converter.

can be read out at the end of the conversion period.   Only 10–50 msec
are required to come to balance for each new conversion, compared to
several hundred milliseconds for the continuous null method.

Voltage-to-Time-to-Digital Converter

An inexpensive approach for obtaining conversion from analog
voltage to a digital word is to go through the time interval domain.   An
A-Δt-D converter based on the generation of a linear ramp was illustrated
in Chapter 1 (Fig. 1-19).   The input voltage e_u was compared with a
reference ramp voltage e_r.   When the ramp voltage starts its linear sweep,
a gate opens.   When e_r reaches a value equal to e_u, the comparator output
reverses polarity and the gate is closed.   The result is a time interval
Δt that is proportional to the input analog voltage.   The time interval Δt
is readily converted to a digital word by sending clock pulses into a
counter for the duration of the time interval.

It takes relatively little electronic equipment to produce a time interval pulse by the ramp technique.   This pulse can also be conveniently
transferred from one instrument to another over long transmission lines.
Therefore conversion of the analog voltage to the time interval domain
can be accomplished near the transducer, and the time pulse can then be
transmitted to a data processor where it can be converted to the digital
domain by counting.   The disadvantages of this ramp technique are the

relatively low accuracy (typically 0.1–1%) and the possibility of large errors with noise on the input signal e_u. Input filters are usually required with this type of converter.

Experiment 7-23 *Ramp-Type A-D Converter*

The conversion of an analog voltage to a digital word is performed by an A-Δt-D technique using a ramp generator and digital timer.

The Dual-Slope Integrator A-D Converter

Another A-Δt-D converter is called the *dual slope* or *up-down integration* technique. The inaccuracies inherent in the single-slope linear ramp technique can be greatly minimized by a dual-slope integration method. A time interval pulse that is proportional to the analog input voltage is produced by making a time comparison between two integrations. The input voltage e_u is first integrated for a fixed time interval t_1, as shown in Fig. 7-49. The input to the integrator is then switched to a stable fixed reference voltage $-e_r$. The capacitor then discharges to

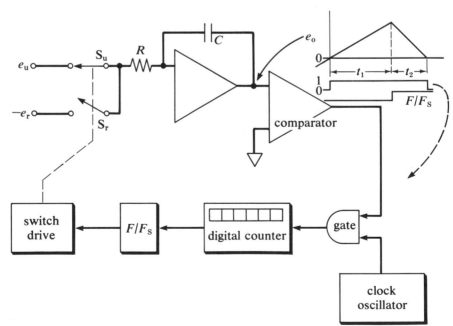

Fig. 7-49 Dual-slope integrator A-D converter.

zero, and the time required for discharge is proportional to the capacitor charge which is proportional to the input voltage e_u.

At the start, a pulse resets the counter and F/F_S to zero, and S_u is closed and S_r is open so the integrator capacitor C begins to charge at a rate that depends on the magnitude of e_u. As the integrator output exceeds zero slightly the comparator output voltage changes state. This opens the gate so that clock pulses are fed to the counter. The counter runs through its total capacity (e.g., 9999), and this represents a fixed time because time increments from the clock are being counted. On the next clock pulse, all digits go to zero and a spillover pulse triggers F/F_S to logic level 1. This causes switch S_r to close and S_u to open so that e_u is disconnected and $-e_r$ is connected to the integrator. The integrator output voltage now decreases linearly to 0 V where the comparator output state again changes and closes the gate, thus locking out the clock pulses.

The number of clock pulses in the counter depends on t_2, and t_2 is proportional to e_u. If e_u is doubled the capacitor charge is doubled in the fixed time required to run through the counter capacity. Therefore, it takes twice as long for the capacitor to discharge to 0 V and the counter will have twice the number of counts.

In the time t_1 the integrator output voltage e_0 is

$$-e_0 = \frac{1}{RC} \int_0^{t_1} e_u dt = \frac{e_u t_1}{RC} + K \tag{7-22}$$

The constant $K = 0$ if there is an initial small negative integrator offset and the comparator detects the crossover points during charge and discharge at exactly zero.

The output voltage e_0 is decreased to zero by discharging capacitor C with a constant current $-e_r/R$, so that

$$e_0 = \frac{1}{RC} \int_0^{t_2} -e_r \, dt = -\frac{e_r t_2}{RC} \tag{7-23}$$

Combining Eqs. 7-22 and 7-23,

$$e_0 - e_0 = \frac{e_u t_1}{RC} - \frac{e_r t_2}{RC} = 0 \tag{7-24}$$

and, simplifying,

$$e_u = \frac{t_2}{t_1} e_r \tag{7-25}$$

It is apparent from Eq. (7-25) that the accuracy of the measured voltage is independent of the accuracy of the integrator time constant. The times t_1 and t_2 are obtained by counting the numbers n_1 and n_2, respectively, of the clock oscillator time increments Δt, thus,

$$e_u = \frac{n_2 \Delta t}{n_1 \Delta t} e_r = \frac{n_2}{n_1} e_r = k n_2 \tag{7-26}$$

Equation 7-26 shows that the method is also independent of the absolute value of the oscillator frequency.

The dual-slope technique has excellent noise rejection because it is an integration technique. The speed and accuracy are readily varied according to specific requirements, and accuracies of 0.05% in 100 msec are feasible.

Voltage-to-Frequency-to-Digital Converter

The conversion of the input voltage into a proportional frequency and then to a digital word (A-Δt-D) is frequently used for DVM's. The output of the V-F converter can be fed directly to any digital frequency meter so that it is a convenient method.

The technique is illustrated in Fig. 7-50. The analog input voltage produces a charging current e_u/R_1 that charges a capacitor C to a refer-

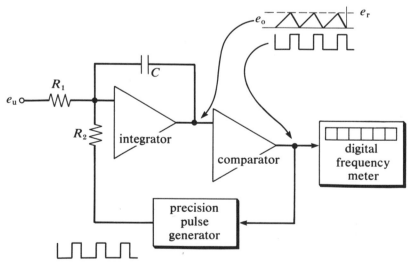

Fig. 7-50 Voltage-to-frequency converter.

ence voltage e_r. When the reference level is reached the comparator changes state so as to trigger a pulse generator. The pulse generator produces a pulse of precise charge content that rapidly discharges the capacitor. The rate of charging and discharging the capacitor provides a signal frequency that is directly proportional to e_u.

The V-F method could also be considered a dual slope method. However, instead of holding e_r/t_1 constant the product t_2e_r of Eq. (7-25) is held constant by providing feedback pulses of fixed charge content. Therefore,

$$e_u = t_2 e_r \left(\frac{1}{t_1}\right) = k f_u \qquad (7\text{-}27)$$

where the output frequency f_u is directly proportional to the input voltage e_u. The proportionality constant k can be adjusted by changing the coulomb content of the feedback pulses ($e_r t_2$), so that the frequency can be related directly to volts or other desired units of pressure, temperature, pH, etc. This integration method also has a high noise rejection and can be made very accurate (0.005%).

Another type of V-F converter utilizes a unity gain inverter in front of the integrator. The input voltage is switched alternately to the inverter-integrator combination and then directly to the integrator. The switching occurs at two preset voltage levels (e.g., 0 and 5 V). Therefore, the input signal alternately charges and discharges the integrator capacitor C. The slope of charge and discharge, and thus the output switching frequency from the comparator, are directly proportional to the input voltage.

Experiment 7-24 *Digital Voltmeter*

A DVM is constructed using the V-F converter card and the frequency meter constructed in Chapter 6. Characteristics are determined and an OA follower is used at the input to provide high input impedance.

7-6 Digital Computation

We have all come to know the digital computer as a powerful mathematical tool. Its application to model formulating, data reduction, statistical evaluation, curve fitting, and similar problems have had a tremendous impact on all areas of science. Experiments that yield data requiring extensive analysis are now practical and more information is being extracted from the traditional measurement techniques. Many of the recent developments in scientific instrumentation have come about from the application of circuits and techniques developed for digital computers. The purpose of this text has been to describe these circuits and techniques and to demonstrate their usefulness for measurement and control instrumentation.

The development of integrated circuits and improvements in computer design are bringing about significant advances in computer speed and versatility and, at the same time, substantial reductions in cost. As instruments and accessories are developed that use increasingly complex digital circuits and as versatile basic computers decrease in cost, a new interaction between the digital computer and scientific instruments has come about. The basic digital computer is beginning to be used as a part of the instrument. If a mass-produced general-purpose computer can functionally and economically perform the tasks required of the special digital circuitry it replaces, there is much to be gained. The computer can perform many other tasks such as linearization, signal averaging, conversion, normalization, storage, data transmission, and correlation. In addition, there is the possibility of computer control of instrument functions and system parameters to increase measurement efficiency and convenience.

A basic computer needs to perform only a few operations such as fetch a word from storage, add, subtract, input, output, test and jump, and put a word into storage. The power of the computer comes from performing these operations many times very rapidly. This section will describe how these operations are performed and give a general description of the organization and control of the computation and storage circuits.

Computer Arithmetic

The arithmetic operations performed directly by the computer are simply addition and subtraction. More complex operations such as multiplication, division, integration, or taking a log or a power are performed by repeated addition or subtraction. Circuits that perform these operations will be described in this section.

The addition of binary numbers was discussed in Chapter 4, Section 4-3, and adder circuits were given in Fig. 4-14. Half-adder and full-adder circuits are shown again in Fig. 7-51 to illustrate what circuits are in each block. In a computer, the binary numbers are stored and handled in groups of 8, 12, 16, 18, 24, or 32 bits called a *word*. For a given

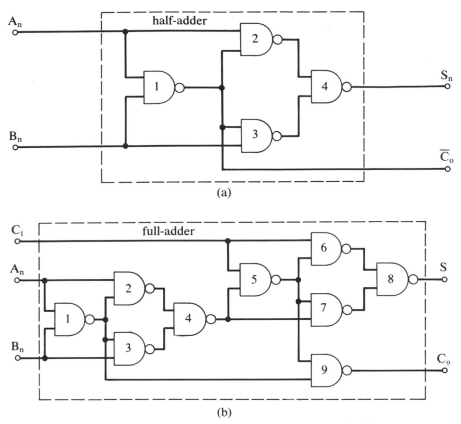

Fig. 7-51   Circuit of (a) half-adder and (b) full-adder.

computer, all numbers have a fixed word length with **0**'s for the unused most significant figures. Thus computer addition circuits are designed to add numbers as stored in these fixed-length words. There are two ways to do this: use a separate full-adder for each bit in the word and add all bits simultaneously (parallel addition) or use one full-adder and add the bits one at a time starting with the least significant bit (serial addition).

Parallel addition.   An 8-bit parallel adder is shown in Fig. 7-52. The numbers in the addend and augend registers are summed by the eight full-adder circuits. The carry output, C_o, from each full-adder is connected to the carry input of the full-adder for the next most significant bit position. The sum outputs of the full-adders are connected to D inputs of the sum register. When the sum is available at the sum outputs, a clock pulse is applied to the sum register. Note that there are nine flip-flops

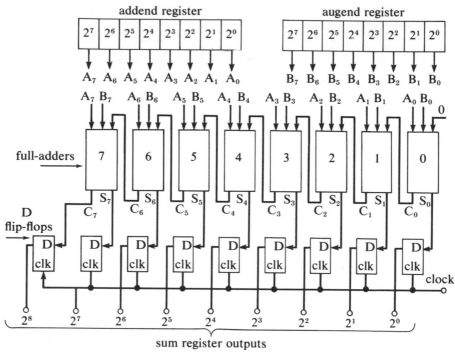

Fig. 7-52   **A parallel adder.**

in the sum register to accommodate the carry output of the eight full-adder.

The addition of all 8 bits in the addend and augend registers is not performed simultaneously. The carry input information for each full-adder circuit does not appear until the previous full-adder has generated its carry output signal. For the full-adder circuit of Fig. 7-51, the input information passes through a series of five gates (1, 2 or 3, 4, 5, and 9) before appearing at the carry output C_o. If the gate delay time is 15 nsec, C_o will appear 75 nsec after the information from the A and B registers is applied to the adders. When the C_o signal is then applied to full-adder 1, A_1 and B_1 have been applied for 75 nsec. The delay between C_i and C_o is, therefore, just two gate delay times (gates 5 and 9). The signal appears at C_1 30 nsec after C_o and at C_2 30 nsec after C_1, and so forth. This is called a *ripple-carry* adder since the carry information "ripples" through the adder circuits. The total delay between the application of the A and B register information and the appearance of the correct level at C_7 is 75 nsec plus 7×30 nsec equals 285 nsec for the circuit illustrated. In parallel adders designed for larger words, the delay of the ripple-carry circuit will be excessive. Integrated circuit adders are available with additional logic gates that reduce the carry time.

Sums of more than two numbers are obtained by adding two numbers and then adding a third number to the sum of the first two, and so on. After the first sum, the sum register output of Fig. 7-52 is transferred to the augend register where it can be added to the next addend number. It is quite common for the sum and augend registers to be one and the same, i.e., the sum register outputs are connected to the corresponding B inputs of the full-adder circuit. The combination sum-augend register is called the *accumulator*. An addition is now performed: first, clear the accumulation register; second, enter the first number in the addend register—the full-adder outputs are now the sum of the addend and the accumulator (addend + 0 = addend); third, clock the accumulator to enter the addend into the accumulator; fourth, enter the second number into the addend register—the sum of the first two numbers now appears at the adder outputs; fifth, clock the accumulator to enter the sum into the accumulator; sixth, continue to add numbers by entering into the addend register and clocking the accumulator.

Serial addition. In a serial adder, numbers are added by applying them one bit at a time to a single full-adder circuit. The numbers to be added and the sum are stored in shift registers. Starting with the least significant bits the numbers are shifted out of the addend and augend registers into the adder, and the sum is shifted from the adder into the sum registers. Just as in the parallel adder case, the sum and augend registers are usually combined into a single accumulator register. A serial adder circuit is shown in Fig. 7-53. To perform an addition the accumulator and carry delay flip-flop are cleared. The first number to be added is entered into the addend register by the serial or parallel inputs. Then the shift input is pulsed 8 times. The adder sums the bits at the 2^0 position in each register. After the first shift pulse, the sum of the 2^0 bits is stored in the leftmost position in the accumulator and the carry information is stored in the carry delay flip-flop. Now the 2^1 bits and the 2^0 carry are at the adder inputs. After the shift input is pulsed a second time, the 2^1 sum is stored in the accumulator and the carry information is in the carry delay flip-flop. After eight shift pulses, all 8 bits are added and the sum is in the accumulator (and carry delay flip-flop). Since the accumulator is CLEAR to begin the addition, the first eight shift pulses have the effect of placing the addend in the accumulator. Then the second number is entered into the addend register and the shift input is pulsed 8 times again. This process of entering and shifting may be repeated to add additional numbers to the sum in the accumulator. If

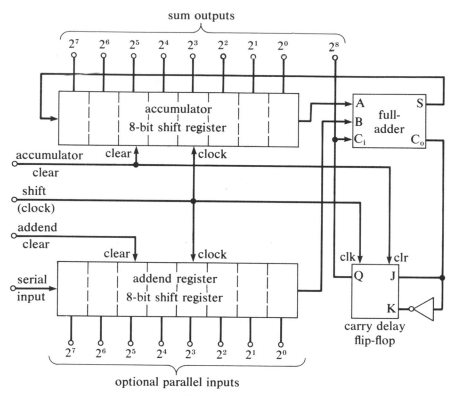

Fig. 7-53 A serial adder.

the serial input to the addend register is used to load the register, the addend register can be loading the next number while the sum is being taken.

The serial adder requires substantially fewer parts than the parallel adder, and the serial form of data transfer can result in great reductions in system interconnections. Its primary disadvantage is that it is much slower than parallel addition. Some computers use combinations of serial and parallel addition and data transfer. For instance, several 8-bit "bytes" or 4-bit "nibbles" can be added by a parallel adder one after another to complete the addition of a 16- or 24-bit word.

Binary subtraction. A gate circuit that will perform the logical operations required for binary subtraction can readily be designed. However, in a small general-purpose computer, it is far more convenient to have a single circuit that will add or subtract on command. The most popular method of satisfying this requirement treats subtraction as the addition of a negative number. A *sign bit* is used to distinguish positive and negative numbers. This sign bit is normally the left-most bit in the word. A **0** indicates a positive number, and a **1** a negative number. When writing the number, an asterisk is sometimes used to separate the sign bit from the magnitude. For instance, 0*0000111 is +7 and 1*0000101 would be −5 if the code for positive and negative numbers were identical. However, as will be shown next, the arithmetic operation of adding positive and negative numbers is greatly aided by representing negative numbers as the complement of the corresponding positive binary number.

The complement of a number is obtained by changing all 1's and 0's and all 0's to 1's, in other words by complementing each bit. The complement of 0*0000101 (+5) is 1*1111010 (−5). When a number and its complement are added, the sum is all 1's. For this reason, the complement is called the 1's complement. If a number and a complement of a smaller number are added, the sum is 1 less than the difference between them as illustrated by the following examples:

$$
\begin{array}{ll}
0*0001101 \ +13 & 0*0010110 \ +22 \\
\underline{1*1111000 \ - \ 7} & \underline{1*1110011 \ -12} \\
10*0000101 \ + \ 5 & 10*0001001 \ + \ 9
\end{array}
$$

The sign bits are added along with the magnitude as though they were the most significant bits. Note that there is a carryover to the left of the sign bit. If a negative number in complement form is added to a smaller positive number, the sum is the difference in complement form since it is a negative number. Note that the difference is accurate and that there is no carryover from the sign bit.

$$
\begin{array}{ll}
0*0001011 \ +11 & 0*0000101 \ + \ 5 \\
\underline{1*1101101 \ -18} & \underline{1*1101010 \ -21} \\
1*1111000 \ - \ 7 & 1*1101111 \ -16
\end{array}
$$

From these examples a general rule and procedure for adding positive and negative numbers in complement form can be formulated. That is, if there is a carryover from the sign bit, add 1 to the result; if not, add nothing to the result. In practice, this is accomplished by applying the sign carryover bit to the carry input of the 2^0 adder in an operation called the *end-around carry*. This is especially convenient for parallel adders. For serial adders, a second pass of the number in the accumulator register through the adder circuit is required. To subtract one number from another, simply enter the minuend in the accumulator, enter the subtrahend in the addend register, complement the addend register, and add the numbers utilizing the end-around carry for parallel adders or the double circulation for serial adders. Note that the complement of a number is very easy to form in the 1's complement number system.

Another form for negative numbers called the 2's complement is also frequently used. The sum of a number and its 2's complement is all 0's (plus the carry). For instance, the 2's complement of 0*0001101 (+13) is 1*1110011.

$$
\begin{array}{ll}
0*0001101 & +13 \\
\underline{1*1110011} & -13 \text{ (2's complement)} \\
10*0000000 &
\end{array}
$$

The 2's complement is obtained by adding 1 to the 1's complement. When negative numbers are represented in 2's complement form, addition of positive and negative numbers yields the correct answer directly as shown in the following examples:

$$
\begin{array}{ll}
0*0011011 & +27 \\
\underline{1*1110010} & -14 \text{ (2's)} \\
10*0001101 & +13
\end{array}
\qquad
\begin{array}{ll}
0*0001001 & +\ 9 \\
\underline{1*1101101} & -19 \text{ (2's complement)} \\
1*1110110 & -10 \text{ (2's complement)}
\end{array}
$$

Using 2's complement arithmetic, subtraction is carried out by complementing and adding 1 to the subtrahend number and then adding. The sign bit carry can be ignored. The magnitude of the 2's complement negative number can be obtained by complementing and adding 1 to the result.

Thus the arithmetic operations are different for 1's complement and 2's complement negative numbers, but either is readily implemented with the standard adder circuits. All that is required for subtraction is the assignment of the left-most bit to the sign and the ability to complement the addend register. A complementing shift register is shown in Fig. 7-54. Gates are used to determine whether the information at the D inputs is derived from the Q of the adjacent flip-flop (shift operation) or from $\overline{Q}$ of the same flip-flop (complement operation). To complement the register, the complement input is held HIGH and the shift input LOW for one clock pulse. Forming the complement of a number in the 2's complement number system will in general require a pass through the adder circuitry.

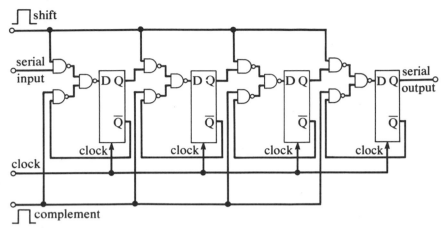

Fig. 7-54   Complementing shift register.

Experiment 7-25   *A Serial Adder*

A four-bit serial adder is wired following the block diagram of Fig. 7-53. Binary numbers are entered into the accumulator and addend registers, and the operation of adding is carried out using a manual clock source. The use of negative numbers in addition and subtraction is also illustrated.

Computer Memory

In even the most basic computers, several different devices are used for the storage of binary information. Flip-flops are used for active storage registers such as those in the adder circuits described above. Punched paper tape or punched cards are used for long-term storage of computing routines or data. Temporary or semipermanent storage is also provided in the electronic part of the computer. Such memory types are magnetic disk or drum memory, circulating shift registers, and ferrite core memory. The core memory is presently almost universally used as the working storage in all sizes of computers and, thus, will be described briefly in this section.

A magnetic core is a very small doughnut-shaped piece of ferromagnetic material of high retentivity. When a current is passed through a wire threading the hole in the core, a magnetization is induced in the core as shown in Fig. 7-55. Because of the high retentivity, the mag-

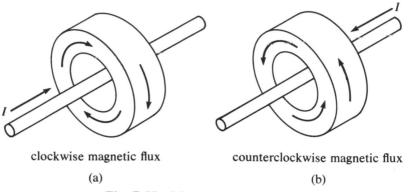

clockwise magnetic flux   counterclockwise magnetic flux

(a)   (b)

Fig. 7-55   Magnetizing the core.

netism remains when the current is turned off. The core is now set in
the **1** state. If the current is now reversed, it is found that the magnetism
of the core does not change appreciably until the magnitude of the current is great enough to counteract the stored magnetic flux in the core.
When the reverse current exceeds this minimum value, the direction of
the magnetic flux is reversed as shown in Fig. 7-55b. This behavior is
illustrated in the B-H curve of Fig. 7-56. A magnetizing force or current
$+I_m$ is applied to put the core in the $+B_r$ or **1** state and a current $-I_m$ is
applied to put the core in the $-B_r$ or **0** state.

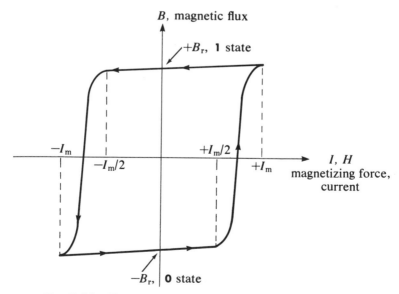

Fig. 7-56   Hysteresis loop for a ferrite magnetic core.

To write information into and read information out of the core, the cores are wired in a matrix called a *bit plane* and shown in Fig. 7-57a. This arrangement takes advantage of the hysteresis of the core to accomplish some addressing logic. If the core at the intersection of X_3 and Y_5 is to be SET, half of the required set current or $+I_m/2$ is applied to the X_3 and Y_5 drive lines. Only the core X_3Y_5 receives the required setting current. The other cores on the X_3 and Y_5 drive lines only receive $I_m/2$ which, as Fig. 7-56 shows, is not sufficient to SET the core. The part-current drive line matrix reduces the number of current drivers for N cores from N to $2N^{1/2}$. Each core has a unique matrix intersection or *address*.

The state of a particular core is read or "sensed" by applying a CLEAR pulse to that core. If the core was in the **1** state, the resulting change in magnetic flux induces a current pulse in a *sense line* that threads through all the cores in the plane. If the core was in the **0** state, a much smaller pulse, induced by the CLEAR pulse, appears in the sense line as shown in Fig. 7-58. Even though the sense line threads through all the cores, the induced current pulse can only come from the core address that received the CLEAR pulse. The reading operation is "destructive" so if it is desired to continue to store the information some provision must be made to write the bit back into the addressed core.

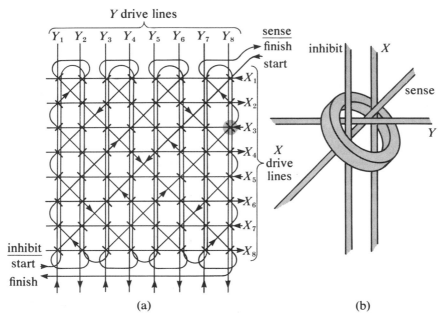

Fig. 7-57 A core memory bit plane.

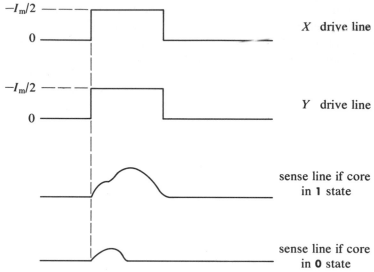

Fig. 7-58   Waveforms in drive and sense lines for stored 1 and 0.

Core memories store words by using as many bit planes as there are bits in the word. One method of organizing the bit planes into words is the "3-D" technique shown in Fig. 7-59. Corresponding X and Y drive lines for each bit plane are wired together so that all bit planes are addressed together. Each bit plane stores one bit in the addressed word. The memory will store as many words as there are cores in each plane. A separate sense amplifier is used for the sense line in each bit plane so that when an address is read (when a $-I_m/2$ pulse is applied to one X and one Y drive line) the outputs of the sense amplifier present the word in parallel digital form. This information is read into a register that provides a stable output for transmission of the word to the desired location and stores the word until it can be applied to the word inputs and written back into the core address.

When a write pulse is applied to an address in a 3-D core stack, a **1** will be written into every bit of the word. To enable a **0** to be written, an *inhibit line* is threaded through every core in a plane. When a $-I_m/2$ current is applied to the inhibit line, the $+I_m$ setting current has a net effect of $I_m/2$ which is insufficient to SET the core. Since the $-I_m/2$ inhibit current is opposite to the two $+I_m/2$ drive line currents, no core receives sufficient magnetizing force to change state. Since the addressed bit was just read, it must be in the **0** state. A word is then written into that address by applying SET currents to the appropriate X and Y drive lines and an inhibit current to the inhibit line of each plane where a **0** bit is to be stored. Figure 7-57b shows the X and Y drive lines and the sense and inhibit lines threading through a typical core.

Another form of memory organization that is increasingly popular is the "$2\frac{1}{2}$-D" memory. In this memory, the corresponding X drive lines of each plane are wired in series or parallel, but the Y drive lines of each plane are driven separately. The read operation is the same as for the 3-D memory, but the write is accomplished simply by not applying a SET pulse to the Y driver line of the planes that are to remain **0**. The $2\frac{1}{2}$-D requires more drivers than the 3-D, but the need for the long inhibit line is eliminated. As attempts are made to design smaller (and consequently faster) core memories, the elimination of the fourth wire can be an important performance and economy factor.

Current core memories are made in packages of 512 words by 8 bits per word up to almost any conceivable size. The switching speed of standard current designs allows a complete read–write cycle in 1–2 μsec

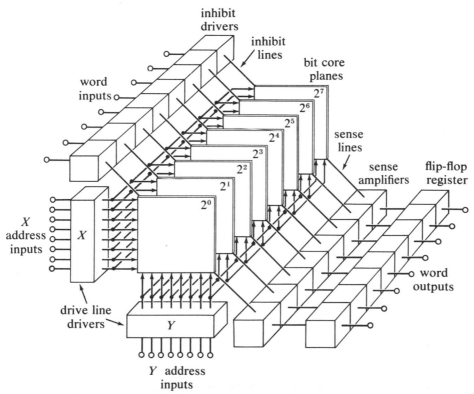

Fig. 7-59   A 3-D core memory stack.

with faster designs available. The size of the core is of the order of 1 mm outside diameter with a complete 32×32 bit plane taking only about 6 cm² of area.

Computer Organization

It was mentioned earlier that the digital computer derives its great capability from being able to perform many very simple operations in rapid succession. It is essential that the sequence of operations be performed automatically since manual control of the sequence would make calculation times prohibitively long. In practice, the desired sequence of operations, commonly called instructions, is stored in the core memory in binary-coded form. The reading and decoding of the operations and the control of their execution is carried out by a group of circuits called the *control unit*. A representation of the relationship between the control unit and the memory unit, arithmetic unit, and input and output devices is shown in Fig. 7-60. It is seen that the control unit controls all the interactions among the various parts of the computer and its accessories.

Though the design of control units differs greatly from one computer to another, all follow the general approach described briefly here. A sequence of instructions to the arithmetic unit, memory, and input and output devices is written which will carry out the desired calculation.

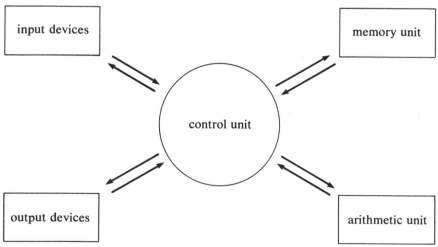

Fig. 7-60 The control unit controls all computer data transfers and operations.

These instructions are translated into a binary code and entered into the core memory in order of their desired execution. A separate memory word is used for each instruction. Any numerical data required are stored in binary form at other memory word locations. The memory address is now set at the location of the first instruction word, e.g., location 1.

When the start button is pushed, the first instruction word is read out, decoded, and used to generate a sequence of pulses that command part or parts of the computer to execute the first instruction. For instance, if the first instruction were "clear the accumulator register and add the contents of memory location 136 to the accumulator," the control unit would send pulses to clear the accumulator register, set the memory address to 136, read the word at location 136 into the memory register, enter the data word into the addend register, write the data word back into memory location 136, and perform the addition of the addend to the accumulator. Now the first operation is complete, but before the computer can begin on the next instruction, the memory address must be set to location 2 where the second instruction is stored. This is accomplished by a *program counter* that is automatically advanced one count (incremented) by the control unit at the completion of each instruction word transfer. The control unit will continue to "fetch" the next instruction, decode the instruction, increment the program counter, generate the required sequences of pulses to execute the instruction, fetch the next instruction, and so on until a stop instruction is received.

A basic stored-instruction computer is shown in block form in Fig. 7-61. Some aspects of computer organization and control will be looked at more closely with respect to this diagram. The sequence of instructions is loaded into the memory one word at a time from an input device such as a Teletype, punched card reader, punched paper tape reader, or magnetic tape. In some computers the input devices can feed the memory register directly while the control unit pulses the memory write circuit and increments the program counter each time a new word appears at the input. Other computers transfer data only between the memory and the accumulator or the addend register. In this case, the input device enters data to one of these registers under the control of the control unit.

Once the instructions and the data have been stored in the core memory the computer is ready to start the desired computation. The start command can be given manually or by an input signal that follows the last datum or instruction word to be entered. The start command sets the program counter to the address of the first instruction. The operation cycle generator sends a pulse to clear the instruction generator, then a pulse to transfer the program count to the memory address regis-

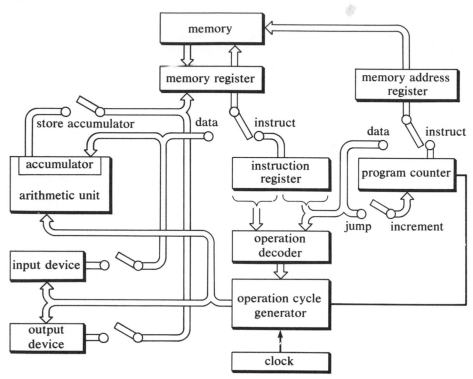

Fig. 7-61 Block diagram of basic computer.

ter, then a pulse to read the memory word into the memory register and write the word back into the memory, then a pulse to increment the program counter, and then a pulse to transfer the contents of the memory register to the instruction register. This succession of five pulses comprises the operation of fetching a new instruction and is common to all instructions. The subsequent pulse sequence depends on the instruction to be executed.

The instruction word will generally have the form shown in Fig. 7-62a. The leftmost bits indicate the operation to be performed and the remaining bits give the memory address where the number to be operated on *(the operand)* is to be found. Such instructions are "clear the accumulator and add the number in memory location 635" and "subtract the number in memory location 125 from the accumulator." In principle, if the computer is capable of following sixty-four different instructions, 6 bits will be required for the operation code. Similarly, if the memory has 4096 addresses (a "4K" core), 12 bits will be required for the memory address. Thus the instruction word would have to be 18 bits long.

Some smaller computers have a word size that is too small to encode all the instructions and addresses. To save space, instructions that do not require a memory location such as "shift accumulator to right," increment accumulator," and "print out accumulator on the Teletype" can use instruction word bits normally saved for the operand address. One or two of the leftmost bits are used to indicate that the entire word is an instruction. Another technique is to provide a few additional temporary storage registers in the arithmetic section to reduce the number of times data has to go in and out of the memory. In that case the movement of data in and out of memory might always be a separate instruction word called a *memory reference* word shown in Fig. 7-62b. The left-most bits identify the instruction as data movement and identify the register that is the source or destination of the word for the memory address indicated. Instruction words would then use operands in registers in the arithmetic unit.

Data words are transferred from the memory register to a register in the arithmetic section. They generally have the form shown in Fig. 7-62c. Data words are never mixed in with the instruction words in the memory, but are always called out of their memory address as the result of an instruction.

Refer again to the block diagram computer of Fig. 7-61. The instruction has just been placed in the instruction register. The operation decoder identifies the instruction and controls the remaining output pulses from the operation cycle generator accordingly. The operation cycle generator is a cyclic sequencer driven by the clock. The switch-tail ring counter and decoder shown in Fig. 7-30 is frequently used for this pur-

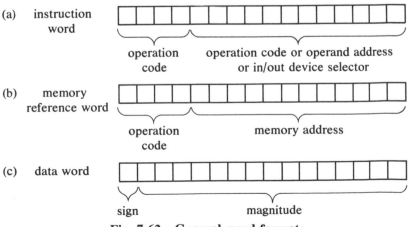

Fig. 7-62 General word formats.

pose. The operation cycle generator can produce a series of pulses that includes all parts of all the instructions in the proper order, and the operation decoder then gates off those pulses that are not needed for the specific instruction. This is a fixed cycle time computer. All instructions will take an equal time to execute. Another approach is for the operation decoder to gate the generated pulses to the appropriate location in succession, resetting the cycle generator when the operation is complete. This type of decoder-cycle generator is more complex but more versatile and results in faster computations. The operation cycle generator is the "heart" of the computer. In addition to the connections shown in Fig. 7-61, the cycle generator controls the memory cycle and all the information transfer gates (shown as switches in the diagram).

Thus far, it has been assumed that the complete list of instructions from beginning to completion of the problem has been written into the memory in sequence. Frequently an instruction or a set of instructions has to be repeated many times. Such is the case in multiplication (repeated shift and add instructions) and division (repeated shift and subtract instructions). Most computers provide a *jump* instruction. The operation code indicates a jump to the instruction address given in the address part of the instruction word. There are also instructions called *conditional* jump instructions such as "jump to XXX if the accumulator is minus." Such instructions allow the programmer to provide alternative instructions for the computer which depend on the results of previous calculations.

It is hoped that this brief introduction to computers will indicate how the basic digital concepts, devices, and circuits described in this book provide the necessary functions for complex computations. Dozens of books have been written about the computer and its organization, various components, and peripheral equipment. Several good books that are specifically devoted to digital computation are listed in the Bibliography. It is, of course, the utilization of the basic concepts and circuits not only in computers, but also for measurement and control, that will provide the completely automated scientific experiments and laboratories of tomorrow.

Bibliography

Bartee, T. C., *Digital Computer Fundamentals,* 2nd ed., McGraw-Hill, New York, 1966. Computer arithmetic, circuits, and organization.

Benrey, R. M., *Understanding Digital Computers,* Rider, New York, 1964. Computer arithmetic and organization on an introductory level.

Caldwell, S., *Switching Circuits and Logical Design,* Wiley, New York, 1958. Boolean algebra, switching logic, and logic minimization techniques.

Chu, Y., *Digital Computer Design Fundamentals,* McGraw-Hill, New York, 1962. Extensive treatment of binary arithmetic, Boolean algebra, and logic circuit design.

Delholm, L., *Design and Application of Transistor Switching Circuits,* McGraw-Hill, New York, 1968. A practical guide to the design of component switching circuits and flip-flops.

Gibbons, J. F., *Semiconductor Electronics,* McGraw-Hill, New York, 1966. A good basic text in general electronics.

Giles, J. N., *Linear Integrated Circuits Applications Handbook,* Fairchild Semiconductor, Mountain View, California, 1967. A valuable guide to the applications of integrated circuit operational amplifiers and comparators.

Hakim, S. S., and R. Barrett, *Transistor Circuits in Electronics,* Haydon, New York, 1964. A basic book on transistory electronics including switching and logic applications.

Harris, J. N., P. E. Gray, and C. L. Searle, *Digital Transistor Circuits,* Wiley, New York, 1966. An introduction to digital gates.

Hoeschele, D., Jr., *Analog-to-Digital-to-Analog Conversion Techniques,* Wiley,

New York, 1968. A recent and extensive treatment of A-D and D-A techniques.

Hughes, J., *Digital Computer Laboratory Workbook,* Digital Equipment Corp., Maynard, Massachusetts, 1968. A book of basic experiments in gating, counting, and adding using TTL gates and flip-flops.

Littauer, R., *Pulse Electronics,* McGraw-Hill, New York, 1965. A description of switching devices and component gate and flip-flop circuits.

Lo, A. W., *Introduction to Digital Electronics,* Addison-Wesley, Reading, Massachusetts, 1967. Descriptions of the switching devices and basic circuits used for digital gates and memories.

Maley, G. A., and J. Earle, *The Logic Design of Transistor Digital Computers,* Prentice-Hall, Englewood Cliffs, New Jersey, 1963. Heavy emphasis on Boolean algebra and minimization techniques in logic circuit design.

Malmstadt, H. V., C. G. Enke, and E. C. Toren, Jr., *Electronics for Scientists,* W. A. Benjamin, New York, 1962. An earlier book that can provide a companion study of basic electronics and linear circuits and techniques.

Marcus, M. P., *Switching Circuits for Engineers,* Prentice-Hall, Englewood Cliffs, New Jersey, 1962. A text on logic and minimization techniques in logic design of switching networks and sequential circuits.

Millman, J. and H. Taub, *Pulse, Digital, and Switching Waveforms,* McGraw-Hill, New York, 1965. The standard source book for switching and wave-shaping and component gate and flip-flop circuits.

Nashelsky, L., *Digital Computer Theory,* Wiley, New York, 1966. An introduction text on digital circuitry and computations.

Philbrick Researches, Inc., *Applications Manual for Computing Amplifiers for Modelling, Measuring, Manipulating, and Much Else,* Philbrick Researches, Inc., Dedham, Massachusetts, 1966. An extensive operational amplifier applications manual by a company that has set the pace in educating the scientist in the use of operational amplifiers.

Sevin, L. J., Jr., *Field Effect Transistors,* McGraw-Hill, New York, 1963. An early reference on field-effect transistors and their applications.

Wood, P. E., Jr., *Switching Theory,* McGraw-Hill, New York, 1968. A recent and advanced text on the design of combinational and sequential switching networks.

Experiments

Experiments

General summaries for dozens of experiments were given throughout the text. Specific instructions are presented here on how to perform these experiments efficiently. A system of integrated circuit cards, modules, parts, and instruments is used. Equipment details are presented as they are needed for specific experiments.

A new experimental system and new methods of interconnecting parts and circuit cards were considered essential so that a student could progress rapidly from connecting the most basic circuits to building and experimenting with digital and analog-digital instruments such as frequency meters, timers, counters, and digital voltmeters (DVM's). The requirements, ideas, and concepts of the authors for specific digital circuit cards, parts, modules, and instruments have been implemented by the scientific instruments program of the Heath Co.[1] An analog-digital designer (ADD), a universal digital instrument (UDI), and a rapid-connect bread-boarding system were developed to meet the experiments' requirements.

The circuits and instruments that the student builds are representative of those now used for research-quality scientific instrumentation. Therefore, when the set of experiments is completed the experimenter will have gained a working ability with the electronic digital circuits and instruments that are directly applicable for scientific measurements, computation, and control functions.

In addition to the equipment presented or outlined in the specific experiments, a high-performance dual-beam or dual-trace oscilloscope is recommended. Scope characteristics should include DC-15 MHz (or greater) frequency response for the Y channels, 10 MV/cm sensitivity, sweep speeds from about 2×10^{-1} to 5×10^{-7} sec/cm, and sensitive X-Y capability to at least 100 kHz. Also, a voltage reference source (such as the Heath EU-80A) and a few other items are called for in the experiments.

[1] UDI of Heath Co., Benton Harbor, Michigan.

Accurate measurements of frequency, time interval, period, events counting, voltage and voltage integration, and frequency ratio are important in many scientific laboratory experiments. Many of these same measurements are useful and convenient in characterizing the digital circuits, systems, and instruments to be investigated in subsequent chapters of this book. Therefore, it is appropriate at the start to gain experience in making and using digital measurements.

The general concepts of these digital measurements have been described in Chapter 1 and it is worth while to review these principles so as to have an understanding of the general circuit functions while performing the measurements. Specific instructions are given for use of the Model EU-805A universal digital instrument. This instrument can very accurately perform all of the digital measurements described in this chapter. It is also built on a modular basis so the circuit cards in the instrument which perform the specific electronic functions can be easily removed and investigated. The type of electronic discrete and integrated circuits on these cards will be thoroughly studied through experimentation in subsequent chapters, and specific cards will be removed and utilized for other instrumental applications.

Because instrument cards can be patched in the ADD to make digital measurement systems, an ADD thus equipped and patched could be used in place of the UDI to make many of the digital measurements suggested in these and later experiments.

Experiment 1-1 *Introduction to the Universal Counter*

A familiarity with the physical arrangement of the basic digital measurement instrument is obtained by locating and identifying the controls and connections. Manual time measurements are then made to check the operation of the instrument and to learn the function of several basic controls.

1-1a   Controls

Refer to Fig. E1-1 to locate and identify the front and rear panel connections and controls of the instrument.

1-1b   Instrument Check

Plug the instrument in.   Set the controls as shown in Table E1-1. The instrument is in the time measurement mode.   The 1-MHz crystal oscillator output is scaled down to provide a pulse every 10^{-1} sec which

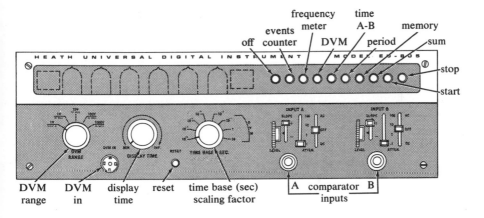

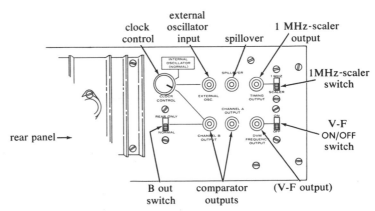

Fig. E1-1   Universal digital instrument controls and connectors. (Courtesy of Heath Co.)

can be counted over the desired time interval and displayed, as shown in Fig. E1-2. Momentarily push the START button to begin the measurement and the STOP button to end it. Observe that the gate light is on during the time measurement.

Table E1-1   Manual Time Interval Measurement

FUNCTION	MEM	SUM	DISPLAY TIME	TIME BASE, SEC	INPUT A			
					LEVEL	SLOPE	ATTN.	AC-OFF-DC
Time A-B	Out	Out	12 O'clock	10^{-1} sec	_____	_____	_____	OFF

INPUT B				REAR PANEL			
LEVEL	SLOPE	ATTN.	AC-OFF-DC	CLOCK CONTROL	1-MHz SCALER	DVM	B OUT
_____	_____	_____	OFF	Normal	_____	OFF	Normal

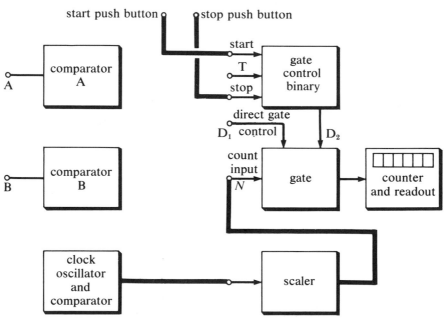

Fig. E1-2   Manual time-interval digital instrument.

1-1c Display Time

Vary the Display Time control and note its effect on the time between the end of the count period and the automatic reset. Note that the next measurement cannot be started until the display time is over. Return the control to the 12 o'clock position.

1-1d Time Base

Use other settings of the Time Base switch to change the precision of the time measurement. Note that the units and decimal point lights change correspondingly. Place the Time Base control in the 10^{-6} sec position to measure time with microsecond resolution. Note that in an interval of 1 sec the counter is filled and the OVER light turns ON indicating an overrange condition. To avoid this overranging reduce the resolution or count the spillover pulses from the SPILLOVER output on the rear panel. One spillover pulse occurs each time the counter goes from 999,990 to 000,000, that is, every million counts.

1-1e Memory

Push the MEMORY button in. Make another time measurement and note that the results of the measurement are not displayed until the measurement is over and that the display remains until the next measurement is complete. Note also that a measurement cannot be started until the display time is over. This delay can be eliminated by turning the Display Time control to MIN. The MEM light is lit whenever the memory is ON.

1-1f Sum

In the sum mode, the counter does not reset after each start–stop cycle. Thus the second measurement interval is added to the first, and so on. This effect is easiest to observe with the memory OFF (button out). The summing function is especially convenient in cumulative time measurements. Turn the Display Time control to MIN so that successive measurements can be started anytime. The SUM light is ON whenever the SUM button is in.

Experiment 1-2 *Input Comparator*

The characteristics and response of the input comparator and its setting for reliable measurements, including slope, level, and attenuator controls, are studied by using an oscilloscope to display the input signal and comparator output.

1-2a AUTO Level Position

Set the controls on the UDI as summarized in Table E1-2.

Connect Input A of the UDI to the sine-wave output of a sine-square generator as shown in Fig. E1-3. Use a dual-trace (or double-beam) oscilloscope if available to observe simultaneously the generator signal and the comparator A output from the rear panel of the UDI. Set the generator amplitude at 10 V rms, and the frequency at 1 kHz.

Note that the comparator output changes state at each zero crossing and that the negative slope of the comparator output occurs on the positive slope of the input signal.

1-2b Slope Control

Switch the slope control to the (−) position and note that the negative slope of the comparator output now occurs during the negative slope of the input signal.

Table E1-2 Input Comparator

				TIME BASE, SEC		INPUT A			
			DISPLAY	SCALING					
FUNCTION	MEM	SUM	TIME	FACTOR		LEVEL	SLOPE	ATTN.	AC-OFF-DC
Events counter	___	___	___	___		AUTO	+	100	Ac

INPUT B				REAR PANEL			
LEVEL	SLOPE	ATTN.	AC-OFF-DC	CLOCK CONTROL	1-MHz SCALER	DVM	B OUT
___	___	___	OFF	___	___	OFF	___

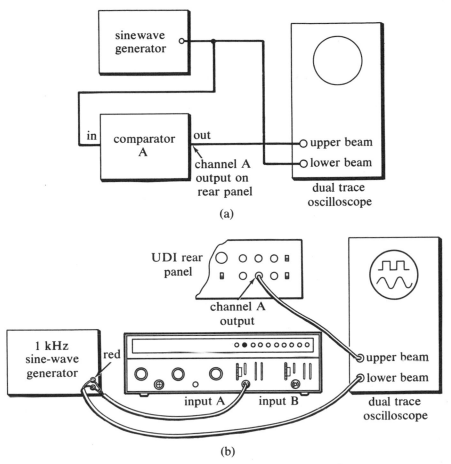

Fig. E1-3 **Input comparator test connections.**

1-2c   Level Control

Switch the LEVEL control out of the AUTO position and adjust it so as to set the comparator trigger point. Note that the trigger point on the input waveform changes as a function of the LEVEL control setting. Note also that when the trigger level is set too near the input waveform extremes, erratic triggering can result.

1-2d   Attenuator

Lower the input signal voltage to about 100 mV. The attenuator setting must be lowered to maintain sufficient comparator sensitivity.

Use output waveform to check for proper attenuator and level adjustment.

1-2e Alternating Current and Direct Current Controls

Some input signals might have a dc component added to the waveform. In the ac position the dc component is blocked and the AUTO (zero-crossing) level setting can be used. For slowly changing signals or for dc level inputs, the dc position must be used. The LEVEL control must then be set for the desired trigger point. To illustrate this point, connect a unidirection pulse such as the square-wave output from a square-wave generator or the square-wave or ramp signal from the ADD timing module to the comparator A input. Observe the effect of ac-dc and level controls.

Experiment 1-3 *Events Counting*

Counting measurements are made in order to become familiar with the counting measurement and to experiment with manual and remote start and stop and with the scaling of input events.

1-3a Push-Button Control of Start and Stop

Adjust the controls of the UDI as outlined in Table E1-3 and connect a signal generator to the comparator A input as shown in Fig. E1-4. Set the signal generator to provide a 5-V, 20-Hz sine or square wave. The block diagram for this measurement is shown in Fig. E1-5. Push

Table E1-3 Events Counting

| | | | | TIME BASE, SEC | INPUT A | | | |
FUNCTION	MEM	SUM	DISPLAY TIME	SCALING FACTOR	LEVEL	SLOPE	ATTN.	AC-OFF-DC
Events counter	Out	Out	12 O'clock	1	AUTO	+	10	Ac

| INPUT B | | | | REAR PANEL | | | |
LEVEL	SLOPE	ATTN.	AC-OFF-DC	CLOCK CONTROL	1-MHz SCALER	DVM	B OUT
———	———	———	OFF	———	———	OFF	Normal

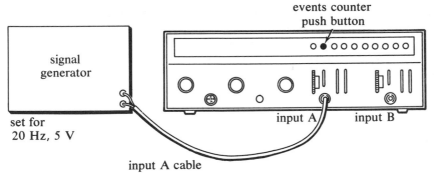

Fig. E1-4   Events counting.

the START button and observe the count of the input cycles on the read-out.   Push the STOP button at the end of the desired count interval. Note that the MEMORY, SUM, and Display Time controls have the same functions as in Experiment 1-1.

1-3b   Scaling and Overrange

Turn the sum and memory OFF.   Increase the signal frequency to about 100 kHz.   Start the count and notice that the count register is

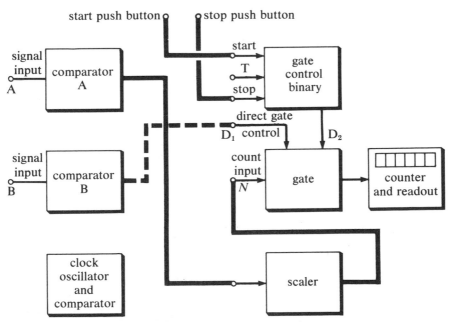

Fig. E1-5   Block diagram for events counter mode.

quickly filled. When this happens the OVER range light comes ON to indicate that the most significant figure(s) of the measurement has been lost. High-frequency input signals can be scaled down by as much as 10^7 before they enter the counter. Increase the scaling factor one step at a time and observe the effect on the count rate. The check (✔) light comes on to indicate that the reading must be multiplied by the scaling factor to obtain the actual count.

To count more than 999,999 events without losing any significant figures, it is necessary to count how many times the counter counted to 1 million. The output of the last DCU is connected to the SPILLOVER output at the rear panel. This may be connected to an electromechanical or electronic counter to increase the counting range of the instrument.

1-3c   Electronic Counting Gate

The events count at input A will be started and stopped (gated) by the output from comparator B, if it is on. Connect the comparator B input to a 1-V signal from a voltage reference source. Turn input B to the dc × 10 position and adjust the level control so that depressing and releasing the zero button on the voltage reference source (VRS) causes the gate light to go ON and OFF. Set the slope switch so that depressing the zero button turns the gate ON.

Using the zero button of the VRS, turn the gate ON for 10 sec while counting a 100-kHz sine-wave signal.

Note: When the instrument is in the EVENTS COUNT mode, the A input is always the *count* input, and the B input is always the *gate control* input.

Experiment 1-4   *Frequency Measurement*

The frequency of a signal generator output is measured. The effects of input frequency and time base on precision are studied. The effect of the comparator setting on measurement accuracy is observed. The frequency of the crystal oscillator and scaler outputs is measured.

1-4a   Frequency Measurement

It is clear from the electronic gate counting experiment above, that if the B input signal remained above the trigger level for precisely 1 sec, the count would be exactly equal to the average *frequency* of the input signal for that second. In the FREQUENCY mode of operation, the counting gate is held open for a precise time interval derived from a high-precision crystal oscillator and decade scaler. The block diagram for this measurement is shown in Fig. E1-6.

Connect about a 100-kHz, 5-V signal to the UDI as shown in Fig. E1-7. Set the controls as summarized in Table E1-4. Observe the frequency. Then change the Time Base control and note the effect on the number of significant figures and the measurement repetition rate. On the 10-sec time base, the time between measurements is shortened considerably when the rear panel scaler output switch is in the 1-MHz position.

Using the 1-sec time base, measure a frequency in each range of the generator. Note the decrease in significant figures as frequency is decreased, even though the accuracy is constant to ±1 cycle.

1-4b Comparator

For accurate measurement it is essential that the comparator trigger accurately and reliably for each event or cycle. All comparators have sensitivity and frequency limitations. In addition, a faulty setting of the level control can yield unreliable counts. Using the sine-square generator and the UDI, notice how the frequency measurement is affected by

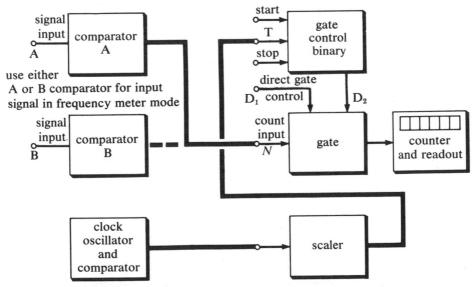

Fig. E1-6   Block diagram of frequency meter mode.

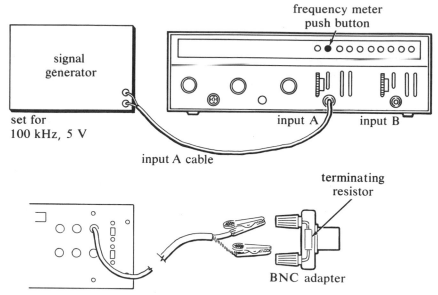

Fig. E1-7 Frequency measurement. (a) Connection of signal source to input comparators. (b) Connection of 1-MHz clock output of UDI to input comparator. (See Section E1-4c.)

Table E1-4   Frequency Measurement

FUNCTION	MEM	SUM	DISPLAY TIME	TIME BASE, SEC	INPUT A			
					LEVEL	SLOPE	ATTN.	AC-OFF-DC
Freq.	In	Out	Min	1 sec	AUTO	+	×10	Ac

INPUT B				REAR PANEL			
LEVEL	SLOPE	ATTN.	AC-OFF-DC	CLOCK CONTROL	1-MHz SCALER	DVM	B OUT
——	——	——	OFF	Normal	1 MHz	OFF	Normal

the amplitude of the input signal and the setting of the attenuator and level controls, including the automatic level position. If a generator with an output frequency greater than 10 MHz is available, the triggering sensitivity and upper frequency limitations may be studied.

1-4c Crystal Oscillator and Scaler Output

The Timing Output connector on the rear panel provides a 1-MHz output from the crystal oscillator or the output of the decade scaler depending upon the position of the adjacent switch. With this switch in the 1-MHz position, connect the Timing Output to the A input.

Note: Whenever a low-impedance signal source, such as the 1-MHz output of the UDI, is connected to a high-impedance input, such as the comparator of the UDI, it is necessary to terminate the cable connection at the high impedance end with the characteristic impedance of the cable. One method for terminating the cable connection is shown in Fig. E1-7b.

"Measure" the frequency of this 1-MHz signal using various time base settings. Since the same oscillator is providing the measurement pulses and the gate interval, the display would read 1.00000 MHz even if the oscillator were not that accurate. This connection and measurement is useful, however, because it checks almost all the circuits of the instrument.

Set the time base at 10^{-2} sec and make sure the MEMORY is in. Note that by varying the Display Time control the repetition rate of the measurement can be varied (observe the gate light). In some positions the measurements recur so frequently that the gate light appears to be ON continuously.

Put the Timing Output switch in the scaler position and measure the scaler output frequency on the oscilloscope. Note that the scaler output frequency is determined by the front panel Time Base switch.

Experiment 1-5 *Period Measurements*

Period measurements of a sine-wave source are made for a variety of input frequencies and time base settings. The effect of these variables and the comparator setting on the measurement precision and accuracy are studied.

In the period mode, the output from the time base is counted for one period of the input signal. If the time base is set for 10^{-6} sec, a 100-Hz signal would display a period of 10,000 μsec. Note that even though the measurement requires only 0.01 sec, there are two more significant figures to the measurement than the frequency measurement using the 1-sec time base. The block diagram for the period measurement is shown in Fig. E1-8.

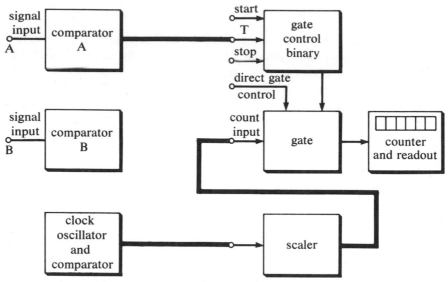

Fig. E1-8 Block diagram for period measurements.

Connect a signal generator to the A input as shown in Fig. E1-9 and set the UDI controls as given in Table E1-5.

Measure the periods of signals over a wide frequency range using several different time bases. Observe how the number of significant figures changes with time base (numerator N) and signal frequency (denominator D).

For the period measurement to be accurate and reproducible the counting of the time base must begin and end at precisely the same point on the signal waveform. For this to be absolutely true, the comparator

Table E1-5 Period Measurement

FUNCTION	MEM	SUM	DISPLAY TIME	TIME BASE, SEC	INPUT A			
					LEVEL	SLOPE	ATTN.	AC-OFF-DC
Period	In	Out	Min	Vary	AUTO	+ OR −	×10	Ac

INPUT B				REAR PANEL			
LEVEL	SLOPE	ATTN.	AC-OFF-DC	CLOCK CONTROL	1-MHz SCALER	DVM	B OUT
___	___	___	OFF	Normal	1 MHz	OFF	Normal

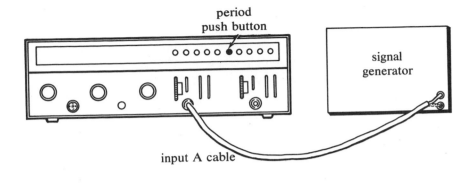

period
push button

signal
generator

input A cable

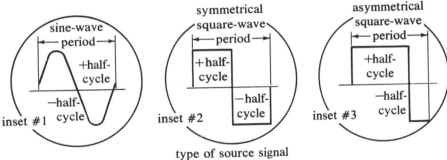

sine-wave
←period→
+half-
cycle
−half-
cycle
inset #1

symmetrical
square-wave
←period→
+half-
cycle
−half-
cycle
inset #2

asymmetrical
square-wave
←period→
+half-
cycle
−half-
cycle
inset #3

type of source signal

Fig. E1-9   Period measurement.

and signal source must be free of drift, jitter, and noise. To the extent that they are not, there will be a random error in the elapsed time measurement. Measurement errors from this source can be reduced by (a) triggering at the point of greatest rate of change of the input signal, (b) averaging a number of period measurements, or (c) measuring the duration of a known number of successive periods.

Measure the period of a signal 100 cycles or less in frequency and notice the effect of level setting and signal amplitude on the reproducibility of the measurement. (Vary the amplitude in the AUTO level position and then vary the trigger level with the amplitude constant.)

Experiment 1-6   *Multiple Period Average*

This experiment demonstrates how the comparator B input signal can be used in place of the internal crystal oscillator and how this is used to make multiple period measurements with 1-μsec resolution. The advantages of the multiple period measurement for certain signals are determined.

The multiple period average measurement is the number of microseconds in F periods of the input signal. Thus the instrument needs to be set up to count microseconds while the count gate stays open for F periods. As shown in the block diagram for this measurement, Fig. E1-10, the B output is connected to the scaler which is connected to the toggle input of the gate control binary. The 1-MHz output from the crystal oscillator is connected to the A input which is counted. This is essentially the same as the FREQUENCY mode; the number of "A" counts per scaler output period.

When the Clock Control switch on the rear panel is in the B OUTPUT position, the output of comparator B is used in place of the internal crystal oscillator for any mode. In the case of the FREQUENCY mode, the clock (now B output) is connected to the scaler as desired. To count microseconds at A, the 1-MHz rear panel output is connected to the A input.

Thus, to make a multiple period measurement, connect the signal to the B input, the 1-MHz output to the A input (with properly terminated cable), push the FREQUENCY mode selector button, and turn the rear panel Clock Control switch to B OUTPUT shown in Fig. E1-11 and Table E1-6.

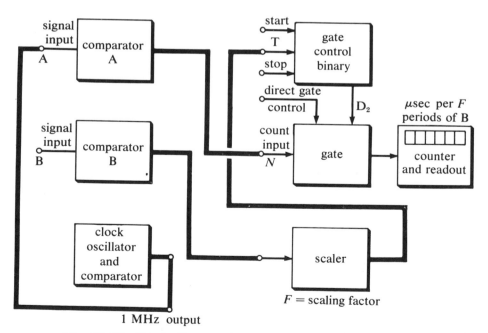

Fig. E1-10   Block diagram for multiple period measurements.

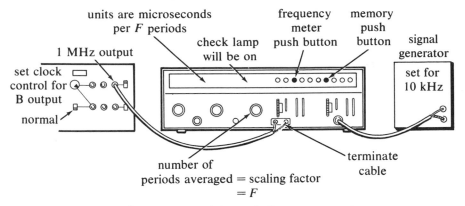

Fig. E1-11   Multiple period measurement.

Table E1-6   Multiple Period Measurement

					INPUT A			
FUNCTION	MEM	SUM	DISPLAY TIME	SCALING FACTOR	LEVEL	SLOPE	ATTN.	AC-OFF-DC
Freq. Meter	In	Out	Min	$1, 10, 10^2, 10^3, 10^4$	AUTO	+ 00 −	×1	Ac

INPUT B				REAR PANEL			
LEVEL	SLOPE	ATTN.	AC-OFF-DC	CLOCK CONTROL	1-MHz SCALER	DVM	B OUT
AUTO	Same as A	×10	Ac	Ch. B output	1 MHz	OFF	Normal

Measure the duration in microseconds of 1, 10, 100, 1000, and 10,000 periods of the input signal. The number of periods measured is equal to the scaling factor F. Note the increase in resolution as F is increased.

Experiment 1-7   *Frequency Ratio*

The frequency ratio measurement is seen to be the same as the multiple period except that both signals to the comparator inputs are from external sources. The line frequency is compared with the frequency scale markings on a signal generator.

measurement is $F \times \dfrac{A\ (Hz)}{B\ (Hz)} = F\left(\dfrac{f_A}{f_B}\right)$

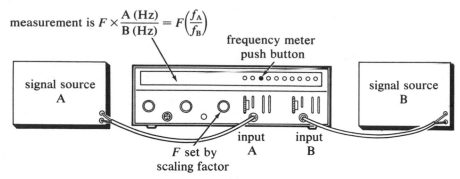

Fig. E1-12   Frequency ratio measurement.

The measurement of frequency ratio is very similar to the multiple period measurement above. The signal at input B is used as the clock so that in the Frequency Meter mode the instrument measures the number of cycles at input A during the number of B periods by the scaling factor switch. Therefore, the measurement is identical to the Multiple Period measurement except that input A is connected to another signal source instead of the 1-MHz output shown in Fig. E1-12. The measured quantity is $F \times A$ Hz/B Hz or $F\ (f_A/f_B)$, where F is the scaling factor.

The calibration of the sine-square generator is checked at 60 kHz. Connect the generator output to input A, connect a low-voltage 60-Hz signal (stepped down from the line) to input B, and switch the Clock Control input to B OUTPUT at the rear panel. A suitable 60-Hz source would be a 6-V filament supply or the 5-V, half-wave rectified 60-Hz signal in the EU-801-11 Digital Power Supply. In the latter case, use #20 solid hook-up wire to connect to the terminal strip behind the front panel. Use terminal numbers 10 (common) and 18 (60-Hz signal output). Use a scaling factor of 1 for the initial adjustment. Then switch to larger scaling factors for more significant figures in the ratio measurement.

Experiment 1-8   *Time A-B Measurements*

The Time A-B mode which was introduced in Experiment 1-1 is also used with electrical start and stop signals to measure positive and negative half-cycles of a square wave to check for symmetry.

Connect the output of a square-wave generator to both A and B inputs and set the controls as shown in Fig. E1-13 and Table E1-7.

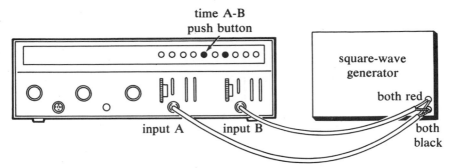

Fig. E1-13   Time A-B measurement.

Consider the input waveform as shown in Figure E1-14. If comparator A is set to trigger on a positive (+) slope and B on a negative (−) slope, the time between the A trigger and B trigger, that is, the + half-cycle will be measured. Measure the duration of the positive half-cycle of a 100-cycle signal from the square-wave generator. Set the time base output to give four significant figures. Reverse the A and B comparator slope settings to measure the duration of the negative half-cycle of the same signal. Measure the total period of the signal to see if it agrees with the sum of the durations of the two half-cycles. Calculate the relative asymmetry of the square-wave signal $(|t_+ - t_-|/t_+ + t_-)$.

Optional: Devise and test a digital technique for the measurement of the phase angle between two signals of equal frequency.

Table E1-7   Time A-B Measurement

FUNCTION	MEM	SUM	DISPLAY TIME	TIME BASE, SEC	LEVEL	SLOPE	ATTN.	AC-OFF-DC
						INPUT A		
Time A-B	In	Out	Min	10^{-6}	AUTO	+	×10	Ac
						−		

INPUT B				REAR PANEL			
LEVEL	SLOPE	ATTN.	AC-OFF-DC	CLOCK CONTROL	1-MHz SCALER	DVM	B OUT
AUTO	−	×10	Ac	Normal	1-MHz	OFF	Normal
	+						

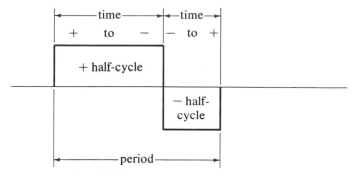

Fig. E1-14   Asymmetrical square wave.

Experiment 1-9   *Voltage Measurement with the DVM*

The DVM is used to check the calibration of a voltage reference source. The effect of source resistance on the accuracy of a voltage measurement is studied. To measure voltage, the UDI uses a very accurate voltage-to-frequency (V-F) converter. Since the output of the V-F converter is a frequency proportional to the voltage, a digital frequency measurement of this output is made for known input voltages, and the frequency-to-voltage ratio of the voltage-to-frequency converter is determined.

The block diagram for the DVM mode is shown in Fig. E1-15.

Connect the DVM cable to a VRS as shown in Fig. E1-16. Switch the DVM range switch to the 1-V, 10-MΩ range, and push the function switch to the DVM mode.

Allow the instrument to warm up for several minutes after turning the DVM mode ON. If you are going to switch modes (i.e., DVM to FREQUENCY to DVM) turn the V-F converter output ON at the rear panel so that the V-F analog circuitry remains ON while switching modes. Then it will not be necessary to wait for warmup each time you switch back to the DVM mode.

Measure a 1-V signal from the VRS and recalibrate the VRS on the 1-V range if necessary. Now check the VRS calibration on the other ranges. Is the VRS within the accuracy specifications?

Measure a 1-V output from the VRS, noting the exact value. Now insert a 1-MΩ resistor in series with the VRS output. Measure the voltage again using the 1-V 10-MΩ, and 1-V Hi Z ranges. Note the effect of the source impedance and compare it with the expected effect based on the known input impedance. What is the largest possible

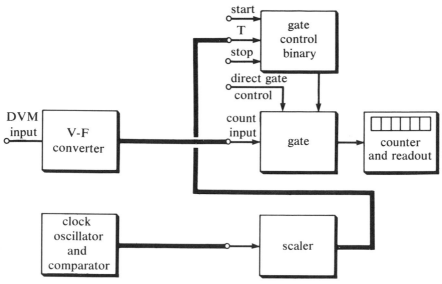

Fig. E1-15   Block diagram for digital voltmeter.

source impedance for a 9.5-V signal which will keep the loading error within the 0.05% accuracy specification of the DVM?

Connect from the V-F output on the rear panel (with a 50-Ω terminated cable) to the input of comparator A. Switch to the FREQUENCY mode and observe the frequency of the V-F converter as a function of input voltage. What is the megacycle/volt ratio of the V-F converter? Note that the above connection is made internally when switched to the DVM mode. Also note that the V-F converter can be used independently when the UDI is being used on modes other than DVM (such as FREQUENCY, PERIOD, and TIME INTERVAL). The multiple period measurement of the V-F output can be used to accurately measure the reciprocal of the input voltage (1/V).

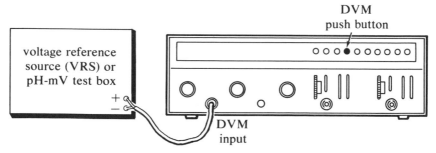

Fig. E1-16   Voltage measurement.

Optional: Check the calibration of the pH/mV test box on several millivolt ranges. Determine relative error of several ranges. Is the pH/mV test box within the 1% accuracy specification. With the pH/mV test box set at 500 mV note the effect of changing the output resistance of the pH/mV test box from 100 kΩ to 1000 MΩ. From your measurements estimate the input impedance of the DVM on both 1-V ranges.

The experiments in this chapter are designed to provide an introduction to switching devices and to specific diode parameters and circuits. Current-voltage (I-V) characteristic curves are obtained from a curve tracer (that the experimenter builds), and important device parameters are determined. Diode rectifier circuits, including half-wave, full-wave, bridge, and voltage-doubler circuits are constructed and used in conjunction with power supply filters. A Zener diode voltage regulator circuit is also built and tested. Diode clipping and clamping circuits are investigated so as to demonstrate their important functions in electronic circuits. The set of experiments is completed by the construction of a diode switching circuit for control of signals to the output of a system.

Analog-Digital Designer

Part of the analog-digital designer (ADD) will be used in this set of experiments and other parts of the ADD will be required in subsequent chapters. There are several basic modules and many circuit cards that

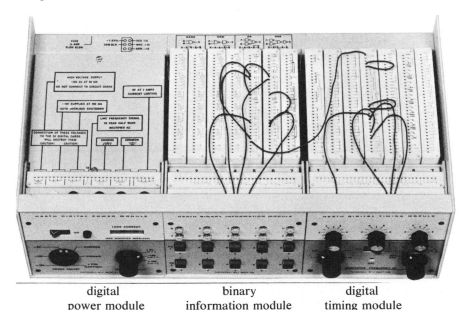

| | digital
power module | binary
information module | digital
timing module |

Fig. E2-1 Analog-digital designer. (Courtesy of Heath Co.)

make up the ADD. Two of the modules (power supply module and digital timing module) and one circuit card (an operational amplifier card) will be used in the Chapter 2 experiments and are introduced here together with some general information and characteristics of the complete system.

The standard cabinet holds three modules as illustrated in Fig. E2-1. The arrangement of modules depends on the specific application, but generally the power supply module will occupy the left-hand position in a 3-module cabinet. Additional modules can be operated outside the cabinet.

Module Installation

Module installation is illustrated by the rear view of the chassis in Fig. E2-2. The module is pushed in from the front of the cabinet and a spring catch at the rear of the chassis snaps onto the catch pin on the base of the cabinet. A thumbscrew at the bottom front of the cabinet will lock the chassis in place.

Power Connections and Card Installation

The 6-pin connectors and cables at the rear of each module (Fig. E2-2) connect power between the power supply (PS) module and other modules. The rear cable on the module adjacent to the power supply

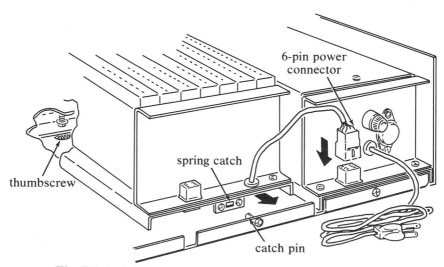

Fig. E2-2   Module installation. (Courtesy of Heath Co.)

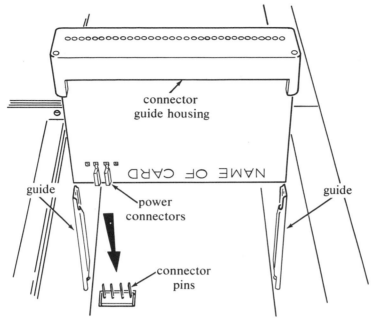

connector
guide housing

guide

NAME OF CARD

power
connectors

guide

connector
pins

Fig. E2-3   Card installation in analog-digital designer.   (Courtesy of Heath Co.)

connects to the PS module and each successive module connects to its adjacent module so that all modules are powered.   The maximum number of modules that can be connected to one PS module depends on the current required by the circuit cards inserted in the modules.

The circuit cards are installed in the module as illustrated in Fig. E2-3.   The ends of the card are inserted in the front and rear guides that cause the power connectors in the base of the module to mate with the connectors on the circuit card.   It is good to check to see that the connectors mate properly before pressing the card firmly into place.

Circuit Connections

The connectors below the marked holes in the circuit cards can be interconnected quickly by inserting lengths of #22 solid hook-up wire. The bare ends of the hookup wire are inserted in the proper holes.   *Do not insert more than one wire in a connector at one time!*   For multiple connections to one circuit point, there are usually several connectors linked together on the circuit card.   These are indicated by joined lines on the card label.   If enough common connectors are not available use a multiple connector card.

Precautions

Do not use other than #22 wire for patch connections, and use only components with #22 or smaller gauge leads. The stripped portion of the connector wires should be straight and smooth and the end cut clean to avoid damage to the connectors. Connector wires that have been soldered or kinked should be discarded. The ±15-V power must not be patched to any integrated circuit cards unless specified. Circuits will be destroyed by excessive voltage. Also when introducing external signals into the circuit cards, check to be sure of the correct interfacing voltages so as to prevent damage to the circuit cards.

Power Supply Module

This module is completely self-contained and can be used in or out of the ADD cabinet with any circuit requiring the voltages it supplies.

As shown in the block and card diagrams of Fig. E2-4, there are five different voltages provided to four different types of connectors. The +5, +15, and −15-V dc are for the plug-in circuit cards in the ADD. The 5-V half-wave rectified ac at 60-Hz power line frequency is for timing purposes, and the +170-V dc is for operating neon lamps and cold-cathode display tubes.

The four types of connectors are an octal socket, a patch-wire connector card, banana jacks, and a 6-pin intermodule chassis connector. The many types of connectors provide maximum convenience for general purpose applications.

The meter on the front panel can be switch selected for the +5, +15, and −15-V supplies so as to indicate the fraction of current capacity being utilized. If there is a short across the output, the power supply will current limit and the meter will indicate zero.

Note that most circuit cards are automatically powered with the required voltages when inserted into the guide rails and pressed into place. Use caution when patching in auxiliary voltages. Excess voltages can destroy the circuit components.

Always turn the supply OFF when making connections.

Digital Timing Module

The operational part of this module is primarily a function generator that provides timing pulses and other repetitive waveforms for use with the digital and analog circuit cards.

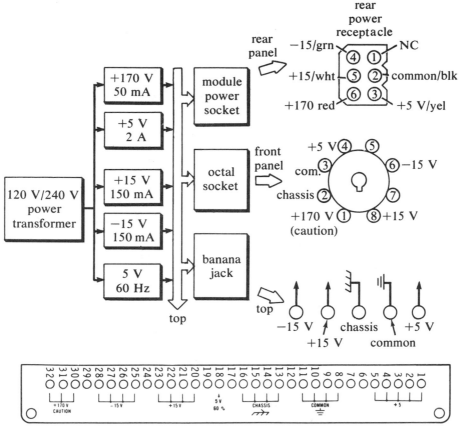

Fig. E2-4 Digital power module—block diagram and patch connector label. (Courtesy of Heath Co.)

Repetitive pulses (complementary), square waves (complementary), and sawtooth (ramp) voltage signals are all available simultaneously over a wide range of frequencies. The block diagrams and patch card are illustrated in Fig. E2-5.

The repetition rate can be varied by front panel controls from less than 1 Hz to 10 kHz. External capacitors can be patched into the circuit to extend the output frequency to 100 kHz.

Three 100-kΩ potentiometers can also be controlled from the front panel. The three contacts for each potentiometer are available at the patch card at the top front of the module (see top label in Fig. E2-5). These are very useful as timing and level controls, and will be used in this set of experiments.

The module provides space and power connectors for up to seven

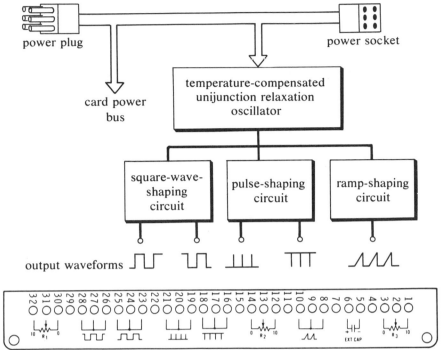

Fig. E2-5   Digital timing module — block diagram and card label showing connectors.   (Courtesy of Heath Co.)

circuit cards.   Insert only those cards in the module that are required for the experiment or connected instrument.   Store the other cards carefully to prevent damage.

Operational Amplifier Card (EU-900NA)

Although the operational amplifier (OA) circuits will not be studied until later, the circuit card will be used in the Chapter 2 and subsequent experiments to provide well-defined functions.   At this time, therefore, it is only necessary to know the connections on the card so that it can be connected to perform the required function, and later the specific OA circuits will be investigated.

The card contains a high-quality general purpose OA, an independent digital logic driver, four precision resistors, and a high-resolution balance potentiometer.   All important component contacts are brought to the top connectors as illustrated by Fig. E2-6.   Note that the plastic top cover can be removed from the card if it becomes necessary to repair a contact.

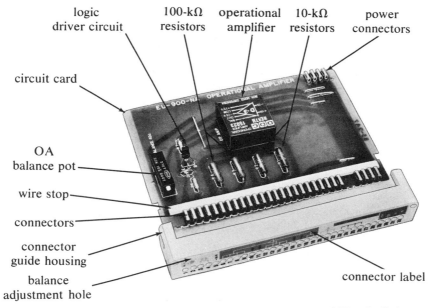

logic driver circuit   100-kΩ resistors   operational amplifier   10-kΩ resistors   power connectors

circuit card

OA balance pot

wire stop

connectors

connector guide housing

balance adjustment hole

connector label

Fig. E2-6   Operational amplifier card.   (Courtesy of Heath Co.)

Experiment 2-1   *Characteristic Curves and Load Line for an "Ideal" Switch*

A curve tracer is constructed so as to display the characteristic curves of an ideal switch on the screen of an oscilloscope. The same curve tracer will be especially useful in studying diode and transistor characteristics in subsequent experiments.

2-1a   Construction of Curve Tracer

The characteristic curves of a device can be readily displayed on an oscilloscope by using a system that is connected as shown in Fig. E2-7. The output voltage of a sweep generator is applied across a voltage divider made up of a resistance R_L in series with the device under test. The voltage drop across R_L is a measure of the current i through the device and is connected to the vertical input of an X-Y oscilloscope, and the voltage applied across the device is connected to the horizontal input. If the level and amplitude controls of the generator are adjusted to provide voltages over the applicable range of the device a current-voltage (I-V) characteristic curve is repeatedly plotted on the scope face for every sweep.

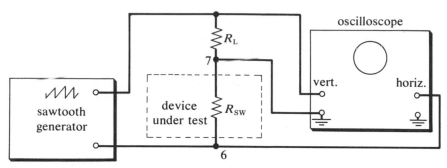

Fig. E2-7 Block diagram of curve tracer for obtaining characteristic curves of devices.

The magnitude of the <u>output ramp voltage</u> from the signal generator of the analog-digital designer (ADD) is not sufficient to cover the operating ranges of several devices to be studied so it is amplified with an OA circuit as illustrated by the diagrams in Fig. E2-8.

The operational amplifier is wired as an inverting summing amplifier. The output voltage is the inverted sum of the level and sawtooth input voltages ($e_{sawtooth}$ and e_{level}, respectively) according to the relationship

$$-e_{out} = \left[e_{sawtooth} \cdot \frac{100\ k\Omega}{R_{in}} + e_{level} \right]$$

where R_{in} in the resistance of the amplitude control. However, the output voltage of the operational amplifier is limited to about 10 to 13 V plus or minus. Therefore, if the sawtooth amplitude and the level control are set so that the sum exceeds the amplifier limit, the output waveform will appear truncated at the limit value. The amplitude and level controls can be adjusted to provide sweep outputs of any desired span in the range $+10$ V to -10 V. A diagram showing how to make connections to the OA circuit card after plugging into the ADD is shown in Fig. E2-8b. Connect the OA sweep circuit as shown in Fig. E2-8b.

2-1b   Use of Curve Tracer

Make the following adjustments while observing the output of the circuit of Fig. E2-8.

Set the generator at 100 Hz.

Set the Amplitude Control at about the mid-point.

Adjust the Level Control to be just short of the positive limit.

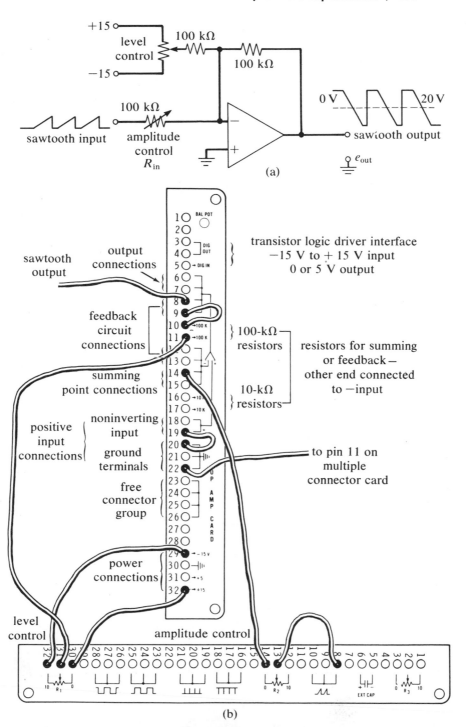

Fig. E2-8 Operational amplifier (OA) circuit for curve tracer sweeps: (a) Schematic diagram; (b) OA card connections.

Adjust Amplitude Control for maximum excursion with only slight limiting at negative extreme.

Check to see that there is no CHASSIS-COMMON jumper on the power card of the ADD for this experiment if the scope is grounded. (Possibly through the three-wire line cord.)

Breadboard connection system (EU-51). The three-card breadboard chassis used in the next experiment is shown in Fig. E2-9 with three cards inserted in the guides. A power patch card with a power cable and octal plug is used to supply voltages from the PS module to the patch strip. These voltages are then readily connected by #22 wire to the cards in the other two slots. A multiple connector card is used to provide multiple common connections, and the breadboard card is inserted so that components can be readily interconnected as illustrated in Fig. E2-9.

The breadboard card has a connector under each hole in the plastic

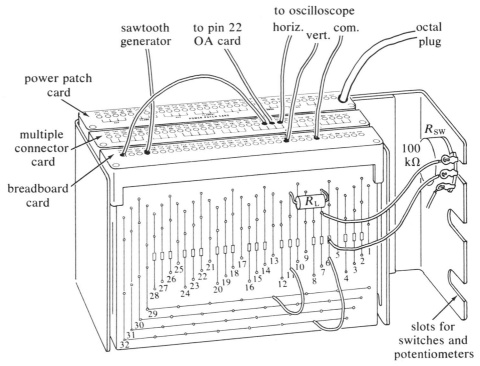

Fig. E2-9   Pictorial of breadboard connection system (EU-51).   Example shown is for Fig. E2-7.

top and side covers. On the side cover there are 32 rows of multiple contacts, and each of these rows leads to one of the 32 top connectors. The contacts in each row are connected together and are identified by a number. For example, the 4 contacts in row 15 of the side cover are connected to contact 15 at the top cover. Rows 29 to 32 each interconnect 10 or 11 contacts and are especially useful for interconnecting a large number of leads to the circuit common and power supply voltages. The rectangular holes in the side cover are large enough to accept special pin connectors or certain adapters used in the EU-51 system, and are in-line for convenience in connecting transistors. When components are used that have solder lugs, a length of #22 hook-up wire should be soldered to the lugs as shown for the potentiometer in Fig. E2-9.

Again, *do not* connect more than one wire in each hole.

2-1c Resistor Current-Voltage (I-V) Curves

Connect the ramp (sawtooth) output of the OA to the series circuit of R_L and the device under test (R_{SW}) and make the oscilloscope connections, as shown in Fig. E2-9. Although this simple circuit could be connected without using the EU-51 breadboard cards and chassis illustrated in Fig. E2-9, the example provides an introduction to the system. This will expedite the construction and testing of dozens of subsequent circuits.

Note: In the diagrams E2-7 and E2-9, use a 10-kΩ resistor for R_L, and a 100-kΩ potentiometer (pot) for R_{SW}.

Observe the current-voltage curve on an X-Y scope. Vary R_{SW} over its range of resistance values and note the slopes of the I-V curves.

2-1d Load Lines

While varying R_{SW} rapidly, note the first and third quadrant load lines. Record.

2-1e "Ideal" Switch

Replace the pot R_{SW} with a wire jumper contact acting as an "ideal" switch. Note the ideal I-V characteristic for each state and observe operating points for maximum output voltage from the sweep generator.

Note: Keep your curve tracer set up for obtaining the diode characteristic curves in Experiment 2-3.

Experiment 2-2 *Charge and Discharge Curves for an RC Circuit*

A square wave is applied across an RC circuit and the output signals across the resistor and capacitor are observed with an oscilloscope. The frequency, resistance, and capacitance values are varied to show the effects of period and time constant on the output signals.

2-2a RC Circuit

Connect the RC circuit of Fig. E2-10 using the resistor and capacitor substitution boxes for R and C.

2-2b RC Differentiator

Set the square-wave generator for an output of about 1 kHz and 1 V amplitude. Start with a value of $C = 0.1$ μF and vary R in steps from about 100 Ω to 100 kΩ. Observe the waveforms on the scope and record (sketch) the cases where the RC time constant relative to the period of the input signal is such that the output across the resistor is a series of (1) spikes, (2) undistorted square waves, (3) square waves with a slight tilt on the top. Note the conditions for each case.

For the case where the output is a series of spikes, measure the time it takes for the voltage to decay to $1/e$ times the initial value as compared to the time constant RC.

2-2c RC Integrator

Reverse R and C and connect the scope leads across the capacitor.

Record (sketch) the curves for the same values of RC and signal period for which the curves were sketched in Experiment 2-2b. Compre the curves.

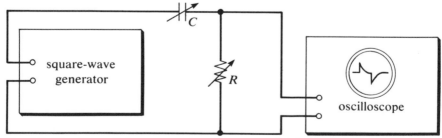

Fig. E2-10   Circuit for measuring response of RC circuit to square waves.

Experiment 2-3 *Diode Characteristic Curves*

A silicon diode characteristic curve is displayed utilizing the oscilloscopic curve tracer of Experiment 2-1, and significant diode parameters are measured from the curve. The experiment may be repeated for a germanium diode.

Change R_L of Fig. E2-9 to a 1-kΩ resistor, and replace R_{SW} with a junction diode, with the cathode (banded end) connected to the ADD common and the horizontal input of the scope.

The usual orientation for *I-V* curves may require the use of inverting inputs on the horizontal amplifier, but if not available the forward-bias characteristics will appear in the second quadrant as a result of voltage sweep direction.

Draw or photograph the characteristic curve.

Experiment 2-4 *Resistance of a Diode*

Values obtained from the forward-bias characteristic diode curve are used to plot log I vs E so as to demonstrate the logarithmic relationship over a few decades. The dynamic resistance is measured and compared with the static resistance at a current of 1 mA. The problems of using an ohmmeter to measure diode resistance are determined.

2-4a   Forward-Bias Characteristics

Put the HORIZ SENS at 0.1 V/division and, using a range of vertical sensitivities, measure 5 points over a two-decade range of current on the *I-V* curve.

Plot log I vs E and check the linearity of the curve, calculate the slope, and compare with theoretical and empirical values. Determine the dynamic resistance and the forward voltage drop for 1 mA of current. Use the DVM or calibrated scope for voltage measurements.

2-4b   Reverse-Bias Characteristics

Observe the reverse breakdown voltage for the above diode, and estimate the reverse current I_r.

2-4c   Measurements with an Ohmmeter

Measure the forward and reverse resistances of the above diode using the ohmmeter function of a VTVM or VOM. Use several resistance ranges and record the measured resistance values. Explain why the ohmmeter readings are greatly different for each meter range.

2-4d   Germanium Diode

(Optional) Repeat Experiments 2-3 and 2-4 for a germanium signal diode.

Experiment 2-5   *Reverse Breakdown and Zener Diodes*

The reverse characteristic curve for a Zener diode is observed and the significant parameters of the Zener diode are measured from the curve.

2-5a   Zener Diode

Repeat Experiment 2-3 for a Zener diode.

2-5b   Reverse Characteristic Curve

Observe the reverse characteristic curve of the Zener diode on the scope and record the signal.

2-5c   Dynamic Resistance of a Zener Diode

Measure the reverse voltage drop for 10 and 20 mA of reverse current using the circuit in Fig. E2-11 and a high precision DVM. Calculate the average dynamic resistance in the 10–20 mA reverse current region.

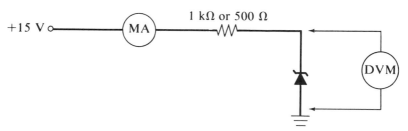

Fig. E2-11   Measurement of dynamic resistance of Zener connected as a shunt regulator.

Experiment 2-6 *Measurement of Transformer Voltage*

The rms and peak voltages of a low-voltage transformer are measured and the turns ratio is determined.

Connect the oscilloscope leads to the secondary terminals of a step-down transformer (e.g., 115 V primary, 6-3 V secondary). See Fig. E2-12. Measure the peak-to-peak secondary voltage and calculate the rms voltage.

Because the power supply patch card will be used in subsequent experiments, it is recommended that this card be inserted into the breadboard chassis and the power cable connected into one of the supplies that provides 6.3-V output, such as the EUW-15 and EU-40 supplies.

The 6.3-V ac secondary terminals of the transformer of the power supply go to the power patch card (PPC) terminals marked FILAMENT (25–28 for one and 17–20 for the other), and the center tap goes to the

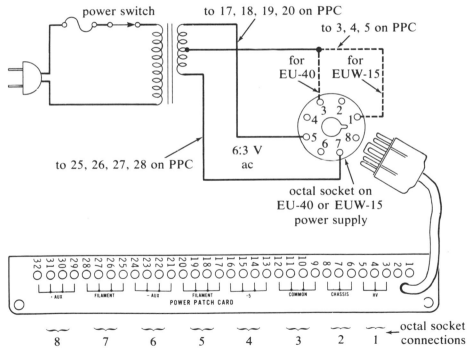

Fig. E2-12   Connections from octal socket of EUW-15 supply to power patch card (PPC).

terminals marked CT COMMON, numbers 9–12, except for the EUW-15 power supply that has the center tap connected to connectors 3, 4, and 5, marked HV on the power patch card. These connections are shown in Fig. E2-6.

Caution: When using the EUW-15 supply, disconnect·the high-voltage (HV) jumper (if connected) so that the high voltage will not be connected to the power patch card for this set of experiments, and keep HV switch OFF. Always turn the complete power supply OFF when connecting circuits or replacing power card. Note that for the EUW-15 supply the HV goes to terminal 3 of the octal socket which goes to PPC terminals 9–12 marked COMMON, and that the power supply common goes to the HV terminals of the PPC.

Assuming that the input voltage on the primary terminals is the average line voltage (115 V, 60 Hz in many United States locations) determine the turns ratio of the transformer.

Note: If you decide to measure the line voltage, connect the probes only when voltage is OFF and be careful that the probes are insulated and not shorted to the chassis or other parts of the circuit. Use only *one hand* in making connections if supply is ON.

Experiment 2-7 *Diode Rectifier Circuits*

All of the rectifier circuits commonly found in power supplies are connected and their characteristics observed, including half-wave, full-wave, bridge, and voltage-doubler rectifier circuits. The ac and dc components are measured and compared.

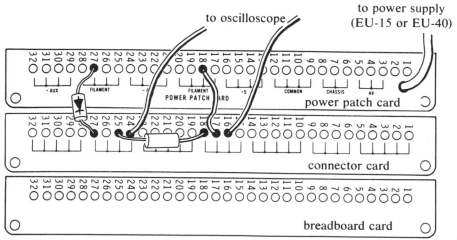

Fig. E2-13 Arrangement of cards in breadboard chassis.

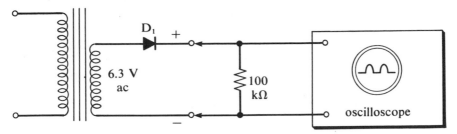

Fig. E2-14   Half-wave rectifier.

2-7a   Half-Wave Rectifier

Insert the power card, multiple connector card, and the breadboard card in the 3-card chassis (EU-51), as illustrated in Fig. E2-13, and connect the half-wave rectifier circuit shown in the schematic diagram in Fig. E2-14.

Observe the output waveform for the half-wave rectifier circuit and measure the amplitude on the oscilloscope.   Reverse the connections of the diode and again observe the waveform and compare the two cases.

2-7b   Full-Wave Rectifier

Wire the full-wave rectifier shown in Fig. E2-15.   Observe the full-wave rectifier output waveform.   Measure the peak value of the rectified voltage and compare with the half-wave rectifier output.   Mount the 10-kΩ potentiometer as shown in Fig. E2-15 and connect the components using the connectors on the top or on the side.

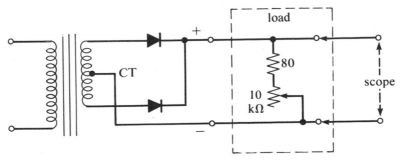

Fig. E2-15   Full-wave rectifier.

2-7c Full-Wave Bridge Rectifier

Wire the full-wave bridge rectifier as shown in the schematic diagram of Fig. E2-16a and pictorial of Fig. E2-16b. Observe the output waveform and magnitude and compare with the output of the rectifier in Fig. E2-15.

Note: Keep this rectifier circuit connected for Experiments 2-8 and 2-9.

2-7d Voltage-Doubler Rectifier

This circuit will not be connected until Experiments 2-9, after investigation of capacitor filters and completion of experiments with the bridge rectifier.

Experiment 2-8 *Power Supply Filters*

The ability of capacitance and *RC* filters to reduce ac ripple on the dc output voltage of a power supply is investigated, and the effectiveness of different filter values under various conditions is determined.

To characterize the filter, the *output voltage* and *ripple voltage* are noted as a function of the *output current*.

2-8a Capacitor Filter for Full-Wave Rectifier

Connect a 20-μF capacitor across the output of the bridge rectifier and observe on the oscilloscope the effect of changing the load current from 1 to 50 mA. Measure the ripple frequency and ripple and dc voltages and determine the ripple factor.

Change the filter capacitor to 100 μF and compare its effectiveness to the 20-μF capacitor.

2-8b Capacitor Filter for Half-Wave Rectifier

Rewire the half-wave rectifier (*but do not disturb* the bridge rectifier which will be used in subsequent experiments) and observe the effect of changing the load current from 1 to 50 mA with the 20-μF filter capacitor at the output. Measure the ripple frequency and ripple and dc voltages and determine the ripple factor.

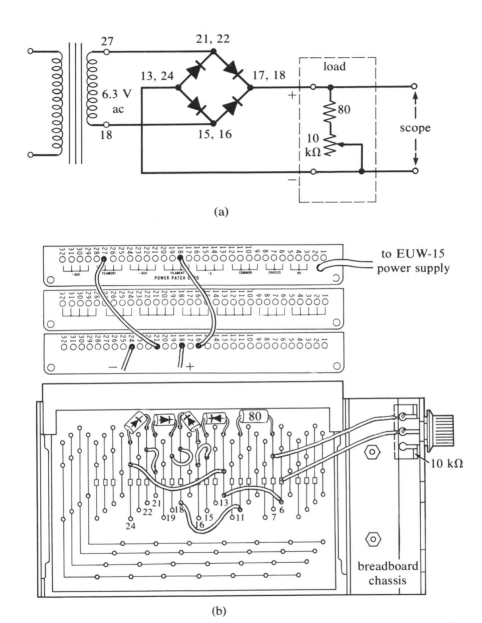

(a)

(b)

Fig. E2-16 Bridge full-wave rectifier. (a) **Schematic diagram; (b) pictorial of connections to breadboard card.**

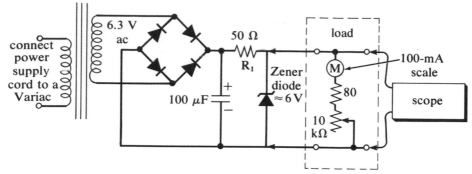

Fig. E2-17 **Zener regulator at the output of bridge rectifier with capacitor filter.**

Experiment 2-9 *Zener Diode Regulator*

The effectiveness of a Zener diode to regulate a power supply against changes of input voltage and load is measured. Conditions for optimum regulation are determined.

2-9a Zener Diode Regulator

Connect the Zener diode voltage regulator at the output of the bridge rectifier as shown in Fig. E2-17. Observe the effect of the Zener regulator to changes of load and input voltage, and the reduction of ripple. Determine the range of load for which the regulation is within 2% of the average output voltage.

If a Variac is available, plug the supply into the Variac and determine the range of line voltage variation for which the output of the Zener supply remains within 2% of the average dc output.

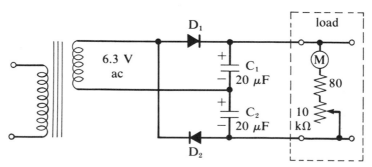

Fig. E2-18 **Voltage-doubler circuit.**

2-9b   The Voltage Doubler

Wire the voltage-doubler circuit shown in Fig. E2-18, which necessarily includes the input-filter capacitances C_1 and C_2. Measure the output voltage as a function of load current and observe the ripple voltage. Note that the ripple frequency is twice the line frequency and that the output voltage is 4 times the full-wave rectifier of Fig. E2-15.

Experiment 2-10   *Diode Clipping*

The use of diodes to clip or slice input signals is investigated.

2-10a   Zener Clipping Circuit

Clip the sine-wave output signal of the sine-square generator between -0.6 and $+4.7$ V using the Zener diode in the circuit shown in Fig. E2-19. Use an input signal of 10-V peak or greater.

Observe and record input and output signals, including voltage levels.

2-10b   Silicon Diode Clipper

Design and test a clipping circuit which will clip between $+0.6$ and -1.2 V using silicon signal diodes in place of the Zener diode.

Experiment 2-11   *Diode Clamps*

A diode clamp is used to restore the dc component of an input square wave after it has been coupled through an RC circuit.

Couple the output signal from a square-wave generator which is a

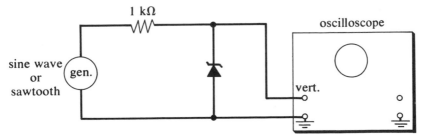

Fig. E2-19   Zener clipping circuit.

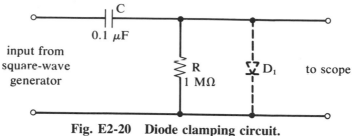

Fig. E2-20   Diode clamping circuit.

pulsating dc signal at a frequency of 1 kHz through an RC network as shown in Fig. E2-20 without the diode shown in dotted lines.

Observe the input and output waveforms, measure the magnitudes, and note that the dc level of the input signal is lost at the output of the RC coupling network.

Now connect the diode across R and observe, measure, and record output signal and the dc level.   Reverse the diode and again record the output signal.

Experiment 2-12   *Diode Switching Circuit*

The diode switching circuit of Fig. 2-44 is constructed and the appearance of the input signal at the output is controlled with an actuating signal from a suitable generator.

Connect the diode switching circuit of Fig. E2-21.   Use silicon signal diodes for D_1 and D_2.   Switch the output pulses from the digital timing module (set at about 10 kHz) to be ON for 0.01 sec and OFF for 0.01 sec. Use the sine-square generator to provide the switch-actuating signal.

Note: Connect output of sine-square generator to EXT SYNC on the oscilloscope and operate in EXT SYNC switch position.

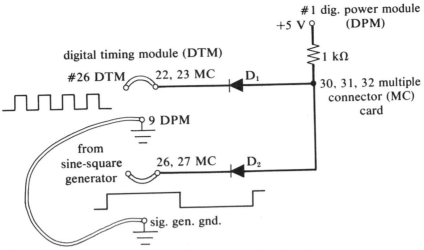

Fig. E2-21   Diode switching circuit.

After observing and recording the characteristic curves for a junction transistor, the ON-OFF voltage levels and inverting characteristics of the common-emitter transistor switch are measured. Then the switching times of the saturated switch are determined and later compared with the switching times for a nonsaturating switch. The junction transistor is also connected in an emitter-follower circuit and important characteristics of this circuit are observed.

The characteristic curves for a field-effect transistor (FET) are determined and familiarity gained with its characteristics as compared to the junction transistor.

A FET switch with transistor driver is constructed and tested. A silicon-controlled rectifier (SCR) circuit and a unijunction transistor time delay and trigger circuit are constructed and used for the control of power in a load. An optoelectronic chopper is also built and tested.

The experiments in this chapter are completed by investigating the characteristics of the electromechanical relay and reed relay, with emphasis on the pull-in and drop-out currents and contact bounce and transfer times.

Experiment 3-1   *Collector Characteristic Curves for Transistors*

The collector characteristic curves for one of the npn-type transistors connected in a common-emitter circuit are obtained using the same oscilloscopic curve tracer as in Experiment 2-1. These curves are recorded so they can be used to evaluate several dc parameters as described in Section 3-2.

Connection of Transistors

Transistors can be wired on the breadboard card in several ways. Their leads can be inserted directly into the connections on the side of the breadboard card, generally into a set of three in-line connectors. When this technique is used, the transistor leads should be bent to conform to the connector spacing *before* insertion. Three small holes can be drilled near the front edge of the desk-top chassis for this purpose. The transistor leads should be inserted into these holes prior to insertion into the breadboard connectors. An alternative approach is to solder #22 hook-up wire to the lugs of a transistor socket. A high-quality "test-type" socket is recommended. Then the other end of the hook-up wire is inserted into the desired connectors.

Keep in mind that the transistor connections are generally shown as *base* diagrams (the leads as viewed looking toward the bottom of the transistor). However, when used on the breadboard card, they are viewed from the top. *Check to see that your transistor connections are not reversed* left for right. Improper connections to the transistor can destroy it.

3-1a Characteristic Curves

Connect the circuit shown in Fig. E3-1 using a 2N3393 npn transistor or its equivalent. Set the horizontal and vertical sensitivities each at 10 V/division, and the voltage reference source (VRS) output at 5 V. Use the same sweep generator as in Fig. E2-2.

Adjust the *amplitude* of the sweep generator so that only the forward characteristic section and a short portion of the reverse current characteristic is observed, as shown in Fig. E3-2. Increase scope sensitivity to 1 V/division on each axis and locate the zero points for convenient observation of the forward characteristics. Vary the VRS output from 0 to 10 V to obtain (plot or photograph) a set of transistor collector characteristic curves.

Experiment 3-2 *The Collector-Emitter Voltage for a Transistor Switch* ON *(in saturation) and* OFF

The values of $V_{CE(sat)}$ and $V_{CE(OFF)}$ are measured and compared with the expected values.

3-2a Load Line

With the VRS set at 9 V, adjust the sweep generator *dc level* so the maximum collector current is 6 mA. Vary the base current from -5 to $+100$ μA with the VRS and observe the cut-off and saturation regions and the load line. Using the momentary ZERO button switch the VRS output between 0 and 10 V to observe the ON and OFF states of the transistor switch.

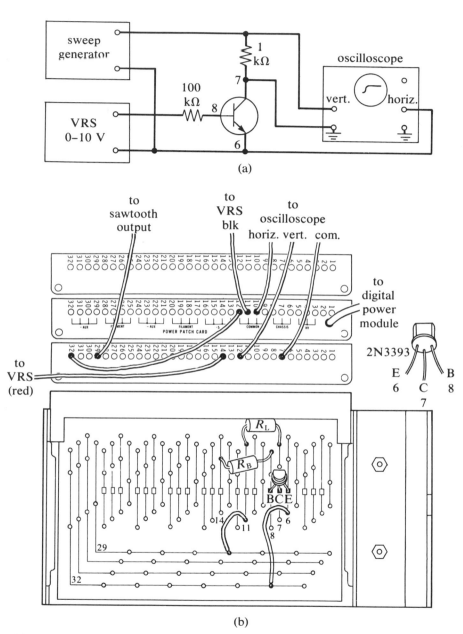

(a)

(b)

**Fig. E3-1   Circuit for obtaining transistor characteristics.   (a) Schematic;
(b) pictorial using breadboard cards.**

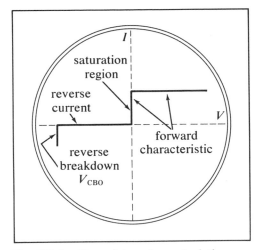

Fig. E3-2   Transistor characteristic curve.

3-2b   ON and OFF Characteristics

Connect the 1-kΩ collector load resistor of Fig. E3-1 to +5 V *instead* of to the sweep generator output.   Switch the base current from 0 to 100 μA to cause the transistor to switch from saturation to cut-off states as shown by the position of the dot on the oscilloscope.

When the transistor is ON, measure $V_{CE(sat)}$ and calculate the equivalent ON resistance.   Also measure $V_{CE(OFF)}$.   Compare the measured values of $V_{CE(sat)}$ and $V_{CE(OFF)}$ with the expected values.

3-2c   OFF Collector Current

(Optional)   Use an operational amplifier (OA) to measure the OFF current.   With the transistor in the OFF state (base current zero), remove the transistor emitter from ground and connect it to the summing point of an OA which has a 10-MΩ feedback resistor.   Use the VTVM or digital voltmeter (DVM) to measure the OA output voltage.   Calculate the equivalent OFF resistance.   If the current is too small to measure, calculate the minimum OFF resistance, and then increase the base-emitter voltage until the emitter current becomes measureable.

Experiment 3-3   *The Direct Current Gain*

The dc β_N or h_{FE} is determined from the characteristic curves and compared with typical values for the specific transistor type.

From the characteristic curves obtained in Experiment 3-1, measure h_{FE} and compare with typical values for the 2N3393 npn transistor or other types of transistor that you might be testing.

If you have several transistors of the same type, note that the h_{FE} values vary considerably from unit to unit.

From the measured value of h_{FE} calculate what the base resistance should be to ensure saturation if V_B and V_{CC} are both $+5$ V and $R_L = 1$ kΩ.

Experiment 3-4   *The Transistor Inverter Circuit*

The ability of a typical transistor switch circuit to provide an output voltage change of opposite direction to the input voltage change is observed.

3-4a   The Inverting Switch

Connect the switching circuit of Fig. E3-3. Set the generator frequency at about 10 kHz and measure the amplitude of the square-wave output.

Observe the shape of the output and compare with the input signal (on dual-trace or dual-beam scope). Note the signal inversion. Explain the rounding of the transitions between the ON and OFF states.

Keep this circuit connected for the next experiment.

Experiment 3-5   *Transistor Switching Times*

The delay, rise, storage, and fall times for a transistor are measured and the net turn-on and turn-off times determined.

3-5a   Saturation

Replace the 100-kΩ base resistor with a 100-kΩ potentiometer. Start with a value of 100 kΩ on the pot and slowly reduce R_B to 10 kΩ while observing the transition from cutoff to saturation.

Note the decrease in the transition time as R_B is decreased (base current is increased).

Repeat the variation of R_B and observe the increase in storage time as I_B is increased beyond the value just sufficient for saturation.

With some transistors the storage time is too short to measure easily. In such cases, another transistor or a slower transistor type may be substituted for these experiments.

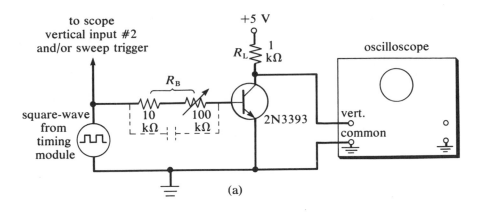

(a)

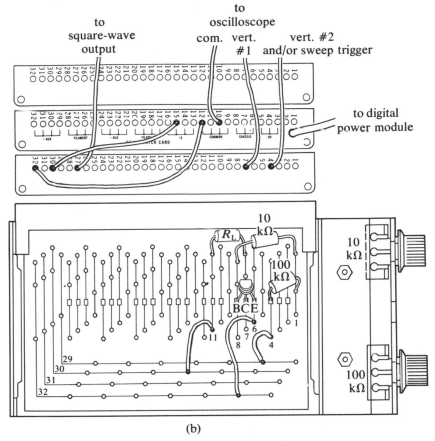

(b)

Fig. E3-3   Transistor switching circuit.   (a) Schematic;   (b) pictorial.

3-5b The "Speed-up" Capacitor

Place a 56-pF capacitor in parallel with R_B and notice the greatly reduced transition times. Vary R_B again and notice and explain the effect on the output waveshape. Try other values of capacitance and note the effect. Remove the speed-up capacitor for the next part.

3-5c Effect of Switch Capacitance

Adjust R_B for minimum storage time (but still saturating) and note that the transition time from cutoff to saturation is shorter than from saturation to cutoff. Why?

Substitute a 10-kΩ potentiometer for the 1-kΩ load resistor in the above circuit. Explain the increased saturation to cut-off transition time at higher R_L values and the departure from saturation at the lower values.

Measure the turn-off time for the minimum value of R_L that maintains saturation. Reduce R_L until the ON state output voltage is about 2-V and decrease R_B to regain saturation. Measure the turn-off time again and explain the difference in values.

Experiment 3-6 *Connection of Loads to Transistor Switch Circuits*

A light is connected as a load in a transistor driver circuit.

3-6a Light Driver

The light-driver circuit in the binary information module is shown in Fig. E3-4. Connect this circuit on the breadboard card and turn the light ON and OFF by applying the square-wave from the digital timing module set at about 1 Hz. Observe that the state of the light follows the input signal as the frequency is changed.

Caution: Be careful to connect the transistor correctly or you can destroy it! If you substitute another type of high-voltage transistor, be sure and check the terminal configuration for the specific type before connecting.

Determine the base current required to turn the light ON. Use a variable voltage applied to the base input.

Binary information module. The binary information module has front panel switches and lights that can be used to provide information to, and

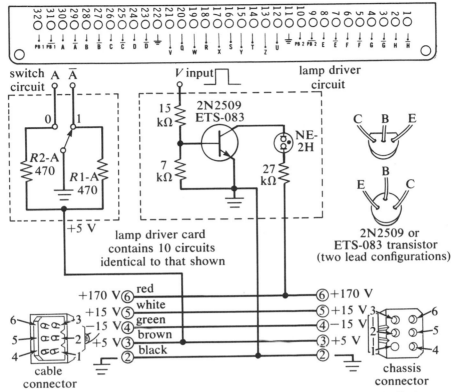

Fig. E3-4 Binary information module. (Courtesy of Heath Co.)

indicate information from, digital logic circuits. Also, the module provides space and power connections for up to seven plug-in circuit cards that are used in the analog-digital designer (ADD).

Ten single-pole, double-throw front panel rocker-type switches are provided. Two of these switches (PB1 and PB2) have momentary contact spring return action. Each of the ten switches is similar to switch A in the diagram of Fig. E3-4. When switch A is in the "0" position, contact A (contact #30 on plastic top cover) is at 0 V, and in the "1" position contact A is at +5 V. Note that contact $\overline{A}$ of switch A is just opposite (the complement) of contact A. The pairs of contacts B and $\overline{B}$, C and $\overline{C}$, etc., are similar to A and $\overline{A}$ but for switches B, C, etc.

The ten neon lamp circuits contain a driver transistor that acts as an electronic switch, similar to the driver circuit shown in Fig. E3-4. Note that the input connection to provide the base current goes to one of the card connectors, contact V (#21) in this example. This circuit controls light V on the front panel. Likewise input contacts Q, W, R, etc., are the inputs to the driver transistors that control the state of lamps Q, W, R, etc. With an open circuit or 0 V on the contact, the light is OFF. More than 1 V should be required to turn the light ON.

3-6b   Logic Level Indicator

Connect a wire from switch contact A to the input of light driver V and observe the operation by switching A back and forth from **0 to 1**. Now connect the input contact V to the square-wave output from the digital timing module and observe the maximum switching speed that can be visually resolved on the light. Note similar operation of other switches and light drivers.

Experiment 3-7   *Emitter-Follower Circuit*

The voltage and current gains of an emitter-follower circuit are measured and checked with expected values. The input impedance is also determined.

3-7a   Voltage Gain

Connect the emitter-follower circuit of Fig. E3-5. Set the total value of $R_E = 1$ kΩ. Measure the output voltage V_o for several values of the input signal V_S. Calculate the voltage gain.

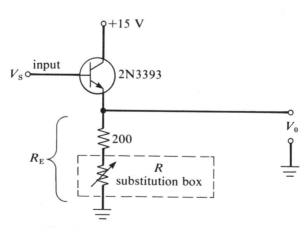

Fig. E3-5   Emitter-follower circuit.

3-7b   Current Gain

Measure the base current and emitter current for several input signal voltages.   Calculate the current gain of the emitter follower.

3-7c   Input Impedance

From the measured values of I_B and the input signal V_S, calculate R_{in}. From the measured value of h_{FE} for your transistor (Experiment 3-3) and the value of R_E, calculate R_{in} and compare with the measured value of R_{in}.

3-7d   Output Impedance

Vary the resistance of R_E in steps from 200 Ω to 10 kΩ and measure the output voltage.   What does the data indicate about the output impedance of the emitter follower?   Determine the effective output impedance from the measured values.

Experiment 3-8   *Nonsaturating Switching Circuit*

A nonsaturating transistor switch circuit is constructed and its characteristics compared with a saturated switch.

Connect the nonsaturating switch of Fig. E3-6.
Apply a 100-kHz, 5-V square wave from the ADD timing module to the input and compare the storage times with and without the diode.

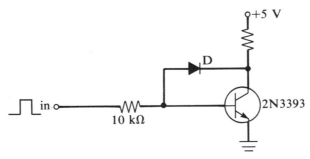

Fig. E3-6   Nonsaturating switch.

Experiment 3-9 *Field Effect Transistor (FET) Switch*

The characteristic curves for a field-effect transistor are obtained from the curve tracer and the load line and ON-OFF characteristics are measured. A FET switch is connected.

3-9a FET Characteristic Curves

Obtain the characteristic curves, load line, and ON-OFF character-istics for a field-effect transistor using the circuit of Fig. E3-7.

3-9b FET Switch Circuit

Connect the FET series switch and pnp transistor driver circuit shown in Fig. E3-8. Connect the 100 MΩ resistor between source and gate if there is excessive noise pickup when making the subsequent meas-urements.

Connect 0 V to the SWITCH CONTROL input and note that the FET switch is closed (ON) so that the input voltage (use about 1 V, and do not exceed 5 V) is transmitted to the output.

Measure the input and output voltages accurately with the FET

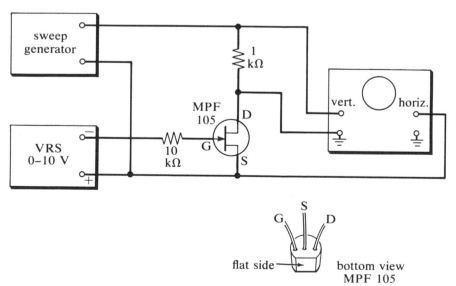

Fig. E3-7   Circuit for obtaining field-effect transistor characteristics.

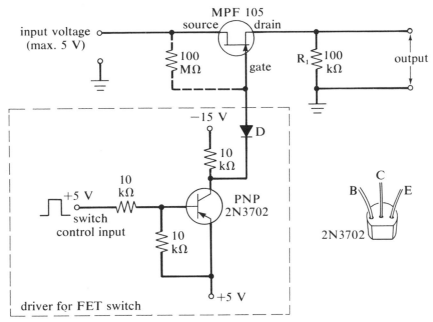

Fig. E3-8 *N*-**Channel JFET switch.**

switch ON. Determine the effective ON resistance and the effect on the output voltage relative to the input. Change R_1 to 10 kΩ and observe the output relative to the input voltage.

Apply +5 V to the control input. Measure the collector voltage of the pnp transistor and note that it is negative with respect to ground. Observe that the FET switch is now OFF, and measure the output voltage. Calculate the OFF resistance.

Connect the square-wave output (0 to +5 V) from the digital timing module to the SWITCH CONTROL input and switch the input signal voltage at rates up to 10 kHz or more. Record the output signal as observed on an oscilloscope. Note the effect of decreasing the value of R_1 on the maximum switching rate.

Experiment 3-10 *Power Control with a Silicon-Controlled Rectifier (SCR)*

An SCR circuit is to be constructed and investigated for control of power in a light bulb.

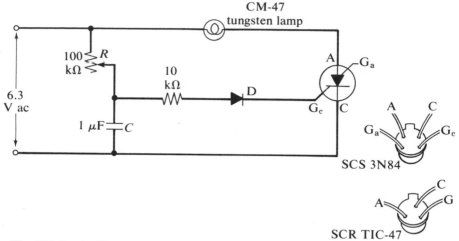

Fig. E3-9   A silicon-controlled rectifier circuit for control of light intensity.

3-10a   SCR Power Controlled Circuit

Construct the power control circuit of Fig. E3-9. Use an SCR or an SCS connected as an SCR. The components R and C provide a phase shift network that controls the ON time of the gate. The diode D prevents the gate from becoming negative with respect to the cathode, and the 10-kΩ resistor limits the gate current to a safe value. The terminal G_a is not connected if the SCS is used.

Observe the effect on light intensity of changing the value of resistor R. Record the signals at the input, across the capacitor, and across the lamp as observed with the oscilloscope. Remember that the scope common is ground so care must be taken to avoid shorting the 6.3-V ac supply. Note the time relationships between the signals, and the effect of R on the observed waveforms.

Experiment 3-11 *Unijunction Transistor Trigger Circuit*

A unijunction transistor time delay and trigger circuit is to be constructed and its characteristics investigated, and then used to trigger an SCR power control circuit.

E3-11a   Unijunction Delay Trigger Generator

Construct the circuit of Fig. E3-10. Observe the effect of changing the RC time constant and the supply voltage. Observe signals across

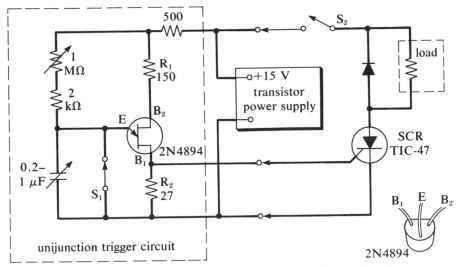

Fig. E3-10   Unijunction time delay and trigger circuit.

the capacitor and at the output (across R_2).   Use a mechanical or an FET switch for S_1.   Set the transistor power supply at about 8 V.

E3-11b   Unijunction SCR Trigger

(Optional) Connect the output trigger to an SCR power control circuit.   Use a load that will not overload the power supply.

Trigger the load ON at a preset time by use of the unijunction transistor time delay and trigger circuit.   If FET switches are used, determine the rate at which power to the load could be turned ON and OFF with this circuit.

Experiment 3-12   *Optoelectronic Switch Circuit*

A photoconductive light chopper is investigated as a modulator for converting a dc signal to an ac signal for subsequent amplification.

3-12a   Optoelectronic Series Chopper

Connect the series chopper of Fig. E3-11a using either of the arrangements shown in Fig. E3-11b and c for the light-detector pair.   If the pair is assembled in the tubing, also connect a light driver similar to

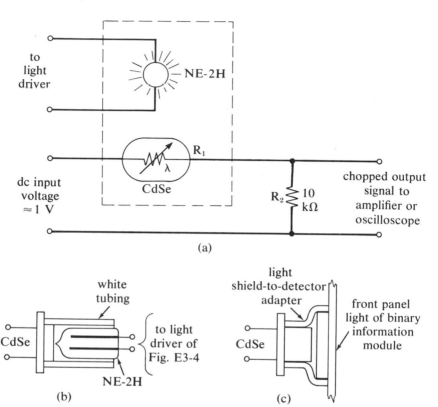

Fig. E3-11   Optoelectronic chopper. (a) Schematic; (b) pictorial showing a neon light and CdSe photoconductive cell mounted in white rubber tubing; (c) pictorial showing use of front panel from binary information and special adapter to accommodate light shield and CdSe detector.

the one in Experiment 3-4 to drive the light.   If the front panel light from the binary information module (BIM) is used, it will be necessary to have a special adapter to make a light-tight coupling between detector and source, but it will not be necessary to connect light drivers because these are already in the BIM.

Observe the output signal from the chopper on an oscilloscope and determine the output amplitude relative to the dc input voltage.   Record the effect on the amplitude of changing the frequency of chopping.

3-12b   Optoelectronic Shunt Chopper

(Optional) Connect a shunt chopper by reversing R_1 and R_2 in Fig. E3-11.   Compare chopping characteristics with a series chopper.

3-12c   Optoelectronic Series-Shunt Chopper

(Optional) Connect a series-shunt chopper by using CdSe detectors for both R_1 and R_2, and drive the two lights out of phase. Compare chopping characteristics with the series chopper.

Experiment 3-13   *Relay Characteristics*

The operate and release and contact-bounce times can be critical for proper operation of some electronic circuits, and it is the purpose of this experiment to measure these values. Also, the pull-in and drop-out currents are determined for one of the relays.

3-13a   The Relay Card

Examine the relays mounted on the printed-circuit relay card. Note the type of relay and the type of contacts. Follow the three contacts from one of the relays to the connectors accessible at the top of the card, and note how they are labeled. Note that the coil of the relay is not connected directly to the top terminal labeled "coil."   The coil is driven by a transistor switch circuit, as shown in Fig. E3-12.

A +5-V signal applied to the input switches the transistor from OFF to ON to provide a current exceeding the "pull-in" current for operating the relay.

Why is it useful to have a transistor drive circuit for the relay?

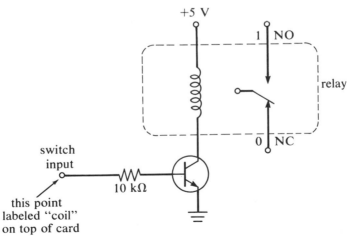

Fig. E3-12   Transistor drive circuit for relay.   (Courtesy of Heath Co.)

3-13b "Pull-in" and "Drop-out" Currents

With the +5-V supply disconnected, measure the resistance of the relay coil by connecting the VTVM between the collector of the transistor drive and the +5-V terminal, as illustrated in Fig. E3-13, and switch to a suitable ohms range.

Now connect the +5-V supply to the relay card, and a variable voltage supply to the input resistor, as shown in Fig. E3-13. Also keep the VTVM connected across the relay coil, but switch to the +5-V dc range.

Vary the input voltage gradually until the relay just pulls in and record the voltage across the relay coil. Now decrease the input voltage until the relay armature just drops out and record the voltage across the relay coil.

Using Ohm's law and the measured voltages and coil resistance, calculate the relay pull-in and drop-out currents.

What would be the current through the coil when the transistor is switched into saturation?

3-13c Measurement of Operate Time

The pictorial diagram in Fig. E3-14 shows one possible procedure for connecting the relay at the top of the card into a circuit utilizing the "patch" capability of the relay card, multiple connector card, and timing module contacts. Note that the OFF-ON characteristics of the square-wave generator are used for the switch S. That is, 0 and +5 V are alter-

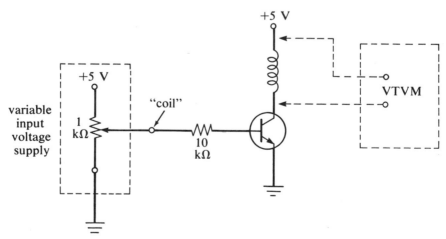

Fig. E3-13   Circuit for measuring pull-in and drop-out currents.

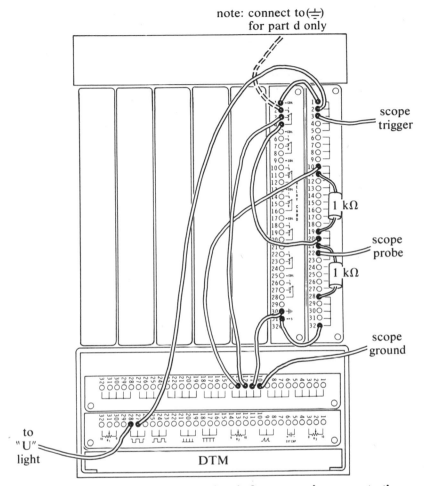

note: connect to (⏚)
for part d only

scope
trigger

1 kΩ

scope
probe

1 kΩ

scope
ground

to
"U"
light

DTM

Fig. E3-14 Pictorial of relay circuit for measuring operate time.

nately applied to the relay coil. The repetition rate of OFF-ON operation can be varied by the Square-Wave Generator Period controls on the front panel of the digital timing module. This switching method will enable the triggered scope to be adjusted more easily and the trace to be observed continuously.

Connect a circuit similar to Figs. E3-14 and E3-15 and measure the *operate time* required to break the normally closed contacts. Vary the period and observe whether there is any change in *operate time*. Note the limits of operation. Explain!

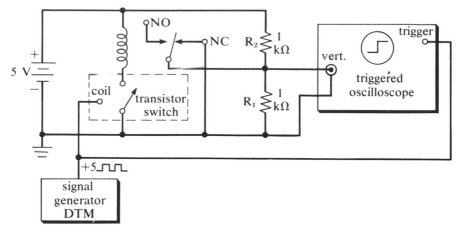

Fig. E3-15   Schematic of relay circuit for measuring operate time.

Note: Suggested scope settings at the start: (1) vertical gain, 2 V/cm; (2) Sweep, 2 msec/cm.

Vary the scope settings to obtain the most accurate readings of the measured times.

3-13d   Operate, Transfer, and Bounce Times

After measuring the operate time as illustrated in Fig. E3-15, measure the operate, transfer, and bounce times as follows.

Connect the NC contact to COMMON as shown in Fig. E3-15. However, connect the NO contact of the relay to +5V instead of to the common. This should give the waveforms on the dual-beam scope, as shown in Fig. E3-16.

Notice that as the square-wave frequency is increased, a point is reached when the movable relay contact never reaches the NO contact before returning. Also, note the effect of the higher frequencies on the operate, transfer, and bounce times.

Measure both the transfer and bounce times. Vary the OFF-ON period and observe if there is any change in times.

Note that, if the function of the circuit requires the closing of the normally open contacts, then the *operate time* is the sum of the operate time and the transfer time, but should not include bounce time.

Why not include *bounce time* in the operate time?

Vary the switching period and determine if the period changes the measured times.

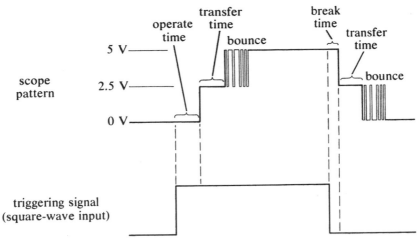

Fig. E3-16 **Scope waveform, showing operate, transfer, and bounce times.**

3-13e   Reed Relay Characteristics

(Optional) Perform the same measurements as in Sections E3-12a to d for one of the reed relays. Wire a transistor switch as the relay driver and use the required suppiy voltages for the specific reed relay coil.

The experiments in this chapter illustrate the characteristics and use of logic gates. A sequence of logic gate circuits is constructed beginning with relay gates and diode gates, then progressing into forms of transistor, diode transistor, and integrated circuit gates. As the gate circuits are studied, they are also used in representative logic circuits. In the last experiment, a binary adder is made with integrated circuit gates.

During these experiments, extensive use will be made of the binary information module that was described in Chapter 3 experiments. The switch circuits shown in Fig. E3-4 are used to provide logic level signals for the gate circuit inputs, and the lights and drivers are used to indicate the logic level of the gate outputs. The set states of the switches thus correspond to the input states or logical conditions and the observed states of the lights indicate the logical solution. The logic level output at contact C, for instance, corresponds to the position of switch C. If switch C is in the 1 position, output C (connector number 25) will be at the 1 logic level ($+5$ V). If switch C is in the 0 position, output C is logic 0 (0 V, a short to ground). The C connection provides logic level outputs that are always opposite to the logic levels at C. The switches and outputs for PB1 and PB2 act the same way except that the switch action is momentary. It is suggested that Fig. E3-4 be studied, that switch outputs be connected to light inputs, and that the logic levels be observed while switching until the action of both circuits is understood.

Other equipment used in this sequence of experiments includes the relay card, the breadboard card, and the NAND gate card.

Experiment 4-1   *Relay AND Circuit*

The AND circuit of Fig. 4-1 is wired using relay switches and a light driver for the load. The truth table for $A \cdot B = T$ is verified.

Wire the AND switch circuit using the relay card and the binary information module as shown in Fig. E4-1. The relay contacts 26 to 28 will be CLOSED when $A = 1$ and OPEN when $A = 0$. The state of switch B,

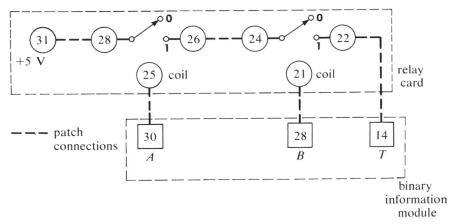

Fig. E4-1 AND circuit using the relay card.

similarly, controls the other relay. Try both positions of *A* when *B* is **1** and again when *B* is **0,** in order to verify that the light *T* is ON only when *A* AND *B* are both **1.** Thus the truth table is confirmed. Additional relay contacts and control switches can be added in series to demonstrate that the AND function $A \cdot B \cdot C \cdot D \cdots = T$ requires all the contacts to be CLOSED for *T* to be ON.

Experiment 4-2 *Relay OR Circuit*

The truth table for the OR function given in Table 4-3 is verified using relay switches in the circuit of Fig. 4-2. An indicator light driver is used as the load.

Change the relay contacts wired in Experiment E4-1 from the series connection to the parallel connection shown in Fig. E4-2. Verify that this circuit performs the OR function by using all combinations of the states of *A* and *B* and noting that *T* is ON when either *A* OR *B* OR both are **1.**

Experiment 4-3 *Relay Logic NOT Circuits*

The truth table for $\overline{A} \cdot \overline{B} = T$ given in Table 4-4 is verified using relay switches and an indicator light driver. Circuits can also be wired to light the light when *A* is TRUE and *B* is FALSE or when *A* is FALSE and *B* is TRUE to prove that there is only one 1 in the *T* column of the truth table for any AND circuit.

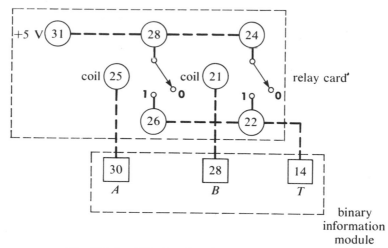

Fig. E4-2   OR circuit using the relay card.

The wiring of the function $\overline{A} \cdot \overline{B} = T$ requires a switch closure when $\overline{A} = 1$ ($A = 0$) and a similar switch closure for B. This can be accomplished by using the **0** (normally closed) contacts on the relay card with the coil connected to $\overline{A}$, or by using the **1** relay contact and the $\overline{A}$ output from the switch. The former method is illustrated in Fig. E4-3.

Use all combinations of A and B and verify that T is only ON when $\overline{A}$ and $\overline{B}$ are both **1** (A AND B are **0**).

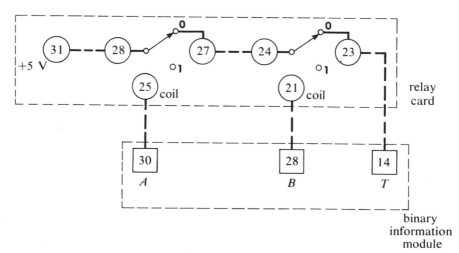

Fig. E4-3   Relay circuit for $\overline{A} \cdot \overline{B} = T$.

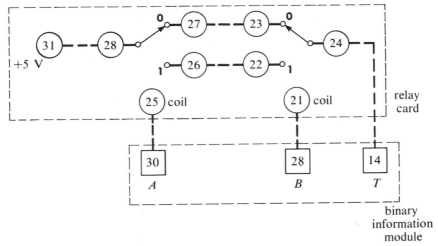

Fig. E4-4 **Relay circuit for** $AB + \overline{A}\overline{B} = T$.

Experiment 4-4 *Relay Equality Detector or EXCLUSIVE-OR Circuits*

A relay circuit to provide the function $T = A \cdot B + \overline{A} \cdot \overline{B}$ is wired and tested to verify the expected response. An EXCLUSIVE-OR circuit $T = \overline{A} \cdot B + A \cdot \overline{B}$ may be wired instead. Either circuit, it will be noted, is an implementation of the upstairs-downstairs light switch problem.

Wire the comparator circuit of Fig. 4-4 (p. 142) using only two SPDT relays as shown in Fig. E4-4. The light T will light only when A AND B are in the same state. Obtain the EXCLUSIVE-OR function $A \cdot \overline{B} + A \cdot B = T$ by interchanging the **1** and **0** contact connections at one of the relays. Verify that T will be ON only when A OR B BUT NOT BOTH are **1** for the EXCLUSIVE-OR circuit.

Experiment 4-5 *Diode AND Gate*

A three-input diode AND gate as shown in Fig. 4-15 is wired. Logic signals are applied to the inputs and a light driver is connected to the output to verify the AND function of the circuit.

Wire the diode AND gate of Fig. 4-15 (p. 152) using columns 1 through 4 of the patchboard card as shown in Fig. E4-5. The other

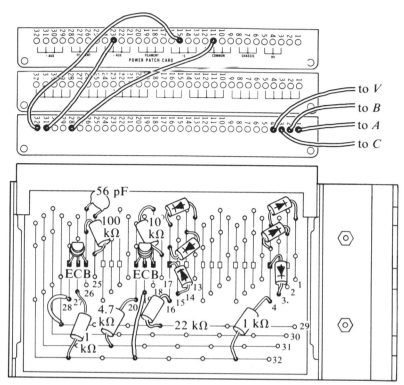

Fig. E4-5   Pictorial for Experiments E4-5 through E4-8.

circuits, shown in Fig. E4-5 will be wired in later experiments.   Connect
A, B, C, and *V* to the corresponding switch and light connections on the
binary information module.   Confirm that the diode AND circuit follows the truth table for an AND gate by trying all eight combinations
of *A, B,* and *C.*

Measure the **0** and **1** level output voltages for the diode AND gate.
Leave this circuit wired because it will be used in a later experiment.

Experiment 4-6   *Diode OR Gate*

The diode OR circuit shown in Fig. 4-16 is wired.   The OR function
is verified using an output level indicator and logic level input
signals.

Wire the diode OR gate of Fig. 4-16 (p. 153) using columns 13–16 of
the patchboard card as shown in Fig. E4-5.   Verify the truth table for

the OR function and measure the output voltages for the **1** and **0** output states. Do not disconnect this circuit because it will be used later.

Experiment 4-7 *The Inverter*

The inverter circuit of Fig. 4-17 is wired and its NOT function is verified.

Wire the transistor inverter circuit of Fig. 4-17 (p. 154) on the patch-board card using columns 25 through 28 as shown in Fig. E4-5. Draw the circuit diagram of the inverter including component values. Test the operation of the inverter and measure the output voltage for each input logic level. Leave this circuit connected for the next experiment.

Experiment 4-8 *Combinations of Gates*

The circuits wired in Experiments 4-5 through 4-8 are used with an emitter-follower and another diode AND circuit to test combinations of gates. Input-output compatibility and expected logic functions are checked. The AND-OR, OR-AND, OR-FOLLOWER-AND, AND-AND, and the half-adder circuit of Fig. 4-13 are wired and tested.

Wire a 2-input diode AND gate, identical to that of Experiment 4-1 using columns 6 and 7 for inputs and 8 for the output. Now wire the emitter-follower amplifier shown in columns 17–20 of Fig. E4-5. Combinations of diode gates will be patched and tested from the top of the card. The gate and amplifier connections as they should appear at the card top are shown in Fig. E4-6. Connect *A* and *B* switch outputs

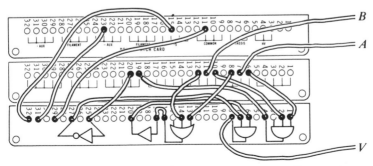

Fig. E4-6 Pictorial for Experiment 4-8.

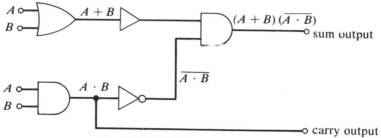

Fig. E4-7   Half-adder circuit.

from the binary information module to the inputs of an AND gate and the AND output and the C switch output to the inputs of the OR gate. Draw the logic diagram and write out the logic function and the expected truth table. Test the logic function and measure the logic level in voltages.

Repeat the above experiment, but reverse the positions of the AND and OR gates to provide the function $(A+B)C$. Note that to make this circuit respond properly, the follower amplifier must be used between the OR gate output and the AND gate input. Why?

Wire the two AND gates to provide the function $(AB)C$ and confirm the response of the circuit. Measure the output logic level voltages particularly noting any deterioration of the **0** level.

Using all the gates and amplifiers, wire the half-adder circuit shown in Fig. E4-7. The necessary connections are shown in Fig. E4-6. Confirm the appropriate circuit response for both SUM and CARRY outputs.

Experiment 4-9   *The DTL NAND Gate*

The DTL NAND gate of Fig. 4-21 is wired. The NAND function of the circuit is verified and the input signal requirements and fan-out are measured.

A patchboard layout for a DTL NAND gate is shown in Fig. E4-8. Draw the circuit diagram, wire the circuit, verify the truth table, and measure the output voltage levels for the gate. Measure and note the input current for a single **0** and **1** input level. Now disconnect the A and B switches from the inputs and connect a resistance substitution box between one input and ground. With the indicator light removed from the output, note the decrease in the **1** level output voltage as the input resistance is increased. If the **0** logic level is to stay between 0 and 2 V, what is the maximum **0** level output resistance you would recommend for

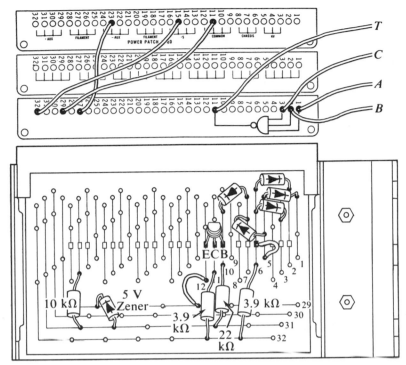

Fig. E4-8   Pictorial of DTL NAND gate.

a signal source to use with this gate? Measure the output current that can be supplied before the **1** output level falls below a desired minimum of 3 V.

Based on the output current capability and input current requirements, estimate the maximum fan-out specification for this gate.

Experiment 4-10   *The RTL NOR Gate*

The RTL NOR gate circuit shown in Fig. 4-22(a) is wired and its NOR function is verified. The **1** and **0** logic level regions and noise immunity are measured.

Wire an RTL NOR gate as shown in Fig. E4-9. Draw the circuit diagram, verify the truth table, and measure the gate output levels. Using a variable voltage source, determine the range of input voltages for the **0** and **1** output levels. Estimate the **0** and **1** logic levels and the noise immunity for this gate on the basis that the transistor amplifiers

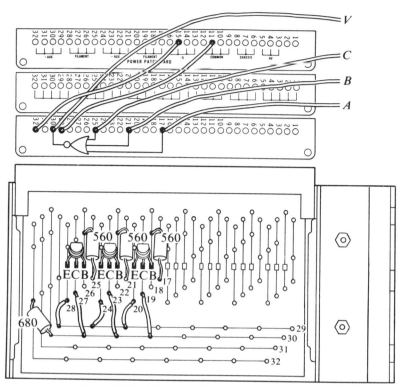

Fig. E4-9   Pictorial for RTL NOR gate.

must be either cut off or saturated. Why? Discuss the basis for your estimations.

Experiment 4-11   *The TTL NAND Gate*

The input and output logic level regions are determined for an integrated circuit TTL NAND gate. The logic function is verified and a gate delay time measurement is suggested.

NAND Gate Card

The model EU-800-JC NAND gate card contains two TTL integrated circuits, each mounted in a standard 14-pin dual in-line socket. One of the integrated circuits contains four 2-input NAND gates and the other contains two 4-input NAND gates. All input and output terminals from these gates are wired by printed circuit to the patch connections along the top edge of the card. Each connection is labeled by showing

the inputs and outputs to the logic gate symbol. A double connection
is provided for each gate output.

To implement a logic function, the gates on this card can be patched
together just as the logic diagram for the function indicates. A fast,
high noise-immunity logic circuit will result. The NAND gate equivalent for the four basic logic functions is shown along the top rear surface of the binary information module.

Although it is true that unused gate inputs represent a **1** level and
will thus not affect the NAND operation on the connected inputs, it is
possible for an unused input to pick up some noise. This presents no
problem for circuit patching, but for more permanent connection, the
unused inputs should be connected to the +5 V supply. In general it
is good practice to minimize the number of unused inputs. (For instance,
use a 2-input gate rather than a 4-input gate for an inverter.)

Connect a variable voltage source to the NAND gate input.

Caution: Always stay within the range of 0 to +5 V!

Measure the output voltage vs the input voltage to determine the **0** and
1 level ranges for input and output signals. Unless the output impedance
of the voltage source is low, be sure to measure the actual source voltage
under the load of the gate input. Measure the **0** level input current and
note the direction of this current.

Verify the logic function for the NAND gate.

Experiment 4-12   *Binary Full-Adder*

Two half-adder circuits are wired, tested, and combined to make a
full-adder. A third half-adder can be wired, if desired, to make an
operating 2-bit binary adding circuit.

Wire the half-adder circuit of Fig. E4-10 using the NAND gate

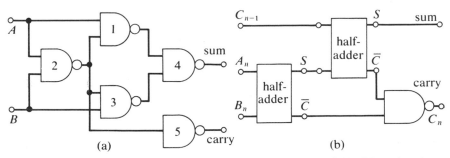

(a)   (b)

Fig. E4-10   Minimized NAND gate half-adder and full-adder circuit.

card. A suggested wiring layout is shown in Fig. E4-11. Wire a second half-adder and combine it with the first to make a full-adder as shown in Fig. E4-10. A third half-adder can be wired, if desired, to make a complete 2-bit binary adding circuit.

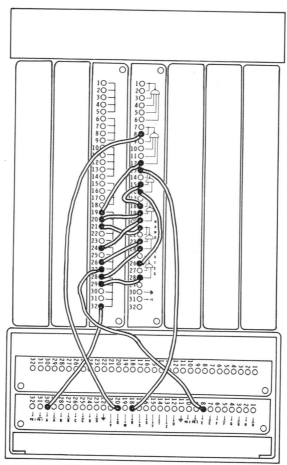

Fig. E4-11 Pictorial for NAND gate half-adder.

Flip-flops, multivibrators, and gates are the basic building blocks of all digital instruments and computers. The experiments in this chapter demonstrate the design and characteristics of the basic memory, RS, T, JK, and D flip-flops, monostable and astable multivibrators (MV's) and the Schmitt trigger. Circuits employing discrete components and those using integrated circuit logic gates are both studied. These two design approaches are integrated in the experimental sequence. For instance, the component flip-flop is wired and tested using the breadboard card and then its logic gate counterparts are wired and studied using NAND gates in the analog-digital designer (ADD). Then the component flip-flop circuit is modified to make a monostable circuit that is compared with a NAND gate monostable multivibrator. In the last experiment several interesting and useful circuits using flip-flops and multivibrators can be built.

Experiment 5-1   *Basic Transistor Bistable Memory*

The circuit of Fig. 5-3 is wired and tested. Voltage measurements are made to confirm the dc analysis, and the set and clear inputs are used to verify their effect. The Q and $\overline{Q}$ states are monitored with a meter or more conveniently with a light and logic-actuated light-driver circuit.

The basic transistor bistable memory is wired and tested in this experiment and then it is modified in later experiments to make an RST flip-flop, a monostable and an astable multivibrator. The basic bistable circuit from Fig. 5-3 is repeated for convenience in Fig. E5-1. This circuit is wired on the breadboard card using the 2N2369 high-speed switching transistors. Figure E5-2 shows one possible layout which can be followed if desired. Connect the set and clear inputs to PB1 and PB2, respectively. Connect the Q and $\overline{Q}$ outputs to two convenient indicator light inputs on the binary information module.

Alternately push PB1 and PB2 to reverse the states of the flip-flop noting the flip-flop state by the indicator lights. Trace the pulses applied

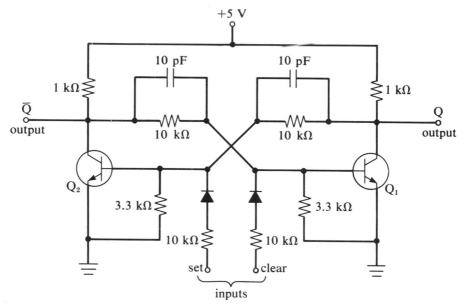

Fig. E5-1   Basic transistor bistable memory.

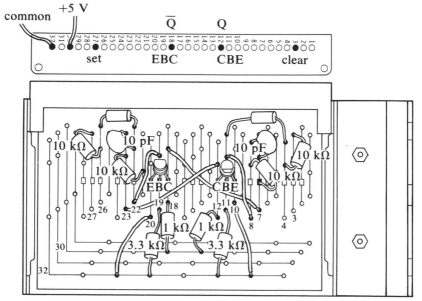

Fig. E5-2   Suggested layout for bistable memory.

by the push buttons to verify that the pulse turns the OFF transistor ON in each case.

Measure the collector and base voltages for ON and OFF transistors using the DVM or VTVM. Disconnect the indicator lights during this measurement because the load they present affects the collector voltages measured.

Save this circuit for modification in Experiment 5-2.

Experiment 5-2 *The Triggered Flip-Flop*

The trigger input and steering networks of Fig. 5-6 are added to the memory circuit of the previous experiment. Q and $\overline{Q}$ are connected to logic level light drivers to observe the flip-flop states. Inputs S and R are connected to logic-level switch outputs and T is connected to a low-frequency square-wave signal or a manually triggered pulse generator. The effect of the R and S levels on the outputs as stated in the truth table is confirmed. The toggling action is observed when S is connected to Q and R to $\overline{Q}$. The frequency and time relationship between the T signal and the Q output is studied using indicator lights or a dual-trace oscilloscope.

Remove the set and clear inputs from the basic bistable circuit of Experiment 5-1 and add the T input and steering networks as shown in Fig. E5-3. Note that the diode to the base is reversed from Experiment 5-1. Two diodes in series are used to provide greater noise immunity in the triggering circuit.

Connect S and R to A and B switch outputs on the binary information module. Connect T to the square-wave output of the timing module. Connect Q, $\overline{Q}$, and the square-wave trigger signal to indicator light

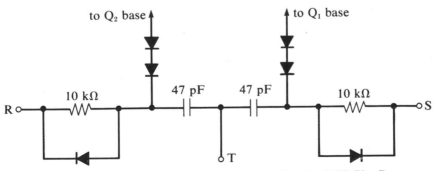

Fig. E5-3   T input and steering networks for the RST flip-flop.

drivers. Refer to the truth table of Fig. 5-6 (p. 186). Only one of the A and B switches should be in the **0** state at any one time to avoid the indeterminate output condition. Note that a **0** applied to a steering input steers the **1–0** transition of the T signal to turn the corresponding transistor OFF.

Now disconnect switches A and B and connect Q and $\overline{Q}$ outputs to the steering inputs S and R, respectively. The flip-flop should now change state on each **1–0** transition of T. Observe the time and frequency relationships between T and Q signals using indicator lights and/or a dual-trace oscilloscope. When lights are used, disconnect the $\overline{Q}$ indicator light and compare Q and T light frequencies. When the oscilloscope is used, set the square-wave trigger source at a frequency of 1 kHz or higher and trigger the oscilloscope sweep from the Q output rather than the T signal. Why should one always trigger the oscilloscope with the lower-frequency signal when two synchronized signals are to be observed?

Leave this circuit wired for later use in Experiment 5-10.

Experiment 5-3 *Logic Gate Memory*

The basic NAND memory circuit of Fig. 5-8 or the NOR circuit of Fig. 5-9 is wired to verify the truth table. It is noted with indicator lights that the outputs are always in different states. Momentary **0** push-button outputs are connected to the set and clear inputs (momentary **1** outputs for the NOR circuit) to observe the alternation of memory state when the set and clear buttons are pushed alternately.

Use the NAND gate card in the ADD to patch the basic memory circuit of Fig. 5-8 (p. 189). Q and $\overline{Q}$ are connected to light drivers in the binary information module. The set and clear inputs are connected to $\overline{PB1}$ and $\overline{PB2}$ switch outputs to provide a momentary **0** when the switches are pushed. Observe that a **0** at the gate with a **0** output changes the state of the memory and that alternating **0** inputs causes the memory to alternate state.

The NOR gates can be used instead in the circuit of Fig. 5-9. This is most easily done using an AND-OR-INVERT (AND-NOR) gate card. When just one input to each AND gate is used, the output is the NOR function for those inputs. Connect PB1 and PB2 to the set and clear inputs and make the comparable observations to those suggested above.

Experiment 5-4 *Gated Memory and Data Latch*

The gated memory and data latch circuits of Figs. 5-10 and 5-11 or 5-12 are wired and their truth tables are verified. A repetitive clock signal is used with the data latch to observe the timing of the data transfer.

Wire the NAND gated memory of Fig. 5-10 (p. 191) or the AND-OR-INVERT gated memory of Fig. 5-12a. Use switch outputs for the set and clear signals and lights to indicate the levels of Q and $\overline{Q}$. Verify the truth table for the circuit wired.

Modify the gated memory to make the data latch circuit of Fig. 5-11 or 5-12b. Connect D to a switch output and the clock input to the square-wave output of the timing module. With the timing module set at a low frequency and with an indicator light connected to the clock signal, observe the time of the data transfer with respect to the clock signal when the D level is changed at various times in the clock signal cycle.

Experiment 5-5 *Triggered Memory*

The ac clocked memory of Fig. 5-13 is wired. The truth table and timing of data transfer are verified. The circuit is wired to toggle and the alternation of the output is observed. Time relationships between high-frequency clock and output signals are observed on a dual-beam oscilloscope and the frequency division is noted.

Wire the triggered memory circuit of Fig. 5-13 (p. 195) using a NAND gate card and a multiple connector card next to it in the ADD. The components and connecting wires can be inserted into appropriate connectors through the plastic connector housings. The circuit could also be wired on the breadboard card by putting the NAND gate card between the breadboard card and the power patch card in the desk-top chassis. Compare the truth table and data-transfer timing with the data latch of Experiment 5-4.

Note: The operation of this circuit depends on the charge stored on the capacitors and the sharpness and magnitude of the T signal transition. For best results, do not connect the T signal source to any other loads. (The complement of T can be observed.) The value of C is also quite critical.

Connect Q to S and $\overline{Q}$ to R to make a toggling flip-flop. Observe the toggling action. Connect the T and Q signals to a dual-trace oscilloscope and observe their time and frequency relationship for T signals of 1 kHz or more.

Experiment 5-6 *RS Master-Slave Flip-Flop*

The RS Master-Slave Flip-Flop of Fig. 5-14 is wired and its operation is verified. It is noted that gates 7 and 8 can accept information at any time during the clock pulse, the state of the master flip-flop being determined by whichever input was **1** last.

Using NAND gate cards, wire the RS master-slave flip-flop shown in Fig. E5-4. Connect one R and one S input to switch outputs and the clock input to the Timing module square-wave output. Indicator lights should be connected to the slave outputs Q and $\overline{Q}$, and the master outputs (outputs of gates 5 and 6). Using a slow clock pulse, verify the truth table shown for Fig. 5-14 (p. 197). What is the effect of changing the R and/or S input levels when the clock pulse is **0**? When it is **1**?

This circuit is used in the next three experiments.

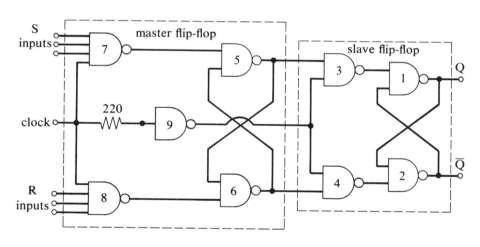

1. gates 3 and 4 close isolating slave from master
2. gates 7 and 8 open connecting master to inputs
3. gates 7 and 8 close isolating master from inputs
4. gates 3 and 4 open connecting master to slave

Fig. E5-4   RS Master-slave flip-flop.

Experiment 5-7 *JK Flip-Flop*

The RS flip-flop of Experiment 5-6 is modified to make the JK flip-flop of Fig. 5-15. The truth table is verified and the toggling operation and frequency division are observed on a dual-trace or double-beam oscilloscope.

Connect Q to R_2 and $\overline{Q}$ to S_2 of the RS master-slave flip-flop wired in the previous experiment. Verify the truth table for the JK flip-flop given on p. 200 and observe the toggling action of the flip-flop when J_1 and K_1 inputs are **1**. Use a dual-trace oscilloscope to observe the time and frequency relationship between T and Q when the flip-flop is toggling.

Experiment 5-8 *Three-Bit Binary Counter*

Two integrated circuit JK master-slave flip-flops are used in conjunction with the NAND gate JK flip-flop wired in Experiment 5-7 to make a simple 3-bit binary counter. The counting action is observed with indicator lights and the frequency division by 2, 4, and 8 is observed on the dual-beam oscilloscope.

The JK Flip-Flop Card

Two integrated circuit JK flip-flops are mounted on a card. The flip-flop circuits used are described on pp. 199–203. All connections including three J and three K inputs for each flip-flop appear at the connector housing. Dual connections are provided for Q, $\overline{Q}$, clear, and T (clock) connections. The connections to the JK flip-flop are summarized in Fig. 5-17.

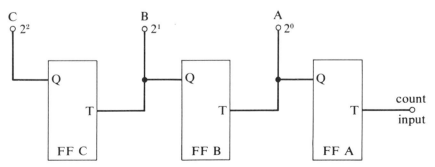

Fig. E5-5   Three-bit binary counter.

Put the JK flip-flop card in the ADD near the circuit wired above and wire the 3-bit binary counter shown in Fig. E5-5. Use the NAND circuit for FF A and the flip-flop card for flip-flops B and C. Connect the outputs A, B, and C to indicator lights in the appropriate sequence (such as S, R, and Q, respectively). With a low-frequency T signal, observe and confirm the binary count sequence. Use a dual-trace oscilloscope and a higher T frequency to observe the frequency relationship between the T signal and A, B, and C outputs.

Experiment 5-9 *D Flip-Flop*

The RS flip-flop from Experiment 5-6 is modified to make the D flip-flop of Fig. 5-18. The data transfer is observed. It is noted that the D input remains active during the clock pulse. When $\overline{Q}$ is connected to D the toggling action is observed.

Wire the D-type master-slave flip-flop shown in Fig. 5-18 (p. 204) by removing the JK connections and adding the inverter to the original RS circuit. Observe the data transfer from master to slave when the clock input goes to **0** and confirm the truth table. Note that D affects the master flip-flop as long as the clock input is **1**. Connect D to $\overline{Q}$ and observe the toggling of the D flip-flop.

Experiment 5-10 *Transistor Monostable*

The transistor bistable circuit of Experiment 5-1 is modified to make the monostable circuit of Fig. 5-21. The effect of C and R on pulse width is measured and compared with the predicted relationship. The maximum duty cycle is determined and related to the rounding of the waveform at the $\overline{Q}$ output. The pulse rise and fall times are measured and the maximum usable range of pulse widths is determined.

Change the circuit of Experiment 5-2 to correspond to that of Fig. 5-21 (p. 208) by removing the R steering input, grounding the S steering input, and changing the cross-coupling network between the Q_2 collector and the Q_1 base as shown. Remove the 3.3-kΩ resistor between the Q_1 base and ground. Change the coupling resistor between the Q_1 collector and the Q_2 base to 6.8 kΩ.

Use a 1.0-μF capacitor for C. With a 1-Hz square-wave trigger source, observe the Q output and the trigger input with indicator lights. Set the generator frequency at 50 Hz and observe the Q and T signals on a dual-trace oscilloscope. Vary R and observe the change in pulse width. Measure the pulse width for $R = 1$ and 11 kΩ and compare with the predicted $\tau \simeq 0.7RC$. Note that the pulse width can be shorter or longer than the trigger pulse. Set R at 11 kΩ and slowly increase the generator frequency. Observe the independence of the pulse width on the trigger frequency. Erratic behavior is observed when the maximum duty cycle is exceeded. Measure the maximum duty cycle. Note that the maximum duty cycle decreases as R is decreased. Why?

Change C to 0.01 μF. Set the generator to 1 kHz and observe the Q and T waveforms. Keep the Q output on the oscilloscope and use the other scope input to observe the Q_2 collector, the Q_1 base, and the Q_2 base. Compare your observations with the waveforms of Fig. 5-22. Now observe the Q_2 collector again while increasing the trigger frequency. The maximum duty cycle will be observed to be related to the rise time of the Q_2 collector waveform.

Decrease the value of C to 100 pF and observe Q and T while triggering at 10 kHz. Measure the pulse width for $R = 11$ kΩ and compare with $0.7RC$ and explain any discrepancy. What is the minimum pulse width available?

Do not dismantle this circuit because it will be used again in Experiment 5-12.

Experiment 5-11 *Logic Gate Monostable*

The pulse width, pulse shape, rise and fall time, and maximum duty cycle characteristics of the gated monostable circuit of Fig. 5-28 or 5-29 are experimentally determined.

Use the NAND gate card in the ADD to wire the gate monostable of Fig. 5-28 (p. 214) or Fig. 5-29. Use 220 Ω for R and a wide range of values for C. Measure the relationship between the pulse width and the time constant RC. Observe the pulse shape and measure the rise and fall times of the pulse and the maximum duty cycle.

Experiment 5-12 *Transistor Astable Circuit*

The transistor monostable circuit of Experiment 5-10 is modified to form the astable circuit of Fig. 5-30. The waveforms and time constants of Fig. 5-31 are confirmed and the useful frequency range is determined.

Modify the circuit from Experiment 5-10 to correspond to the M.V. circuit of Fig. 5-30 (p. 216). Make R_1 and R_2 each 10-kΩ resistors and C_1 and C_2, 0.1-μF capacitors. Observe the output waveform on an oscilloscope and measure the frequency with the oscilloscope or UDI. Compare the collector and base waveforms with those shown in Fig. 5-31 (p. 217) and compare the frequency with the theoretical value of $f = 1/(1.4RC)$.

Change the values of R and C to test the frequency range of the oscillator circuit. Remember that if τ_1 and τ_2 are far from equal, duty cycle limitations can prevent oscillations.

Experiment 5-13 *The Schmitt Trigger*

The characteristics of the Schmitt trigger circuit of Fig. 5-35 are determined experimentally. The measured values of E_{ON} and E_{OFF} are compared with the expected values. The amount of hysteresis and the trigger potential are varied while the response of the circuit to a sine-wave input signal is observed.

Wire the circuit shown in Fig. E5-6.

Connect the sine-wave output from the sine-square generator to e_{in}. Set the generator at 1 kHz and less than 1 V output. Connect the generator and Schmitt trigger outputs to a dual-trace oscilloscope. Increase the generator output until the Schmitt circuit triggers. Carefully measure E_{ON} and E_{OFF} from the oscilloscope waveforms as shown in Fig. 5-36 (p. 222). Compare the measured and expected values of E_{ON} and

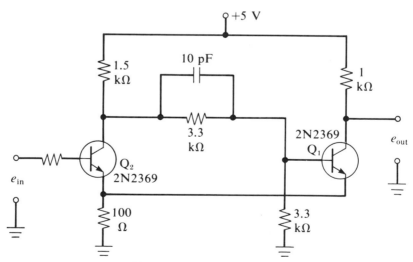

Fig. E5-6 Schmitt trigger.

E_{OFF}. Vary the input frequency to observe the triggering range. Now replace the 100-Ω emitter resistor with the resistance substitution box. Vary the emitter resistance while observing the input and output waveforms. Note the effect of the emitter resistor value on E_{ON}, E_{OFF}, and the hysteresis.

Experiment 5-14 *Waveshaping*

Any of the circuits in this section may be wired and tested. Recommended for particular interest are the oscillator gate, Fig. 5-40; the single pulse generator, Fig. 5-42; and the pulse-width discriminator, Fig. 5-44.

Wire and test any of the circuits in Section 5-7. When a flip-flop is required, use the JK flip-flop card. Similarly, when a monostable MV is needed, use the dual monostable card provided in the ADD. The pulse width of this monostable is determined by an external patch connection that connects one of five values of C on the card into the circuit. A timing resistor R of 0 to 100 kΩ must be connected externally. If continuous variation is desired, the 100-kΩ potentiometers in the digital timing module can be used for R. The monostable MV in the monostable card has a capacitor-coupled input that triggers on a *sharp* **1-0** transition. If the trigger source has too long a fall time to trigger the monostable, it can be improved by going through one or two TTL NAND gates before triggering the monostable. For a description of the actual circuit used in the dual monostable card, refer to the ADD Manual.

The tremendous versatility of the flip-flops studied in the last chapter is clearly demonstrated by this set of experiments. Flip-flops are wired together in a wide variety of counting circuits, memory registers, and shift registers. In addition, decoding circuits and some readout devices are studied. The experimental sequence includes binary counting (up and down, synchronous, and asynchronous) and binary decoding, binary-coded decimal (BCD) counting, decoding, and readout, memory registers, decade counters and scalers, fixed and variable modulus counters with automatic stop or recycle, preset counters, shift registers with serial and parallel inputs, and ring counters.

Complete decade counters are wired together to make a 4-digit counter, a precision time-base generator is connected and, in the last experiment, a functional automatic cycling counter-timer is wired and used.

Experiment 6-1 *Binary Counting*

The 4-bit binary counter of Fig. 6-1 is wired using JK flip-flops. Binary counting is observed by connecting the flip-flop outputs to indicator light drivers and pulsing the input with a push-button pulser. The 16 successive states shown in Fig. 6-2 are verified. Clear connections are wired to a push button for instant clearing of the counting register.

In this experiment, the pulses obtained by pushing PB1 will be counted by a 4-bit binary counter. The output from PB1 cannot be counted directly because the contact bounce of the switch will produce many pulses each time it is pushed. Therefore, a circuit that eliminates contact bounce must be built for the push-button switch. A convenient circuit is the NAND gate basic memory shown in Fig. E6-1.

Wire the basic binary counter of Fig. 6-1 (p. 231). It is desired that the count advance as PB1 is depressed. Should PB1 or $\overline{PB1}$ be used as the pulse source? A momentary **0** is required to clear the flip-flops. Should PB2 or $\overline{PB2}$ be used as the clear input source? Connect outputs

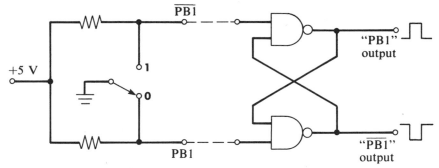

Fig. E6-1   Push-button pulser with contact bounce eliminator.

A, B, C, and D to lights T, S, R, and Q. Clear the register and advance the count verifying the waveforms shown in Fig. 6-2.

Save this circuit for use in the next experiment.

Experiment 6-2   *A Binary Down-Counter*

The circuit of Fig. 6-1 is modified to make a down-counter in two ways. First, by connecting the indicator lights to the $\overline{Q}$ outputs of the flip-flops, then by connecting the $\overline{Q}$ output of the first flip-flop to the T input of the next, and so on, as shown in Fig. 6-4. The response counting and clearing pulses is analyzed and observed.

Change the connections to the lights from Q to $\overline{Q}$ of each flip-flop. Explain the observed count sequence and the effect of the clear button. Another form of down-counter can be obtained by connecting the $\overline{Q}$ output of each flip-flop to the T input of the next flip-flop, as shown in Fig. 6-4, and observing the states of the Q outputs. Why does this circuit count down? What is the state after a clear pulse? What kind of counter would result from observing the $\overline{Q}$ outputs of this circuit?

Experiment 6-3   *A Binary Up-down Counter*

The binary up-down counter of Fig. 6-5 is wired using AND-OR-INVERT gates or the equivalent NAND gate circuit. Counting is observed in the up and down (U/D) counting modes. The necessity of applying a 0 to the count control input when the U/D control input is changed is observed.

Wire the up-down counting circuit of Fig. 6-5. Use switch outputs for the count control and U/D control signals. Obtain the count signal from the square-wave output of the digital timing module set at a low frequency. The outputs A, B, C, and D are observed with indicator lights. Put the U/D and count control inputs at 1 and observe the count.

Experiment 6-4   *A Binary-to-Octal Decoder*

The binary-to-octal decoder of Fig. 6-6 is wired (or at least four elements of the decoder). The A, $\overline{A}$, B, $\overline{B}$, C, and $\overline{C}$ inputs are connected to the Q and $\overline{Q}$ outputs of three flip-flops arranged as a binary counter. The decoder outputs are connected to indicator lights. The 0, 1, 2, 3, etc., lights are observed to come on sequentially as the counter is advanced through its 8 states.

Draw the circuit diagram of a 2-bit (binary to 4-state) decoder using the two-input NAND gates. Wire this 2-bit decoder or the 3-bit decoder shown in Fig. 6-6. Now wire a 2- or 3-bit counter and connect Q and $\overline{Q}$ from FF A to A and $\overline{A}$ of the decoder, Q and $\overline{Q}$ from FF B to B and $\overline{B}$, etc. Connect the decoder outputs to indicator lights and observe the output states as the counter is advanced. If the lights were labeled 1, 2, 3, 4, etc., the flip-flops and decoder would make an easily read counting circuit.

Using this same approach, how many gates would be required to decode a 7-bit counter that can count to 128? How many inputs would be required on each gate?

Experiment 6-5   *Synchronous Binary Counters*

The synchronous binary counter of Fig. 6-8 is wired and tested. Then the circuit is modified to make a synchronous binary down-counter and tested. A high-frequency input signal is used and the frequency at each output is measured and related to the input frequency.

Wire the synchronous binary counter of Fig. 6-8 (p. 239) using the JK flip-flop cards. The dual connectors on the T and clear inputs allow all T and clear inputs to be easily interconnected. Connect the Q outputs of the flip-flops to indicator lights and confirm the binary counting action of this circuit.

Design a synchronous binary down-counter using the waveforms of Fig. 6-4 to help determine the appropriate connections to the J and K inputs. Wire the resulting circuit and confirm the desired count sequence. Use a high-frequency input and an oscilloscope or the Universal Digital Instrument (UDI) to measure the frequency relationship of the input signal and the four flip-flop outputs. Explain why it is no differ-ent than the up-counter.

Experiment 6-6 *Synchronous Up-down Counter*

The synchronous binary up-down counter of Fig. 6-9 is wired and shown to count up and down. It is demonstrated that the Count Con-trol input required by the synchronous up-down counter when chang-ing count direction is not required for this circuit.

Use NAND gates and JK flip-flops to wire the synchronous up-down counter of Fig. 6-9. Show that the control input is not required for the synchronous up-down counter and explain why. Describe possible applications of the up-down counter as a digital adder and subtractor and as digital difference detector and indicator.

Experiment 6-7 *Synchronous Binary-Coded Decimal Counter*

Either the synchronous BCD up-counter of Fig. 6-10 or the BCD down-counter of Fig. 6-11 is wired. The output wave forms of each flip-flop are observed on an oscilloscope and compared to the appro-priate figure. The D flip-flop output frequency is measured and com-pared to the frequency of the input signal.

Wire a synchronous BCD up or down counter as shown in Fig. 6-10 (p. 243) or Fig. 6-11. Study the appropriate waveform chart and justify the J and K input connections shown in the circuit diagram. Using a high-frequency input and an oscilloscope, compare the output waveforms of the circuit with the expected behavior. What outputs can be used to obtain a signal that is one-tenth the input frequency? What other fre-quencies are available and at which outputs?

Experiment 6-8 *Asynchronous Binary-Coded Decimal Counter*

The asynchronous BCD counter of Fig. 6-12 is wired. The outputs of the flip-flop are connected to indicator light drivers so that the count sequence can be observed to follow the binary pattern.

Wire the asynchronous BCD circuit shown in Fig. 6-12 (p. 245). Confirm its operation and output waveforms. How could a similar circuit be made with JK flip-flops with single J and K inputs? How could a circuit be made with RS flip-flops? Save this circuit for use in the next experiment.

Experiment 6-9 *Memory Register*

A memory register of one or more bits is wired using the AND-OR-INVERT (AOI) gate or NAND gate data latch circuit shown in Fig. 6-14. The register inputs are connected to the BCD counter which was wired in Experiment 6-8 and the outputs are connected to indicator lamps. Data hold and transfer operations for the memory register are observed.

Wire the gated memory of data latch circuit of Fig. 6-14 (p. 248) using NAND or AOI gates. If AOI gates are used, the cross-coupling connection can be made to an input of one of the AND gates. Other inputs to that AND gate should be left unconnected or at permanent logic 1. Connect switches to the inputs and lights to the outputs and review the action of the circuit. Wire up to three additional memory circuits to make a memory register that is used to store the output of the BCD counter wired in Experiment 6-8. Connect the memory register outputs to one row of indicator lights in the binary information module and the counter outputs to the other row of lights. Connect the memory transfer input to PB1. While the counter is counting, observe the transfer of data from the counter to the memory registers when PB1 is pushed. Describe one or more possible applications for a memory circuit that will hold information at the outputs while new information is appearing at the inputs.

Experiment 6-10 *A Complete Decade-Counting Unit (DCU)*

A complete decade-counting circuit including the asynchronous BCD counter of Fig. 6-12, the memory register of Fig. 6-13, the decoder-driver circuit of Fig. 6-15, and a neon decimal display tube is studied and operated. Several such circuits are used together to make a 3- or 4-digit decimal counter.

DCU Card

One or more EU-800-DA DCU cards will be used in this experiment. If a separate supply of these cards is not available, they may be obtained from the UDI.

Caution: Take particular care that circuit cards removed from the UDI are returned to their proper location. Follow the instructions in the UDI Manual regarding the removal of the cover and the best technique for removing and replacing circuit cards.

The DCU card contains an input gate, a 1 2 4 8 binary-coded decade counter, a 4-bit memory register, a complete BCD-to-decimal decoder, and ten neon light driver amplifiers. In conjunction with a Nixie-type indicator lamp, the DCU card provides the complete counting, storage, decoding, driving, and indicating function for one decade. The block diagram of the DCU card is shown in Fig. E6-2. The connection numbers are shown in circles on this diagram. Identify these connections on the DCU card and identify the integrated circuit gate, counter, and memory circuits. The decoder-driver circuit might be discrete components or an integrated circuit.

Patch connections to the DCU card for use in counting circuits are shown in Fig. E6-3. A convenient readout device is the EU-800-DC Decimal Readout accessory. It is a neon decimal display lamp mounted on a circuit card that also contains a plug with the appropriate connections to terminals 2 through 12 of the DCU card, the anode-limiting resistor, a decimal point connector and two +170-V connectors for connection to the power supply and/or the other Decimal Readout accessories. Study the Decimal Readout accessory and its connection to the DCU card (make sure it is in the right position!) and to the power supply. The decimal point will be lit if the decimal point connection is grounded.

Connect one to four DCU cards and Decimal Readout accessories to make decimal counter and display in the analog-digital designer (ADD) as shown in Fig. E6-4. Connect the memory control and reset inputs to PB1 and PB2. The gate control input is just one of the inputs to the input gate. Therefore, if the count signal is **1**, any **0–1** transitions of the gate control signal will be counted. Clearly, then, the input gate control signal should be bounce-free. If a switch output is used, a NAND gate bounce-eliminator circuit (see Experiment 6-1) should be used. Note that the input signal is inverted by the NAND gate input. The D output from the counter is inverted by the carry amplifier to provide the carry signal to the next DCU input.

Count pulses from the digital timing module with this circuit. Using a watch, open the input gate for 1 sec to measure the frequency of the timing module output. Keep this circuit for the next experiment in which a precise 1 sec time base will be wired. In the last experiment in this chapter, this counter will be reconstructed and used with an automatic-cycling count gate control to make precise frequency and time measurement.

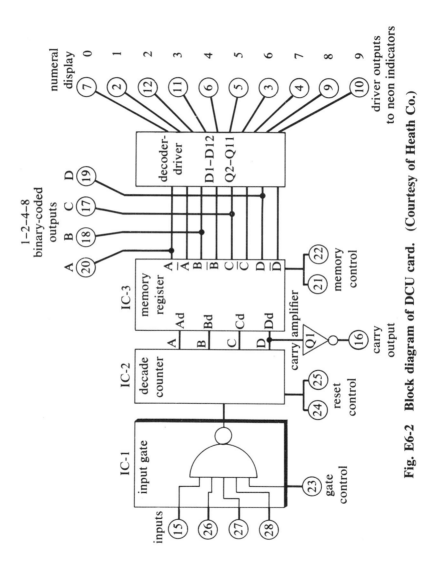

Fig. E6-2 Block diagram of DCU card. (Courtesy of Heath Co.)

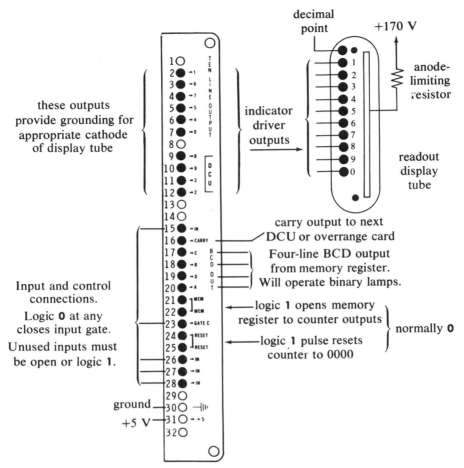

Fig. E6-3 Patch connections to the DCU card. (Courtesy of Heath Co.)

An alternative to the DCU card and Decimal Readout accessories for this experiment is the Quad DCU card and the Decimal Readout module. If these latter devices are used, please look in their manuals for the circuit descriptions and required connections.

Experiment 6-11 *Decade Scaling*

A circuit that contains a series of several integrated BCD circuits is studied and used for scaling. A standard frequency source is connected to the input and precise output frequencies of 1/10, 1/100, etc., of the input frequency are observed.

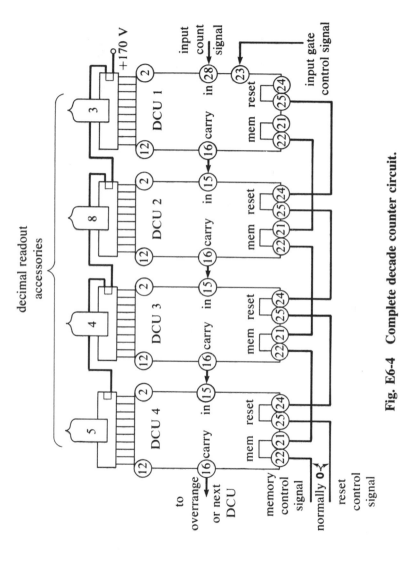

Fig. E6-4 Complete decade counter circuit.

Scaler Card

The EU-800-CA scaler card provides seven decades of scaling in decade steps and gates for input and output signal control. The block diagram is shown in Fig. E6-5. Please refer to the manual for the scaler card or the UDI for the complete circuit diagram and description. Obtain a scaler card from the UDI if extra cards are not available. Study the card and identify the major circuit elements and the patch connector terminals shown in Fig. E6-6. For simple patching applications, the input signal can be applied to terminals 19, 20, or 22 and the desired output taken from terminals 1 through 8 as indicated on the label. Note that it is necessary to apply a **0** (ground connection) to terminals 18 and 26 to avoid activating the reset and preset inputs.

The Crystal Oscillator Card

The EU-800-KA contains a very accurate and stable 4-MHz crystal oscillator with two frequency-divider flip-flops to provide an output of

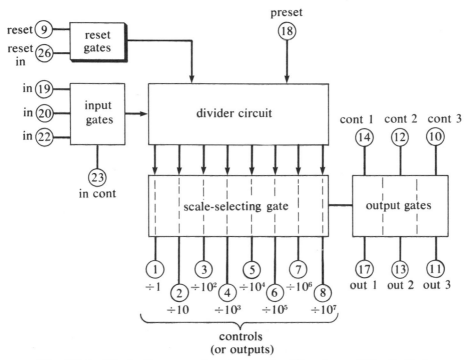

Fig. E6-5 Block diagram of scaler card. (Courtesy of Heath Co.)

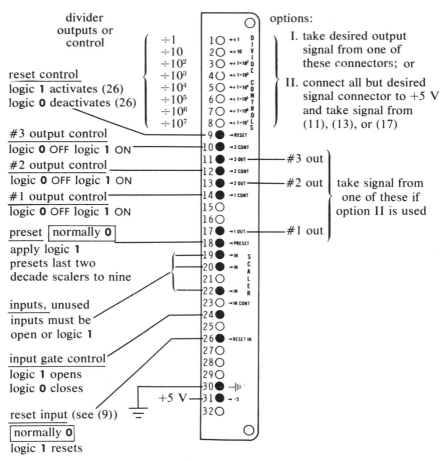

Fig. E6-6   Patch connections to the scaler card.   (Courtesy of Heath Co.)

1 MHz. This signal is available directly at terminal 28 or gated at terminal 22. The Crystal Oscillator card is used in the UDI and the ADD as a precise 1-MHz source. In conjunction with the scaler card, signals of 1 MHz, 100 kHz, 10 kHz, 1 kHz, 100 Hz, 1 Hz, and 0.1 Hz are available for the precise timing of measurements or measurements of time.

Connect the output of the Crystal Oscillator card to the input of the scaler card and observe the signals at the scaler card outputs (terminals 1–8). If possible, use a UDI to measure the frequencies of the output signals. Connect a JK flip-flop to the 1-Hz output to provide a signal that is alternately **1** and **0** for exactly 1 sec each. Use this signal to control the count gate of the counter wired in Experiment 6-10. At the end of the 1-sec gate period, read the counter input signal frequency and manually reset the counter to zero before the next 1-sec gate period appears. Design a circuit to measure the period of the 1-Hz output from the digital timing module to the nearest millisecond.

Experiment 6-12 *Modulo-3 and -6 Counters*

The modulo-3 counter of Fig. 6-17 is wired and the count sequence observed. A modulo-6 counter is made by connecting a T binary before the modulo-3 circuit and then by connecting the T binary after the modulo-3 circuit. The count sequences of the two modulo-6 counters are compared.

Use JK flip-flops to wire the modulo-3 counter of Fig. 6-17 (p. 256). Compare the input signals and the A and B outputs. What is the difference between the A and B outputs? Now connect another JK flip-flop between the count signal and the modulo-3 input to make a modulo-6 counter. Draw the circuit diagram and the output waveforms. Compare these with the circuit and waveforms of a modulo-6 counter made by triggering the third flip-flop from outputs A or B. What advantages might any of these circuits have over the others?

Experiment 6-13 *The Use of Gates in Fixed-Modulus Counter Design*

The modulo-7 counter of Fig. 6-19 is wired and tested. The frequency division (scaling) of this circuit is checked with a digital frequency meter, and, if possible, the maximum input frequency is determined. The design and testing of a modulo-11 or -13 counter is suggested as an exercise.

Wire the modulo-7 counter of Fig. 6-19 (p. 258) using JK flip-flops and NAND gates. Measure the modulus of the A, B, and C outputs. Notice that the modulo-7 output at C is **1** for three and **0** for four out of every seven input pulses. At what point in the circuit is the output level changed for only one pulse every seven counts? Is each set of A, B, and C outputs in the 7-state sequence unique? Must it be?

Design and test a modulo-11 and/or a modulo-13 counter, make a count sequence table and describe the circuit.

Experiment 6-14 *Modulus Control by Direct Clearing*

The direct-clearing modulo-11 counter of Fig. 6-20 is wired and the count sequence is observed. If possible, the maximum input frequency for reliable counting is determined. Counters with several moduli between 1 and 16 are wired by connecting other combinations of flip-flop outputs to the gate input.

Wire the direct-clearing modulo-11 counter shown in Fig. 6-20 (p. 260) and determine the modulus of each of the flip-flop outputs. Can the clear signal at the NAND gate output be observed? Explain how or why not. Describe an experiment to measure the maximum counting frequency for your circuit. What frequency range of generator will probably be required?

Design, describe, and test three similar circuits that count at other moduli between 1 and 16.

The circuit above uses an asynchronous counter. Would a synchronous counter also work and, if so, would there be any advantage?

Experiment 6-15 *Self-Stopping, Variable Modulus Counter*

The counter of Fig. 6-22 is wired. It is observed that the modulus of the counter is equal to the binary number applied to the control inputs. The level of the Count Control signal is observed to follow the automatic stopping action.

A counter that automatically stops at a preset count as shown in Fig. 6-22 (p. 262) is wired and tested. The flip-flop outputs are connected and tested. The flip-flop outputs are connected to indicator lights and the control inputs are connected to switch outputs. The counting is observed to stop at the preset value. To resume counting, the counter is cleared. The Count Control output is 1 during the counting period. Describe a system that could be used to measure how many milliseconds it takes to accumulate fourteen counts in the preset counter.

Keep this circuit for the next experiment.

Experiment 6-16 *Automatic Recycling Variable Modulus Counter*

The circuit of Fig. 6-22 wired in the previous experiment is modified by connecting the Count Control signal to the Clear input instead of to J and K of FF A. A direct-clearing, automatic recycling, variable modulus counter results. The relationship between the binary number set into the control inputs and the counter modulus is noted. The circuit of Fig. 6-23 may also be wired and tested if desired.

Modify the preset counter of Experiment 6-15 to be automatically recycled by connecting the Count Control signal to the clear input rather than to the J and K inputs of FF A. When the counter reaches the preset count, it will thus automatically clear itself and begin counting again. Test this circuit. For a given preset input, is the modulus the same as in the previous experiment?

What are the primary differences between the automatic cycling variable modulus counter wired above and that shown in Fig. 6-23?

Experiment 6-17 *A Preset Binary-Coded Decimal Down-Counter*

The preset BCD down-counter of Fig. 6-24 is wired. The zero detector output is used to stop automatically the counter on the count of zero by connection to the J and K inputs of FF A or to the Clear buss. The circuit is tested for a number of preset input values.

Wire a BCD down-counter with preset inputs as shown in Fig. 6-24 (p. 264). The presetting action is observed and the change in level at the zero detector output is observed when the count reaches zero. Connect the zero detector output to the J and K inputs of FF A so that the counter will automatically stop counting when zero is reached. If it is desired to modify this counter to recycle automatically, the preset control input must be pulsed at the end of each count. Taking careful note of the logic levels required, modify the above circuit to recycle automatically.

Experiment 6-18 *A Serial Input Shift Register*

The basic shift register shown in Fig. 6-26 is wired. Data and clock pulses are applied from manually operated switches to observe the shifting action of the circuit.

Wire four JK flip-flops in the shift register circuits of Fig. 6-26 (p. 267). Connect the outputs to light drivers. Connect a NAND gate bounce eliminator circuit to PB1 to use for the clock signal. The data I and $\bar{I}$ can be obtained from any convenient switch output. Note that the level at I appears at A after one clock pulse and at D after four clock pulses. Enter the binary equivalents of 13, 10, and 5 in the shift register via the serial input. Save this circuit for the next experiment.

Experiment 6-19 *A Parallel Input Shift Register*

Parallel data inputs are added to the circuit of the previous experiment to make the circuit of Fig. 6-28. Data is entered into the register via the parallel input and then observed in serial form at S as the circuit is clocked.

Modify the shift register of Experiment 6-18 to include the gated inputs to the set and clear connections of the JK flip-flops as shown in Fig. 6-28 (p. 269). The inverting gate at each input is not needed if the complements of the input data are available (such as A and $\overline{A}$ from switch A). Connect switch outputs to the parallel data inputs. To set a number into the shift register, enter the number into the switch register and then apply a momentary 1 to the enter input (from PB2). Enter several 4-bit data words into the register and shift them serially out of the register by clocking 4 times. This circuit will be used in the next experiment.

Experiment 6-20 *Circulating Registers and Ring Counters*

The parallel entry shift register of the previous experiment and Fig. 6-28 is made into a circulating register by connecting the S and $\overline{S}$ outputs to the I and $\overline{I}$ inputs. A 4-bit word is entered by the parallel data input and then circulated repeatedly through the register as the circuit is clocked. The word in dc or pulse form (cf. Fig. 6-26) may be observed on an oscilloscope.

The number 0001 is entered into the above circuit to make a ring counter. The sequence of the circuit is observed as pulses are "counted" at the Clock input. Wiring and testing the 5-bit ring counter of Fig. 6-31 is optional.

Connect the serial outputs (S and $\overline{S}$) of the parallel input shift register to the serial inputs (I and $\overline{I}$), respectively, thus making a circulating register. Enter a 4-bit word using the parallel entry, then clock the register and watch as the word is circulated through the register every four clock pulses. By using a higher-frequency clock pulse it is possible to observe the word on an oscilloscope by connecting S to the oscilloscope input. However, with oscilloscopic or other readout from a circulating register, it is important to be able to identify the "start" of the word and to trigger the oscilloscope or readout at this same point each time the word is circulated. Suggest a circuit that could be used for this purpose.

Enter the number 0001 into the circulating register and slowly clock noting that the **1** state passes to each output in turn. If the lights were labeled 0, 1, 2, and 3 the circuit would be a counter and decoder in one. The 5-bit ring counter of Fig. 6-31 may be wired as an optional experiment.

Experiment 6-21 *Count Gate Control and Automatic Cycling*

A complete period or frequency meter is wired as shown in Fig. 6-36. A 1.000-Hz signal is connected to the Gate Time input and a variable frequency source is measured, and the measurement sequence is observed. The counting of the counter can be watched if the memory is defeated by connecting the Memory Transfer input to a logic **1**.

By connecting other signals to the Count and Gate Time inputs and by modifying the count gate control circuit as suggested in Figs. 6-33 and 6-35, the period, duration, or time relationships of a number of signals can be measured. With the addition of a decade scaler this circuit can be used to measure multiple period average and frequency ratio as well.

Wire the complete frequency meter shown in Fig. E6-7 using three or four DCU cards and Decimal Readout accessories, two Dual JK flip-flop cards, a Dual Monostable card, and a NAND card. Apply an accurate 1.000-Hz signal to the Gate Time input and the square-wave output from the digital timing module to the count input. The frequency of the timing module output will be read directly in cycles per second. One source of the 1.000-Hz Gate Time signal is the crystal oscillator and scaler used in Experiment 6-11. Other possible sources are the Clock output at the rear of the UDI, a calibrated 1.000-Hz output from a second timing module, or a separate time-base card that obtains 10 Hz, 1 Hz, and 0.1 Hz signals by dividing down the 50/60-Hz line frequency.

Vary the Display Time control and note its effect. Turn the memory OFF by connecting the memory transfer input to a permanent **1**. Observe the measurement cycle sequence.

Connect other signals to the Count and Gate Time inputs to measure period and frequency ratio. Describe and explain your measurement for each case. Modify the count gate control circuit for the measurement of pulse width or time interval. Describe each measurement made.

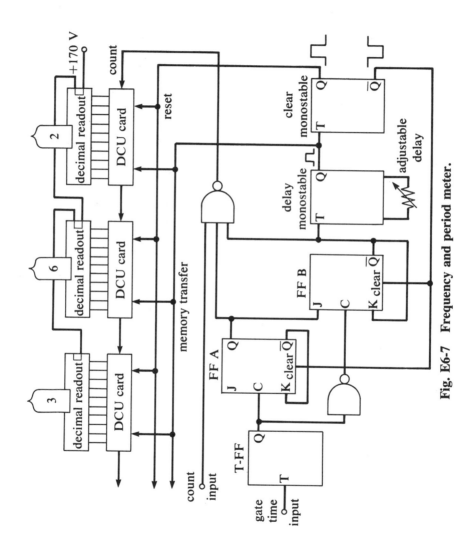

Fig. E6-7 Frequency and period meter.

The experiments in this chapter are designed to provide a familiarization with digital-analog (D-A) and analog-digital (A-D) instrumentation.

The electromechanical servo is investigated both as a potentiometric voltage recorder and as an operational feedback system for performing integration, addition, and other linear and nonlinear analog operations. The features and operations of the all-electronic operational amplifier are tested together with the electromechanical servo, and comparisons are made between the two systems. These operational amplifiers (OA's) are then used in several A-D and D-A systems.

Several digital and analog circuits that are important for system timing, sequencing, and sampling are constructed and tested. These include sample-and-hold circuits, peak samplers, and others. Parallel and serial digital-to-analog converters are built and used in typical applications. Analog-to-digital converters are constructed, and one type of A-D converter is used to make a digital voltmeter (DVM).

Experiment 7-1   *Characteristics of a Servo Potentiometric*
                      Recorder

The damping, dead zone, voltage spans, and other characteristics of a self-balancing potentiometer are investigated. In the potentiometric measurement mode the servo forces a reference voltage to follow the input voltage. The input impedance at balance is extremely high.

The Heath EU-20B and EUW-20A recorders are designed to be utilized in several wiring configurations so that servo experiments can be readily performed. One small modification for these recorders is suggested in Experiments 7-3 and 7-4 so that it will be easy to convert the servo potentiometric recorder to the operational feedback system.

7-1a   Introduction to the Servo Mechanism

Identify the basic components of the servo systems. Remove the top and rear panels and note the major parts by reference to the instrument manual and schematic diagram. The EU-20B input circuit is shown in Fig. E7-1. Note that when the 5-pin plugs labeled A, B, and C are removed there are several convenient terminals exposed. These terminals enable rapid modifications for performance of subsequent experiments.

Always switch the recorder to OFF before making circuit connections!

7-1b Dead Zone

For optimum performance of a servo system the dead zone must be small. The dead zone for a servo recorder is the minimum change of input signal required to move the recorder pen. It is greatly dependent on the amplifier gain (which is changed somewhat with the DAMPING control), the noise that reaches the servo motor drive (which is reduced by the HUM ADJ control and input filter), and the mechanical load (which increases significantly with dirt, etc., in the pen carriage bearing or pulleys or misalignment of potentiometer shaft).

One practical measure of the dead zone is to observe the difference between pen positions for a specific input voltage when the pen approaches the null point from opposite directions.

Set the input SPAN for 100 mV. Turn the Function switch to the RECORD position and apply several successive 5-mV increments from the pH/MV test box, first increasing and then decreasing the voltage. Test the effects of HUM ADJ and DAMPING controls on the dead zone.

Determine whether the dead zone changes when the SPAN is switched to 10 mV. Also, note dead zone at the minimum available span.

Determine the dead zone when a fixed input voltage (for example 50 mV) is used to provide approximately a midscale reading. Turn the recorder to STANDBY and move the pen carriage manually upscale and then turn to RECORD; repeat but manually move the pen downscale. Is there any difference in dead zone by the voltage step or manual displacement methods?

When making measurements the DAMPING control should always be set at the minimum position (counterclockwise) that prevents oscillation of the pen. The most sensitive spans (10 mV, etc.) require least damping.

Damping is an important consideration in all servo systems.[1]

---

[1] For discussion, see H. V. Malmstadt, C. G. Enke, and E. C. Toren, *Electronics for Scientists*, pp. 308-314, Benjamin, New York, 1962.

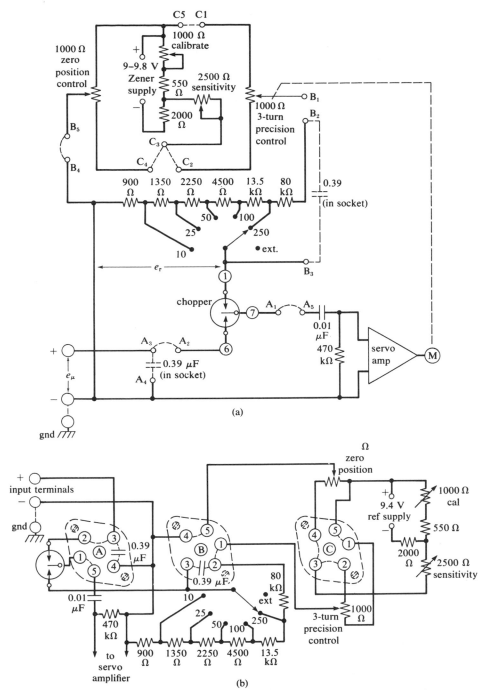

Fig. E7-1 Input circuit of Heath EU-20B recorder. (a) Schematic; (b) arrangement of A, B, and C, 5-pin plugs. (Courtesy of Heath Co.)

The overall gain of the amplifier is about one million so that pickup of noise at the input is greatly amplified and can greatly influence the dead zone and recorder reading. Note that when the back of the recorder is removed, a wave of the hand near the base of the input tube can change the recorder reading. Note that it is important to have the back screwed on to act as a shield while performing tests on recorder performance or calibration.

Caution: When back is off be careful not to touch high-voltage or input line voltage!!

7-1c Voltage Spans

Calibrate the recorder for a 100-mV full-scale span using an accurate low-noise voltage reference source (VRS) (EU-80A). See the recorder manual for the position of the calibration control. Damping should be set so there is negligible dead zone and yet the pen does not oscillate.

Set the zero accurately and then adjust the calibration control for full scale with the VRS set at 100 mV. Check the linearity by increasing the VRS voltage in 10-mV steps from zero to full scale. Repeat in the opposite direction. If the pen position does not reproduce to about the line width of the pen trace, the DAMPING and HUM ADJUST controls probably need to be adjusted.

Observe the effect of the FINE SENSITIVITY control in changing voltage span. Also note in Fig. E7-1 that the SPAN control changes the fraction of reference voltage selected across a voltage divider. Therefore, true potentiometric null-point performance is provided for fixed spans of 10, 25, 50, 100, and 250 mV and any span between 3.5 to 250 mV.

7-1d High-Voltage Ranges

Input voltages greater than a few tenths of a volt can be measured with the self-balancing potentiometer by using a voltage divider at the input to decrease e_u by a known fraction. The disadvantage of this procedure is that current flows from the input source, even at balance. However, the resistance of the divider can be relatively large to decrease the current. Unfortunately, the greater the resistance of the divider the greater the chance of noise pickup at the input.

To make high-voltage measurements, an input voltage divider can be easily connected if the recorder modification shown in Fig. E7-2 has been made. The modification consists of adding a multiple connector card

(EU-50-MD) that was cut off 1-in. from the top of the card. The 1-in. top piece is mounted on an L-shaped bracket. The bracket is slipped under three loosened screws that hold the A, B, and C adapters on the chassis, and then tightened firmly in place. Thus it is easy to connect parts and wires into the multiple connectors as shown in Figs. E7-2 and E7-3. The leads from switch S_4 and B_3 that go to chopper terminal 1 were disconnected at the circuit board and inserted in the multiple connector group labeled Z. One end of a #22 wire lead (about 4 in. long) was soldered at the circuit board in place of the wire removed for switch S_4. The other stripped end is also inserted in the multiple connector group Z. This change is important for subsequent experiments, but for this experiment it just reconnects the reference voltage e_r between chopper contact 1 and ground.

Remove plug A (only), and use #22 wire leads (with a ⅛-in. diameter

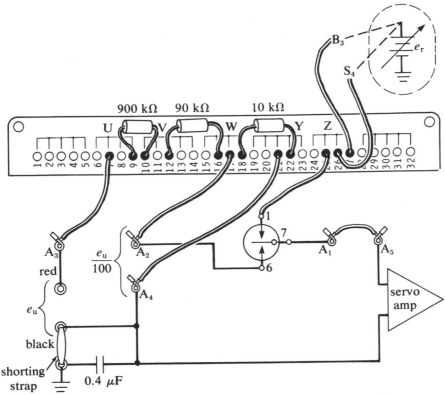

Fig. E7-2 Voltage divider for high input voltages. (Diagram shows suggested modification for EU-20B recorder.)

spring clip on one end and the other end with ½ in. of the insulation stripped off) to connect from A_3 to multiple connector group U, from A_2 to group W, and from A_4 to group Y. Use a lead with spring clips on both ends to connect from A_1 to A_5. Precision resistors are chosen and inserted, as shown in Fig. E7-2. In this case, a divider with a total resistance of 1 MΩ is used so that 1/100 of the input voltage e_u is applied between chopper contact 6 and ground. Therefore, if the SPAN control is set at 100 mV, the full-scale recorder span will now be $100(0.1) = 10$ V.

Check the precision and accuracy by rotating the VRS COARSE control to provide 1-V increments at the attenuator input.

Change the input divider resistors and reference voltage spans so as to provide other desired full-scale input voltage spans.

Disconnect the A-plug connectors, remove the precision resistors, and replace the A plug for normal operation as a potentiometric recorder. Keep the Z group connected.

Always replace the back cover before making tests so as to eliminate noise pickup.

7-1e Zero Suppression

In many laboratory applications it is important to observe small changes superimposed on large signals. For example, the transducer output voltage might be 1 V, but the total change over the operating range might only be 0.01 V (10 mV). Therefore, to spread the useful 10-mV signal across the full scale of the recorder, it is necessary to suppress effectively the scale 0 many chart widths of scale.

The VRS is a good zero-suppress unit because it is very stable (after warm-up), continuously variable over a wide range, and all necessary terminals are built in. By using the VRS in its DIFF position, the recorder becomes, in effect, a linear off-balance indicator for the change of input voltage.

Let a 1-V output from the transistor power supply simulate the output from some transducer, and connect it to the SIGNAL terminals of the VRS. Connect the recorder across the OUTPUT terminals of the VRS. Set the VRS switches in the same positions as used to calibrate the VRS, and adjust the VRS COARSE and FINE controls to the null point, with the recorder switched for the 10-mV span. Observe that a 10-mV variation of the input signal causes full-scale travel of the pen. In other words, a simulated 1% change of input signal causes full-scale deflection. Since changes of input signal of less than 0.1 mV are readily observed, relative signal changes of less than 1/10,000 can be recorded. With the chart ON, determine the stability (in per cent variation) of the transistor power supply over a period of time, assuming the VRS is absolutely stable over the same period.

Experiment 7-2 *Current Measurements Using Electro-*
mechanical Operational Feedback

A servo recorder is modified to the operational feedback mode so as to make a sensitive current recorder with full-scale current spans of 10^{-8} and 10^{-7} A.

Construct the circuit illustrated in Fig. E7-3. Note that the same multiple connector card modification described in Experiment 7-1d is used here. Remove plug A. Connect, as shown, a feedback resistor, $R_f = 1$ MΩ. Connect the unsoldered end of the terminal that goes to chopper contact 1 to the multiple connector group Y. Connect leads between A_3 and group W, between A_1 and A_5, A_2 and A_4, and between groups W and Y. This connects the summing point (chopper contact 1) directly to the red input terminal of the recorder, and the feedback resistor between the summing point and the voltage reference source e_r. Connect a 0.1-μF capacitor C across R_f to average noise pickup.

Connect an input current source. Set the recorder SPAN at e_r (full scale) = 100 mV, so that full-scale current span is $i = 10^{-1}/10^6 = 10^{-7}$ A. Check the current span for linearity. The pH/MV test box can be used as a current source that provides 10^{-8}-A current increments up to 7×10^{-8} A. The internal series resistance is switched to 10 MΩ, and the Function switch set to provide 0–700 mV. The EU-80A VRS with current (resistance) probe can be used to provide any desired current from microamperes to picoamperes.

Observe the full-scale current spans available by changing the span of e_r.

Change R_f to 10 MΩ (0.01 μF in parallel), and set e_r (full scale) = 100 mV. Determine whether the 10^{-8}-A scale is accurate or whether the amplifier input leakage current has become significant.

Use two current sources and show that the two currents can be added and subtracted.

Experiment 7-3 *Voltage Measurements Using Electro-*
mechanical Operational Feedback

A modified servo recorder is used in the operational mode for voltage measurements. A comparison is made between the operational and potentiometric modes for voltage measurements.

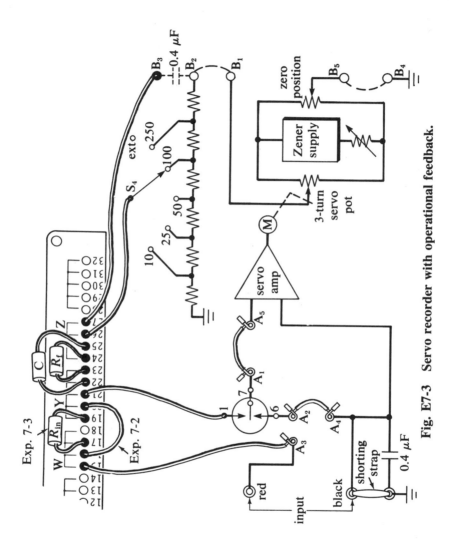

Fig. E7-3 Servo recorder with operational feedback.

The operational feedback system can be used for voltage measurements by inserting an input resistor, R_{in}, between A_3 and summing point as shown in Fig. E7-3. To do this, the jumper lead between connector groups W and Y is removed. Make $R_{in} = 10$ MΩ, $R_f = 1$ MΩ, and $e_r = 100$ mV, and $e_{in} = 0$-1 V. Show that $e_{in} = -e_r R_{in}/R_f$. Change R_{in} to 1 MΩ and check whether $e_{in} = -e_r R_{in}/R_f$. What are the disadvantages of measuring a 100-mV input voltage by the servo operational feedback method as compared with the servo potentiometric method?

Note that voltages could be added or subtracted by using two or more voltage sources and input resistances as is done in Experiment 7-9 with the all-electronic OA.

Keep circuit connected for modification in Experiment 7-10.

Experiment 7-4   *Familiarization with Operational Amplifiers*

The adjustments and connections for a solid-state discrete component OA and an integrated-circuit OA are investigated.

7-4a   Discrete Component Transistor Operational Amplifier

The EU-900-NA Operational Amplifier card was introduced in Chapter 2. A schematic is shown in Fig. E7-4 to illustrate the components and connections on the card. The transistor OA is encapsulated. A multiturn balance potentiometer that is connected to the OA is mounted on the card, and it can be adjusted by a screwdriver inserted through a hole in the plastic top cover.

Four precision resistors each have one end connected to the inverting (summing) input so that they can be easily patched as input and feedback resistors. A transistor Q_1 is connected in a circuit so that its collector (OUT) voltage can only change from 0 to +5 V. When the base (IN) terminal of Q_1 is connected to the output of the OA, the transistor circuit provides an interface to the TTL digital cards. This interface is important because the maximum output voltages from the OA amplifier can destroy the logic circuits on the digital cards if connected directly. An input voltage to the logic driver of more than 0.8 V provides a logic **0** output, and an input voltage less than about 0.7 V provides a logic **1** output.

Connect the wire jumpers indicated by the heavy lines in Fig. E7-5 to make a gain-of-10 amplifier. This amplifies the input offset voltage tenfold so as to provide a more sensitive zero-balance adjustment. Use any type of voltmeter that has a sensitivity of 1 mV or better, and adjust

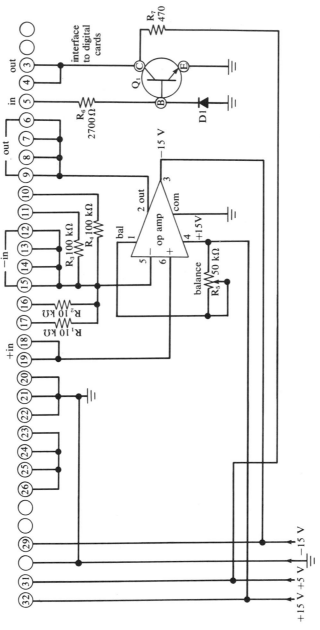

Fig. E7-4 Schematic of the Heath Operational Amplifier card (Model EU-900-NA). (Courtesy of Heath Co.)

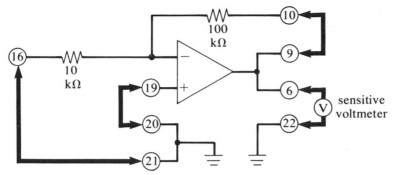

Fig. E7-5 Zero adjustment circuit for operational amplifier.

the BALANCE control for a zero output voltage. Keep the OA card in the analog-digital device (ADD). Most of the remaining experiments in this section will be performed with this general purpose OA.

7-4b   Integrated Circuit Operational Amplifier Card

The EU-800-HB Comparator/V-F card contains a type-709 integrated circuit OA with input protection, high-frequency stabilization, and balance control, as well as other circuits for utilization of the OA as a V-F converter, and as a comparator.

A simplified schematic showing the connector relationships to the numbers on the plastic top cover is shown in Fig. E7-6a.

The circuit for balancing to zero-output voltage with zero-input voltage is illustrated in Fig. E7-6b. A gain-of-100 amplifier is connected, and the OP AMP BALANCE control is adjusted for zero-output voltage as indicated by a sensitive voltmeter.

For all applications except where maximum high-frequency response is desired, it is best to connect the stabilization components by connecting jumper leads from 9 to 10 and from 11 to 12.

This integrated-circuit OA can be used in many of the subsequent experiments, and it will be desirable to use it to compare the characteristics of OA's.

Experiment 7-5   *Operational Amplifier as Current-to-voltage Amplifier*

The basic $e_o = -i_{in}R_f$ relationship is verified for a range of currents and resistances. Limitations of this expression under certain conditions are observed.

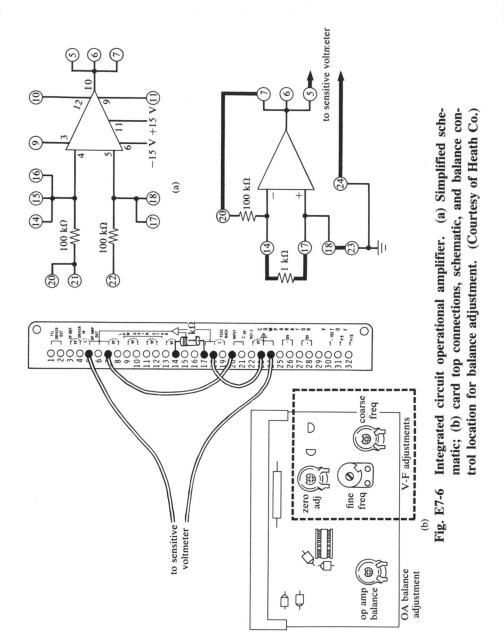

Fig. E7-6 Integrated circuit operational amplifier. (a) Simplified schematic; (b) card top connections, schematic, and balance control location for balance adjustment. (Courtesy of Heath Co.)

Connect the general purpose transistor OA as a current-to-voltage amplifier, as shown in Fig. E7-7. Start with a feedback resistor $R_f = 10$ kΩ and verify that $e_o = -i_{in}R_f$. Note relationship of output voltage polarity with respect to direction of input current.

Increase R_f to 100 kΩ, 1 MΩ, and 10 MΩ and note any deviations from ideal performance. What is the maximum practical sensitivity of this OA as a current measurement device? Compare the performance of the 709 integrated circuit OA with the above.

Experiment 7-6 *Operational Amplifier as Inverting Amplifier*

The basic $e_o = -e_{in}(R_f/R_{in})$ relationship is verified for several ratios of resistances and for a range of input voltages. Voltage and current limits are noted.

Connect on the OA card an input resistor, $R_{in} = 10$ kΩ, and a feedback resistor, $R_f = 100$ kΩ, and accurately measure the output voltage.

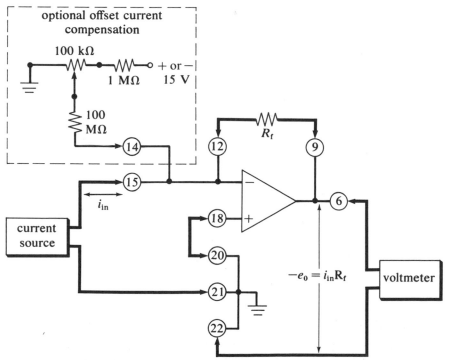

Fig. E7-7   **Current-to-voltage OA amplifier.**

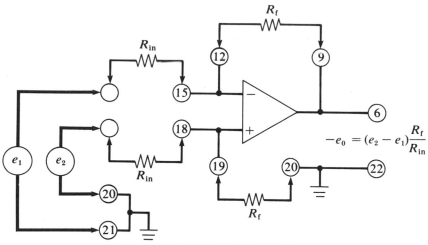

Fig. E7-8   Differential amplifier.

Determine whether $e_o = -e_{in}(R_f/R_{in})$ within the accuracy of the resistors.

Measure e_o for various ratios of R_f/R_{in} and known values of e_{in}. Determine specific limitations.

Connect a 5-V square wave to the input of R_{in} and observe the output voltage signal on the oscilloscope. Start with a low frequency and increase to frequencies where distortion is apparent. Record the OA limitations.

Experiment 7-7   *Operational Amplifier Differential Inputs*

The use of the OA as a differential amplifier is investigated and certain limitations are noted.

Connect input voltages and precision resistors to both inverting and noninverting inputs as shown in Fig. E7-8.

Use stable voltage reference sources for e_1 and e_2 and determine how well the circuit performs as a difference amplifier over a wide range of inputs up to 10 V. What is the CMRR?

Experiment 7-8   *The Operational Amplifier Voltage Follower*

The OA voltage follower is shown to have high-input and low-output impedance. Its use as an impedance transformer over a wide frequency range is tested. Comparisons are made with the servo voltage follower.

Connect a voltage follower amplifier.

Set $e_{in} = 2.000$ V (from the VRS) and measure the input and output voltages accurately with the DVM. Note any difference between input and output voltages.

Connect the resistance substitution box as a load on the follower output and measure the output voltage as the load is increased (resistance decreased). Start with 50 kΩ. When the output voltage drops it is due to the output current limitation of the OA.

The follower input current can be estimated by measuring the voltage dropped across a large input resistor. Carefully balance. Connect a 100-MΩ resistor from the input to ground and measure the output voltage.

Connect a signal generator to the input and determine how well the output follows the input up to high frequencies.

Compare the OA voltage follower and the servo potentiometer of Experiment 7-1.

Experiment 7-9 *Sum and Difference Measurements*

Several voltages are added and subtracted with the OA adder and the servo adder. Comparisons are made between the two systems.

Connect two or more stable voltage sources through precision input resistances to the summing point of the general purpose OA. Ground the noninverting input and start with a precision resistor for $R_f = 10$ kΩ. Show that

$$-e_o = R_f \left(\frac{e_1}{R_1} + \frac{e_2}{R_2} + \cdots + \frac{e_n}{R_n} \right)$$

Observe that either addition or subtraction can be accomplished depending on the polarity of input voltages relative to each other. Note that subtraction can also be accomplished by using the noninverting (+) input as in Experiment 7-7.

Compare the advantages or disadvantages of the all-electronic OA to the servo OA for sum or difference measurements. Perform experiments to illustrate your comparisons. Record the results.

Experiment 7-10 *The Operational Amplifier Integrator*

Servo and OA integrators are connected and their characteristics are observed. Several applications of the integrator are illustrated.

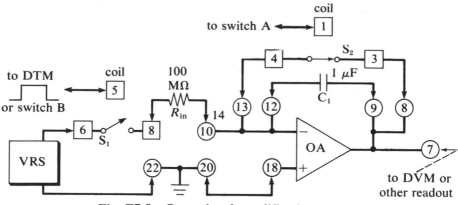

Fig. E7-9   Operational amplifier integrator.

7-10a   Transistor Operational Amplifier Integrator

Check balance on OA with gain-of-10 amplifier.   Connect the voltage integrator of Fig. E7-9.

Start with $C_1 = 1$ μF, $R_{in} = 100$ MΩ, and relay switches S_1 and S_2 controlled by front panel switches A and B on the binary information module.   Set the VRS at about 1 V.

Always keep the capacitor shorted by S_2 except when integrating. Turn OFF the ADD when making connections.

Close switch S_1 for 10 sec (use square wave from digital timing module, DTM, or other clock source, or use switch B and second hand on watch).   Record the ramp output voltage and measure on a chart recorder (with 10-V input divider) the output at the end of 10 sec.   Is the measured voltage equal to calculated voltage output?   Does the output voltage change with time after S_1 is opened?   Discharge the capacitor C_1 before the next integration.

Alternately open and close switch S_1 with a 0.5-Hz square-wave signal from the DTM.   Record the output signal on the chart recorder. Do not exceed 10 V output.   Keep the circuit connected for Experiment 7-11.

7-10b   Servo Integrator

Remove the feedback resistor R_f from the servo operational feedback system used in Experiment 7-2 and illustrated in Fig. E7-3.   Also connect a feedback capacitor $C = 1$ μF or 10 μF.   Insert $R_{in} = 100$ MΩ and be sure the shorting wire across R_{in} in Fig. E7-3 is removed.

Use the VRS set at about 0.1 V and close the input switch S_1 to integrate with the servo integrator. Start with recorder span set at 250 mV and note the limitations imposed by the maximum values of e_r (span) that are selected.

Observe on the recorder chart how the output voltage changes when S_1 is open. Compare performance of the servo integrator with the transistor OA integrator.

Experiment 7-11   *The Operational Amplifier Differentiator*

The output of an integrator circuit is differentiated with an OA differentiator to restore the original input to the integrator and thereby test the circuit over certain conditions.

Connect at the output of the OA integrator from Experiment 7-10, the OA differentiator shown in Fig. E7-10. Connect the square-wave drive signal from the DTM to the coil of S_1 in Fig. E7-9, and the opposite phase square-wave drive signal to the coil of switch S_2. Also change R_{in} to 10 kΩ. Connect the square-wave output from the DTM (across a voltage divider to give 1 V) in place of the VRS, and set the frequency of the generator for 10 Hz.

Observe the original square wave to the input of the integrator and the ramp output of the integrator on the two traces of the oscilloscope. Note the time relationships and sketch. Now observe the output of the differentiator relative to the input of the integrator and compare.

Sketch all three curves showing shapes and time relationships.

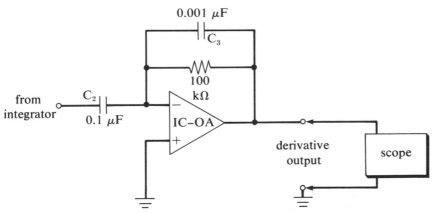

Fig. E7-10   Operational amplifier differentiator.

Experiment 7-12 *Logarithmic Amplifier*

A semiconductor diode or transistor is used as a nonlinear resistance in the feedback path of either the servo or OA system so as to provide logarithmic readout of input currents or voltages.

Connect the logarithmic amplifier shown in Fig. E7-11.

With 0.05 V from the VRS connected at the input, set R_5 to give zero-output voltage as read from a DVM or servo recorder. Switch the VRS to 5 V and set the SENS. control R_8 until the DVM reads exactly 2 V. Now switch the VRS to several voltage-input values (such as 0.1, 0.2, 0.25, 0.5, 1, 2, 2.5, and 4 V) between 0.05–5 V, and determine how closely the readout corresponds to values calculated from log ($e_{in}/0.05$).

Set the limits for only one decade of input voltage and test the accuracy of the logarithmic response for input voltages within the selected decade.

Note that the log element could also be used in the feedback of the servo OA.

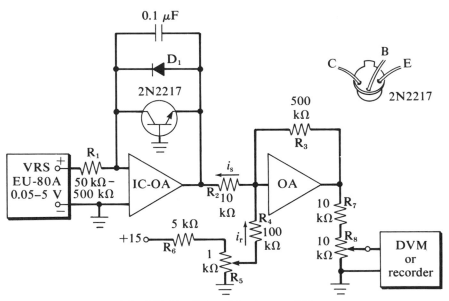

Fig. E7-11   Logarithmic amplifier.

Experiment 7-13   *The Comparator*

The integrated OA comparator on the Comparator/V-F card is investigated and the sensitivity, hysteresis, and other performance characteristics are determined. The use of the servo as a comparator is also noted.

Connect the OA comparator circuit illustrated in Fig. E7-12. Use the Comparator/V-F card (EU-800-HB). Note that a small fraction of the OA output signal is fed back to the noninverting input when connected as shown. This positive feedback helps snap the OA through the transition region when the trigger point is reached, and it results in some trigger voltage hysteresis.

Set the comparison voltage source at +4 V by the 100-kΩ potentiometer on the DTM. Connect the VRS to one of the signal voltage inputs (20 or 21) that has a 100-kΩ resistor between it and the OA summing point. Connect the TTL output (pins 1 or 2) to one of the indicator lights on the binary information module to indicate logic level. Vary the input slowly in the region of the transition voltage. Measure accurately the input voltages at which the transitions from logic **0** to **1** and **1** to **0** occur.

Disconnect the positive feedback signal by removing the wire between pins 18 and 19. Determine the voltages at which the transitions occur with all other conditions the same as above. Explain the differences in measured values with and without the positive feedback.

Reset the comparison voltage source to −3 V and repeat the above measurements.

Show how the servo OA can be used as a comparator (see Experiment 7-1d).

Experiment 7-14   *Analog Gating*

An OA analog gate is constructed and used to gate analog voltage signals.

Connect the OA analog gate of Fig. E7-13.

Use the JFET switches and drivers of Experiment 3-9 and Fig. E3-8. The drivers should be operated so that the two drive signals to the FET's are of opposite logic level at any time. The complementary square waves from the DTM can be used. If the OA is balanced the output should be essentially 0 V when Q_2 is ON and Q_1 is OFF.

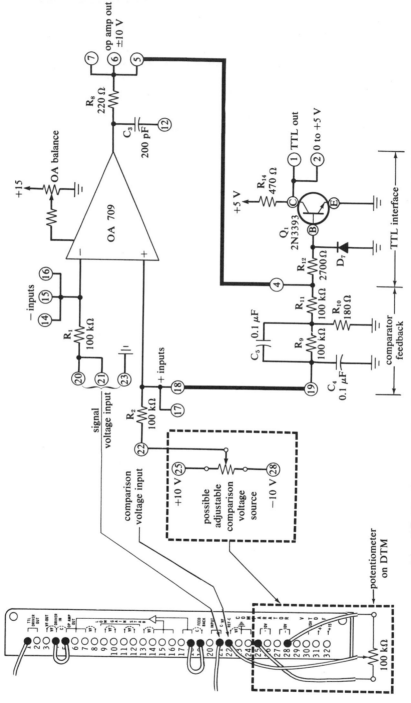

Fig. E7-12 An operational amplifier comparator circuit (EU-800-HB) without input protection or stabilization circuits shown.

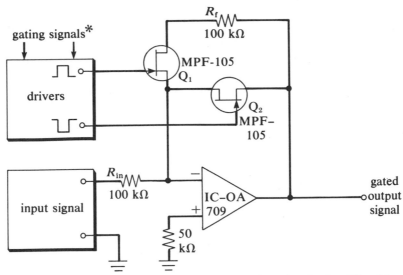

Fig. E7-13 An analog gate. Note that the gating signals should not be applied directly from the DTM to the gate of the FET because a positive voltage can cause excessive current and destroy the FET. Use drivers of Fig. E3-8.

Observe that with this circuit when Q_1 is ON and Q_2 is OFF the OA circuit operates as a normal unity-gain inverting amplifier (within the accuracy of the input and feedback resistors). Apply a load to the output and note that the amplifier provides good isolation of the input signal from the load. Use a 1-V input signal from the VRS.

Measure the maximum rate at which input signals may be accurately gated with this circuit. Use the EXT position and 100 pF external capacitor of DTM to get the frequency output greater than 10 kHz.

Experiment 7-15 *Ramp and Comparator Sequencer*

A ramp generator and one or more comparators are wired to make the sequencer of Fig. 7-28. The adjustability and reproducibility of the sequence are tested.

Wire the circuit of Fig. 7-28 (p. 313) using the Operational Amplifier card or the integrated circuit OA in the Comparator/V-F card for the comparators. The sawtooth signal can be obtained from an OA integrator as shown or from the sawtooth output of the timing module. Use the relay card for the switch across the integrating capacitor, and use the potentiometers in the timing module for the reference voltage dividers. Trigger the oscilloscope at the beginning of the sweep signal and observe the sweep and comparator outputs noting the effect of the reference voltage on the OA outputs.

Use the UDI to measure the reproducibility of the delay times by repeated measurements of the time from the beginning of the sweep to the triggering of the comparator. Readjust the delay time setting and see how close you can come to the original delay time. What is probably limiting the reproducibility of the delay setting and how could it be improved?

Use the digital interface circuits on the OA cards and NAND gates to obtain a pulse signal that begins when OA_1 is triggered and ends when OA_2 is triggered.

Experiment 7-16 *Switch-Tail Counter and Decoder*

An 8-state switch-tail counter and decoder is wired following the design shown in Fig. 7-30. The sequence is observed with indicator lights and a dual-trace oscilloscope. The sequence accuracy is measured with a digital timer.

Wire four flip-flops in a switch-tail counter circuit similar to the five flip-flop design shown in Fig. 7-30 (p. 316). Connect eight decoding gates so that indicator lights are lit one at a time in sequence as the counter is clocked. With a high clock frequency, observe the Clock signal and one or more of the output signals on an oscilloscope. Use the UDI to measure the reproducibility of the delay time between the pulses from decoder gates 1 and 5. What limits the reproducibility of this time and how might it be improved?

Experiment 7-17 *Sample-and-Hold Measurements*

A sample-and-hold circuit is used to measure the amplitude of a periodic signal. Then a differential sample-and-hold circuit is wired and used to measure the difference in potential between two different parts of a signal waveform.

Connect the sample-and-hold circuit of Fig. E7-14. The sampling switches can be relays on the relay card. The sampling switches close

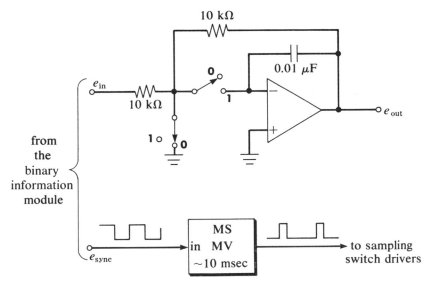

Fig. E7-14   Sample-and-hold experimental circuit.

for the duration of the monostable (MS) pulse immediately following each **1–0** transition of the synchronizing signal.   If e_{in} is the same square-wave output as e_{sync}, e_{in} will be sampled when it is **0** and e_o will be approximately 0 V.   If e_{in} is the complementary square wave of e_{sync}, then e_{in} will be sampled when it is **1** and e_o will be about -4 V.   If e_{in} is the sawtooth output, then e_o will be the inverted sawtooth output voltage at the sampling time.   Use e_{sync} to synchronize the oscilloscope as well and observe $-e_o$ and e_{in} on a dual-beam oscilloscope using the various input signals described above.

Set the oscilloscope sweep at 10 msec/division and observe e_{in} and e_o using the sawtooth input signal and the complementary synchronizing signal (terminal No. 24).   Vary the timing module frequency from 10 to 20 Hz and observe that e_o follows the value that e_{in} is when the sampling switches open.   Now go to very low sampling rates and notice the appearance of a jump in e_o that occurs during the sampling switch closure.   From the repetition rate and the magnitude of this jump, estimate the drift rate of the amplifier in the hold mode.   What minimum repetition rate would you recommend for this circuit?   How might you modify the circuit if lower repetition rates were desired?

If higher-speed sampling is desired, the reed relay or the FET analog switch of Fig. E3-8 can be used for the sampling switches in this experiment.

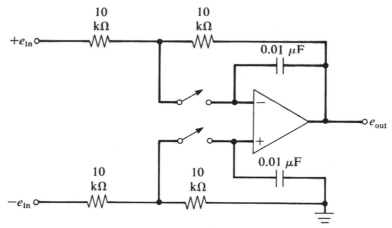

Fig. E7-15 Differential sample-and-hold amplifier.

Differential Sample-and-Hold Circuit

Now connect a differential amplifier in the sample-and-hold mode as shown in Fig. E7-15. Connect the sampling switch drivers to *separate* monostable MV's. First, connect the monostables to the same square-wave-synchronizing signal and confirm the differential nature of the amplifier by measuring the e_o for various values of $+e_{in}$ and $-e_{in}$. Note particularly that when $+e_{in} = -e_{in}$, $e_o = 0$. Connect $+e_{in}$ and $-e_{in}$ to the same and then to complementary square-wave outputs. Note and explain the difference in output.

Now, connect the monostable inputs to the complementary square-wave outputs so that the positive (+) and negative (−) inputs are alternately sampled. Again connect $+e_{in}$ and $-e_{in}$ to the same and then to complementary square-wave outputs and explain the result. When $+e_{in}$ and $-e_{in}$ are connected to the same square wave, reverse the connections to the monostable inputs. Why is e_o inverted? Connect both $+e_{in}$ and $-e_{in}$ to the sawtooth output and explain the e_o observed. This circuit is very convenient for measuring the difference in potential between two points on a waveform or for measuring the difference between two signals that are multiplexed (alternately connected to a common input amplifier).

Experiment 7-18 *Peak Height Measurements*

The peak sampler and memory circuit of Fig. 7-36 is wired and used to sample and store the peak voltage of a sine-wave input signal. The input-output relationship is measured using a dual-trace oscilloscope.

Wire the circuit of Fig. 7-36 (p. 323). Use a relay on the relay card driven by PB1 for the reset contacts. Connect e_2 to -15 V and make $R = 10$ kΩ. Use the sine-wave generator for e_{in} and a 0.01-μF capacitor for C. Observe e_o as the sine-wave amplitude is varied. Note that e_o follows as the amplitude is increased, but that the reset button must be pushed to follow a decrease in input amplitude. Compare the input and output signals on a dual-beam oscilloscope. Try the peak sampler and memory circuit on the sawtooth waveform to see if C is charged to the actual peak value.

Experiment 7-19   *Peak Area Measurements*

A voltage-to-frequency (V-F) converter and a gated counter are used to make a voltage-time integrator. The experimental circuit is part of the digital recorder shown in Fig. 7-38. Integration or peak-area measurements are made on several input waveforms.

Wire a peak area meter by connecting the output of a V-F converter to the input of a gated counter as shown in Fig. E7-16. This can be done by connecting the V-F output of the UDI to the A input with a terminated cable and using the UDI in the Events Counter mode. In this mode,

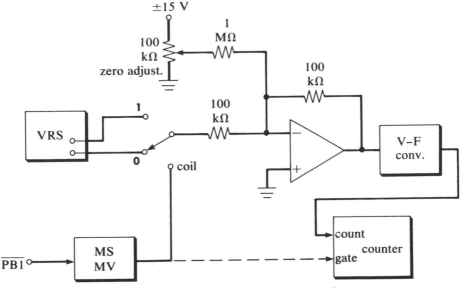

Fig. E7-16   **Peak area measurement.**

the count gate is controlled by input B. An alternative experimental setup is to wire the gated counter of Experiment 6-21 and use the V-F converter that is on the Comparator/V-F card. In this case a gate-time input as shown in Fig. 6-33 should be used for the count gate control.

A pulse signal is made by turning on the output of the VRS for the duration of a monostable pulse. With the relay in the **0** position, open the count gate manually and adjust the zero control so that the count rate is zero. Now set the VRS at 1.00 V, reset the counter, open the count gate, trigger a 1-sec pulse from the monostable, close the gate, and observe the count. Change the VRS output and/or the monostable duration to demonstrate that the counter output is proportional to the pulse area. For convenience in repetitive measurements, the MS output can be used to open the count gate.

Areas of other waveforms can also be measured. For instance, the pulse output from the monostable and the sawtooth output from the timing module can be used. These waveforms have base lines that are significantly different from zero. Therefore it is especially important to "zero" the integrator to compensate for the base-line offset.

Experiment 7-20 *Data Point Sampling*

Two monostable multivibrators and a sample-and-hold circuit are used to measure a point on a repetitive waveform as shown in Fig. 7-39. The sample-and-hold output voltage is plotted as a function of the delay time to reproduce the original waveform with data points. A suggestion is made for a circuit that will automatically sequence the delay and allow the output waveform to be recorded.

Wire the experimental sample-and-hold circuit shown in Fig. E7-14 but include a variable delay MS between the synchronizing signal and the sampling pulse generator as shown in Fig. E7-17a. Connect the square wave from the timing module to e_{sync} and the sawtooth to e_{in}. Observe e_0 and e_{in} for a signal frequency of about 5 Hz. Vary the delay MS over the 10–100 msec range. Note that the sampling point moves along the input waveform as the delay is changed. By adjustment of the delay, any particular point of the input waveform can be sampled.

Successive points on the input waveform can be sampled automatically with the circuit of Fig. E7-17b. A slow sweep signal is started by opening the switch across OA_2. The synchronizing signal triggers the switch driver to open the switch across the fast sweep generator OA_1.

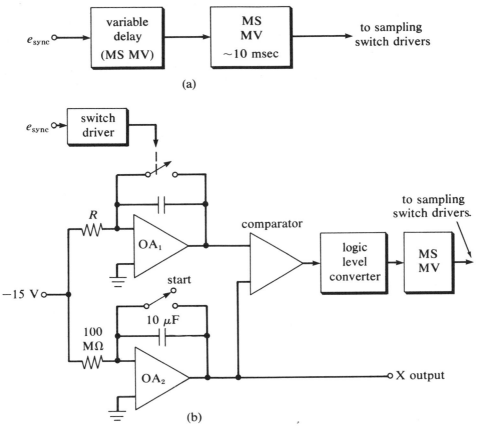

Fig. E7-17 Data point sampling circuits. (a) Single point, (b) automatic sampling advance.

When the OA_1 output equals the OA_2 output, the comparator will trigger the sampling pulse generator. On the next synchronizing trigger, the fast sweep will have to sweep longer to equal the OA_2 output and, thus, the sampling will take place later in the signal cycle. Relays can be used for sampling and for the shorting switch on the fast sweep generator, but FET switches are recommended for better speed. The *RC* time constant of the fast sweep generator is set for the number of samples per slow sweep desired.

An X-Y oscilloscope can be used to observe a sampled and reconstructed waveform by connecting the Y input to e_0 and the X input to the X output. An X-Y or strip chart recorder can also be used to record the signal in equivalent time.

Experiment 7-21 *A Weighted-Resistor Digital-to-Analog Converter*

A 3-decade weighted-resistor D-A converter is constructed. The digital output from the memory of three decade counting units is converted to an analog voltage and recorded on a strip chart recorder.

Connect the D-A converter shown in Fig. 7-41. Use three different sets of weighted resistors. The resistors for the most significant bit n_3 should be 0.1%. Use the VRS set at 1 to 5 V or use the +5 V from the ADD for V_R. Connect a servo chart recorder to measure a fraction of the OA output applied to a voltage divider. Take a fraction of the OA output so the full count of 999 on three Nixie tubes will be 99.9% of full scale on the 100- or 250-mV span of the recorder.

Use the relay card for seven of the switches and a reed relay or FET switch for the eighth switch. The switches for the most significant bit n_3 should be driven from the A, B, C, D memory outputs of the third DCU. The D output from the third DCU card should switch the 12.5-kΩ resistor, the C the 25-kΩ resistor, etc.

Use the 3-digit frequency meter connected in the last experiment of Chapter 6 for the 3-digit readout. Record with the servo recorder the output for various frequencies between 0 and 999 Hz and determine the accuracy of the analog readout compared to the digital input.

Experiment 7-22 *A Serial Digital-to-Analog Converter*

An OA staircase integrator is constructed and used as a serial D-A converter. Also, the circuit is modified and used as a count-rate meter.

7-22a Staircase Integrator A-D Converter

Connect the staircase integrator of Fig. E7-18. Do not connect R_2 or the output filter for this part of the experiment. The square-wave output from the DTM can be used for the input signal, but it is better to use a series FET-switched reference voltage from the VRS as the signal (with the driver controlled by the serial digital information).

Switch a known number N of pulses of amplitude e_{in} at a rate of about 1/sec to the input of the circuit. Record the output voltage e_o on the chart recorder. Start with a recorder span of 250 mV, and use an input divider if necessary. Determine whether $e_o = N\, e_{in}\, C_1/C_2$ within the accuracy of the components.

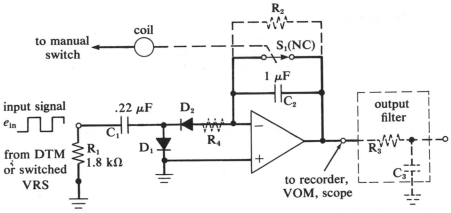

Fig. E7-18 Serial digital-analog staircase integrator and count-rate meter.

Vary the rate of introducing N pulses, and measure whether e_o is the same for a given number N of input pulses. Use a counter to measure N at higher rates. Record the results.

Vary the values of R_1 and C_1 and observe how they affect the maximum rate of introducing input pulses, and also the magnitude of the output voltage.

Select values for C_1, e_{in}, and the fraction of e_o fed to the recorder so that 0–1000 counts will provide 0 full scale on the chart recorder set on the 100-mV span.

7-22b   Count-Rate Meter

The circuit of Fig. E7-18 can be changed to a count rate or frequency meter by adding a resistor R_2 across the feedback capacitor C_2.

Connect a 100-kΩ resistor across C_2, and measure the output voltage e_o as a function of frequency. Plot e_o vs input frequency. Observe the effect of changing the value of R_2.

Select values for C_1, chart span, etc., so that the recorder chart will read 10 kHz full scale and be linear in frequency.

An output filter R_3C_3 with a time constant of about 0.1 sec is also useful for smoothing the signal to the recorder or other readout.

Note that the C_1, D_1, D_2, C_2, and R_2 components could be used in the servo operational feedback system to provide a recording count-rate meter directly. Start with $C_1 = 20$ pF, $C_2 = 10$ μF, $R_2 = 100$ kΩ, the recorder set at 250 mV. Measure the output frequency of the DTM.

Experiment 7-23 *Ramp-Type Analog-to-Digital Converter*

The conversion of an analog voltage to a digital word is performed by two techniques. First the staircase ramp generator of Experiment 7-22 is used in a voltage comparison method, and then a linear ramp is used with a digital timer in an A-Δt-D method.

7-23a Staircase Ramp Analog-to-Digital Converter

A combination of circuits that were connected individually in previous experiments will make an A-D converter as shown in the block diagram of Fig. E7-19. Circuits from specific experiments are indicated on the diagram.

Decide on the type of control desired for your circuit. Then interconnect the circuits to produce an A-D converter so that the digital output will read 1000 counts per 1-V input.

Test the linearity and accuracy of the A-D converter from 0 to 5 V input. Determine the rate at which conversions can be made.

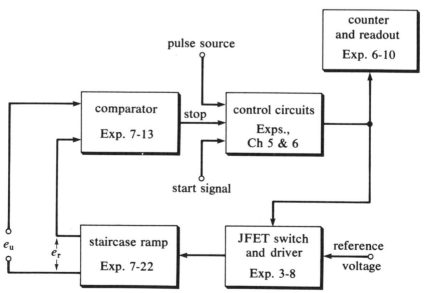

Fig. E7-19 Staircase ramp analog-to-digital converter.

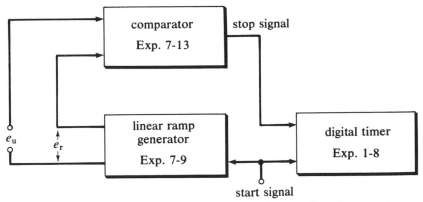

Fig. E7-20   Linear ramp-digital timer analog-to-digital converter.

7-23b   Linear Ramp-Digital Timer Analog-to-Digital Converter

Another combination of circuits that will produce an A-D converter is shown in the block diagram Fig. E7-20.

Choose the circuit values so that 10.00 msec on the digital timer corresponds to 1 V. Test the linearity and accuracy over the range 0–5 V input.

Experiment 7-24   *Digital Voltmeter*

A DVM is constructed using the V-F Converter card and the frequency meter constructed in Chapter 6. Characteristics are determined, and an OA follower is used at the input to provide high input impedance.

A V-F converter can be made by interconnecting the circuits shown in Fig. E7-21, all of which are mounted on the Comparator/V-F card (EU-800-HB). As before, the heavy lines indicate the #22 wire leads that are required between the numbered connectors on the top of the card.

Note that for the V-F circuit the OA is used as an integrator with a 1-μF capacitor (C_6) connected between the inverting input (14-13) and the output (6-8). The noninverting input (18) should be connected to ground (23). The output of the integrator is coupled to a high-gain amplifier composed of transistors Q_2 and Q_3, and its output is connected to the emitter of a unijunction transistor Q_4.

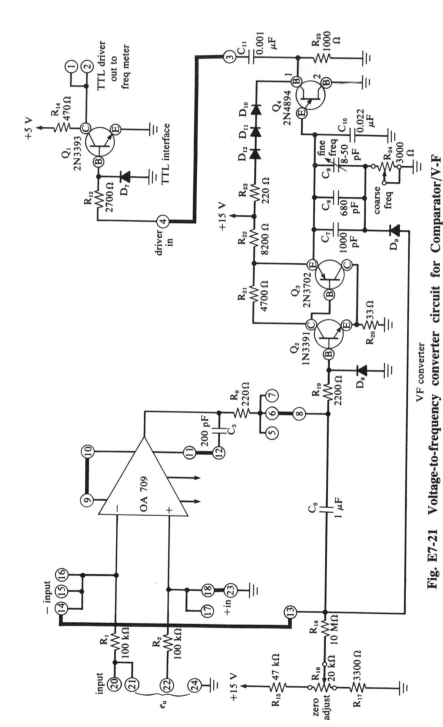

Fig. E7-21 Voltage-to-frequency converter circuit for Comparator/V-F
card. (Courtesy of Heath Co.)

502

When a positive input voltage e_u is applied to the voltage signal input (between 21 and 24) the feedback capacitor C_6 charges at a rate that is proportional to the input voltage. The output of the OA thus decreases and this causes transistors Q_2 and Q_3 to conduct less so the parallel capacitors C_7, C_8, C_9 and C_{10} charge through R_{22}. When the voltage across the capacitors reaches the unijunction firing potential the emitter-base conducts to discharge the capacitors rapidly. The capacitor discharge develops a negative pulse across R_{24}, and this pulse is sent through diode D_9 to the summing point of the integrator OA. The coulombic charge in this pulse causes the OA output to become more positive. Capacitor C_6 is repeatedly charged by the input voltage e_u and discharged a finite amount by the feedback pulses.

Make a complete DVM by using the frequency meter constructed in Experiment 6-21 and the V-F converter.

Insert the necessary circuit cards in the ADD, connect the wire leads, turn ON the ADD and allow the V-F circuit to warm up for at least 5-min.

Check for zero output frequency from the TTL driver with 0 V at the voltage input connectors (between 21 and 24) of the OA.

Check for 1000 Hz output with 1.000 V connected at the input and an integration time of 1 sec. The ZERO and FREQ ADJ controls are shown in Fig. E7-6, but do not adjust without reference to instruction manual and after the following measurements.

With 1 V at the input, observe on the oscilloscope the output signal from the OA integrator (5 or 7). Also, observe the signal at the output of the unijunction transistor (3) and the TTL DRIVER OUT (1 or 2). Note the time relationships and amplitudes of the three signals and record.

If you are not satisfied with the calibration of the V-F converter adjust the FINE adjust. Do not adjust the other V-F controls without reference to the calibration procedure in the instruction manual.

Connect an OA follower at the input to increase input impedance. Determine the linearity with input signals from 10 mV to 10 V.

Experiment 7-25 *A Serial Adder*

A 4-bit serial adder is wired following the block diagram of Fig. 7-53. Binary numbers are entered into the accumulator and addend registers and the operation of adding is carried out using a manual clock source. The use of negative numbers in addition and subtraction is also illustrated.

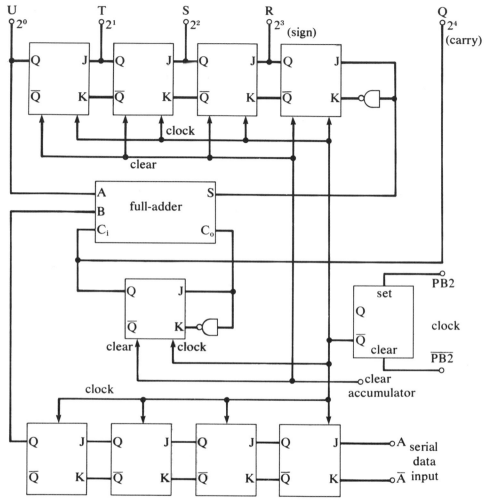

Fig. E7-22   Four-bit serial adder.

Wire the 4-bit serial adder shown in Fig. E7-22. Use the full-adder circuit shown in Fig. E4-10. Check the adder and shift register circuits individually as they are completed. The clock source is a push button with a flip-flop used as a contact bounce eliminator. The circuit can be wired with two NAND Gate cards and 5 Dual JK cards. If available, two Quad JK cards can be used for the registers.

To perform an addition go through the following steps:

1. Push PB1 to clear the accumulator.

2. Enter the least significant bit at A and clock once. Enter the next three bits in order clocking each time.

3. With A on **0,** clock 4 more times to put the number through the adder into the accumulator where it will appear on lights R, S, T, and U.

4. Enter the next 4-bit number into the addend register following step 2.

5. Set A to **0** and clock 4 more times to cycle the numbers in the two registers through the adder. The sum appears in the accumulator. If the sum is more than four bits the carry is indicated by Q.

6. Steps 4 and 5 can be repeated to add more numbers to the number in the accumulator.

Practice the above procedure for several numbers. When you are comfortable with the circuit's operation, steps 3 and 4 can be combined since the number in the addend register is fed out as the next number is entered.

Binary Subtraction

When this circuit is being used for subtraction, there are three number bits and a sign bit. First use the 1's complement form for negative numbers and add a positive and a negative number following the above procedure. If a **1** appears at the carry output, clock 4 more times (with A = **0**) to perform the end-around carry. The answer in positive or 1's complement negative form will appear in the accumulator. Negative numbers in 2's complement form can also be added but the end-around carry is not used in this case.

Appendixes

Decimal Equivalents of Binary Numbers to 2^8

Decimal	Binary	Decimal	Binary	Decimal	Binary	Decimal	Binary
0	00000000	35	00100011	70	01000110	105	01101001
1	00000001	36	00100100	71	01000111	106	01101010
2	00000010	37	00100101	72	01001000	107	01101011
3	00000011	38	00100110	73	01001001	108	01101100
4	00000100	39	00100111	74	01001010	109	01101101
5	00000101	40	00101000	75	01001011	110	01101110
6	00000110	41	00101001	76	01001100	111	01101111
7	00000111	42	00101010	77	01001101	112	01110000
8	00001000	43	00101011	78	01001110	113	01110001
9	00001001	44	00101100	79	01001111	114	01110010
10	00001010	45	00101101	80	01010000	115	01110011
11	00001011	46	00101110	81	01010001	116	01110100
12	00001100	47	00101111	82	01010010	117	01110101
13	00001101	48	00110000	83	01010011	118	01110110
14	00001110	49	00110001	84	01010100	119	01110111
15	00001111	50	00110010	85	01010101	120	01111000
16	00010000	51	00110011	86	01010110	121	01111001
17	00010001	52	00110100	87	01010111	122	01111010
18	00010010	53	00110101	88	01011000	123	01111011
19	00010011	54	00110110	89	01011001	124	01111100
20	00010100	55	00110111	90	01011010	125	01111101
21	00010101	56	00111000	91	01011011	126	01111110
22	00010110	57	00111001	92	01011100	127	01111111
23	00010111	58	00111010	93	01011101	128	10000000
24	00011000	59	00111011	94	01011110	129	10000001
25	00011001	60	00111100	95	01011111	130	10000010
26	00011010	61	00111101	96	01100000	131	10000011
27	00011011	62	00111110	97	01100001	132	10000100
28	00011100	63	00111111	98	01100010	133	10000101
29	00011101	64	01000000	99	01100011	134	10000110
30	00011110	65	01000001	100	01100100	135	10000111
31	00011111	66	01000010	101	01100101	136	10001000
32	00100000	67	01000011	102	01100110	137	10001001
33	00100001	68	01000100	103	01100111	138	10001010
34	00100010	69	01000101	104	01101000	139	10001011

Decimal	Binary	Decimal	Binary	Decimal	Binary	Decimal	Binary
140	10001100	169	10101001	198	11000110	227	11100011
141	10001101	170	10101010	199	11000111	228	11100100
142	10001110	171	10101011	200	11001000	229	11100101
143	10001111	172	10101100	201	11001001	230	11100110
144	10010000	173	10101101	202	11001010	231	11100111
145	10010001	174	10101110	203	11001011	232	11101000
146	10010010	175	10101111	204	11001100	233	11101001
147	10010011	176	10110000	205	11001101	234	11101010
148	10010100	177	10110001	206	11001110	235	11101011
149	10010101	178	10110010	207	11001111	236	11101100
150	10010110	179	10110011	208	11010000	237	11101101
151	10010111	180	10110100	209	11010001	238	11101110
152	10011000	181	10110101	210	11010010	239	11101111
153	10011001	182	10110110	211	11010011	240	11110000
154	10011010	183	10110111	212	11010100	241	11110001
155	10011011	184	10111000	213	11010101	242	11110010
156	10011100	185	10111001	214	11010110	243	11110011
157	10011101	186	10111010	215	11010111	244	11110100
158	10011110	187	10111011	216	11011000	245	11110101
159	10011111	188	10111100	217	11011001	246	11110110
160	10100000	189	10111101	218	11011010	247	11110111
161	10100001	190	10111110	219	11011011	248	11111000
162	10100010	191	10111111	220	11011100	249	11111001
163	10100011	192	11000000	221	11011101	250	11111010
164	10100100	193	11000001	222	11011110	251	11111011
165	10100101	194	11000010	223	11011111	252	11111100
166	10100110	195	11000011	224	11100000	253	11111101
167	10100111	196	11000100	225	11100001	254	11111110
168	10101000	197	11000101	226	11100010	255	11111111

The Powers of 2

2^n	n	2^{-n}
1	0	1.0
2	1	0.5
4	2	0.25
8	3	0.125
16	4	0.062 5
32	5	0.031 25
64	6	0.015 625
128	7	0.007 812 5
256	8	0.003 906 25
512	9	0.001 953 125
1 024	10	0.000 976 562 5
2 048	11	0.000 488 281 25
4 096	12	0.000 244 140 625
8 192	13	0.000 122 070 312 5
16 384	14	0.000 061 035 156 25
32 768	15	0.000 030 517 578 125
65 536	16	0.000 015 258 789 062 5
131 072	17	0.000 007 629 394 531 25
262 144	18	0.000 003 814 697 265 625
524 288	19	0.000 001 907 348 632 812 5
1 048 576	20	0.000 000 953 674 316 406 25
2 097 152	21	0.000 000 476 837 158 203 125
4 194 304	22	0.000 000 238 418 579 101 562 5
8 388 608	23	0.000 000 119 209 289 550 781 25
16 777 216	24	0.000 000 059 604 644 775 390 625
33 554 432	25	0.000 000 029 802 322 387 695 312 5
67 108 864	26	0.000 000 014 901 161 193 847 656 25
134 217 728	27	0.000 000 007 450 580 596 923 828 125
268 435 456	28	0.000 000 003 725 290 298 461 914 062 5
536 870 912	29	0.000 000 001 862 645 149 230 957 031 25
1 073 741 824	30	0.000 000 000 931 322 574 615 478 515 625
2 147 483 648	31	0.000 000 000 465 661 287 307 739 257 812 5
4 294 967 296	32	0.000 000 000 232 830 643 653 869 628 906 25
8 589 934 592	33	0.000 000 000 116 415 321 826 934 814 453 125
17 179 869 184	34	0.000 000 000 058 207 660 913 467 407 226 562 5
34 359 738 368	35	0.000 000 000 029 103 830 456 733 703 613 281 25
68 719 476 736	36	0.000 000 000 014 551 915 228 366 851 806 640 625
137 438 953 472	37	0.000 000 000 007 275 957 614 183 425 903 320 312 5
274 877 906 944	38	0.000 000 000 003 637 978 807 091 712 951 660 156 25
549 755 813 888	39	0.000 000 000 001 818 989 403 545 856 475 830 078 125
1 099 511 627 776	40	0.000 000 000 000 909 494 701 772 928 237 915 039 062 5
2 199 023 255 552	41	0.000 000 000 000 454 747 350 886 464 118 957 519 531 25
4 398 046 511 104	42	0.000 000 000 000 227 373 675 443 232 059 478 759 765 625
8 796 093 022 208	43	0.000 000 000 000 113 686 837 721 616 029 739 379 882 812 5
17 592 186 044 416	44	0.000 000 000 000 056 843 418 860 808 014 869 689 941 406 25
35 184 372 088 832	45	0.000 000 000 000 028 421 709 430 404 007 434 844 970 703 125
70 368 744 177 664	46	0.000 000 000 000 014 210 854 715 202 003 717 422 485 351 562 5
140 737 488 355 328	47	0.000 000 000 000 007 105 427 357 601 001 858 711 242 675 781 25
281 474 976 710 656	48	0.000 000 000 000 003 552 713 678 800 500 929 355 621 337 890 625
562 949 953 421 312	49	0.000 000 000 000 001 776 356 839 400 250 464 677 810 668 945 312 5
1 125 899 906 842 624	50	0.000 000 000 000 000 888 178 419 700 125 232 338 905 334 472 656 25
2 251 799 813 685 248	51	0.000 000 000 000 000 444 089 209 850 062 616 169 452 667 236 328 125
4 503 599 627 370 496	52	0.000 000 000 000 000 222 044 604 925 031 308 084 726 333 618 164 062 5
9 007 199 254 740 992	53	0.000 000 000 000 000 111 022 302 462 515 654 042 363 166 809 082 031 25
18 014 398 509 481 984	54	0.000 000 000 000 000 055 511 151 231 257 827 021 181 583 404 541 015 625
36 028 797 018 963 968	55	0.000 000 000 000 000 027 755 575 615 628 913 510 590 791 702 270 507 812 5
72 057 594 037 927 936	56	0.000 000 000 000 000 013 877 787 807 814 456 755 295 395 851 135 253 906 25
144 115 188 075 855 872	57	0.000 000 000 000 000 006 938 893 903 907 228 377 647 697 925 567 626 953 125
288 230 376 151 711 744	58	0.000 000 000 000 000 003 469 446 951 953 614 188 823 848 962 783 813 476 562 5
576 460 752 303 423 488	59	0.000 000 000 000 000 001 734 723 475 976 807 094 411 924 481 391 906 738 281 25
1 152 921 504 606 846 976	60	0.000 000 000 000 000 000 867 361 737 988 403 547 205 962 240 695 953 369 140 625
2 305 843 009 213 693 952	61	0.000 000 000 000 000 000 433 680 868 994 201 773 602 981 120 347 976 684 570 312 5
4 611 686 018 427 387 904	62	0.000 000 000 000 000 000 216 840 434 497 100 886 801 490 560 173 988 342 285 156 25
9 223 372 036 854 775 808	63	0.000 000 000 000 000 000 108 420 217 248 550 433 400 745 280 086 994 171 142 578 125
18 446 744 073 709 551 616	64	0.000 000 000 000 000 000 054 210 108 624 275 221 700 372 640 043 497 085 571 289 062 5
36 893 488 147 419 103 232	65	0.000 000 000 000 000 000 027 105 054 312 137 610 850 186 320 021 748 542 785 644 531 25
73 786 976 294 838 206 464	66	0.000 000 000 000 000 000 013 552 527 156 068 805 425 093 160 010 874 271 392 822 265 625
147 573 952 589 676 412 928	67	0.000 000 000 000 000 000 006 776 263 578 034 402 712 546 580 005 437 135 696 411 132 812 5
295 147 905 179 352 825 856	68	0.000 000 000 000 000 000 003 388 131 789 017 201 356 273 290 002 718 567 848 205 566 406 25
590 295 810 358 705 651 712	69	0.000 000 000 000 000 000 001 694 065 894 508 600 678 136 645 001 359 283 924 102 783 203 125
1 180 591 620 717 411 303 424	70	0.000 000 000 000 000 000 000 847 032 947 254 300 339 068 322 500 679 641 962 051 391 601 562 5
2 361 183 241 434 822 606 848	71	0.000 000 000 000 000 000 000 423 516 473 627 150 169 534 161 250 339 820 981 025 695 800 781 25
4 722 366 482 869 645 213 696	72	0.000 000 000 000 000 000 000 211 758 236 813 575 084 767 080 625 169 910 490 512 847 900 390 625

appendix C

Summary of Boolean Algebra Postulates and Theorems

Postulates

$X = 1$ or else $X = 0$

$1 \cdot 1 = 1$

$1 \cdot 0 = 0 \cdot 1 = 0$

$0 \cdot 0 = 0$

$\bar{1} = 0$

$0 + 0 = 0$

$0 + 1 = 1 + 0 = 1$

$1 + 1 = 1$

$\bar{0} = 1$

Theorems

1a. $0 \cdot X = 0$

2a. $1 \cdot X = X$

3a. $XX = X$

4a. $X\bar{X} = 0$

5a. $XY = YX$

6a. $XYZ = (XY)Z = X(YZ)$

7a. $\overline{XY \cdots Z} = \bar{X} + \bar{Y} + \cdots + \bar{Z}$

8. $\qquad\qquad \bar{f}(X, Y, \ldots, Z, \cdot, +) = f(\bar{X}, \bar{Y}, \ldots, \bar{Z}, +, \cdot)$

9a. $XY + XZ = X(Y + Z)$

10a. $XY + X\bar{Y} = X$

11a. $X + XY = X$

12a. $X + \bar{X}Y = X + Y$

12a'. $ZX + Z\bar{X}Y = ZX + ZY$

12b'. $(Z + X)(Z + \bar{X} + Y) = (Z + X)(Z + Y)$

13a. $XY + \bar{X}Z + YZ = XY + \bar{X}Z$

13b. $(X + Y)(\bar{X} + Z)(Y + Z) = (X + Y)(\bar{X} + Z)$

14a. $XY + \bar{X}Z = (X + Z)(\bar{X} + Y)$

14b. $(X + Y)(\bar{X} + Z) = XZ + \bar{X}Y$

15a. $X \cdot f(X, \bar{X}, Y, \ldots, Z) = X \cdot f(1, 0, Y, \ldots, Z)$

15b. $X + f(X, \bar{X}, Y, \ldots, Z) = X + f(0, 1, Y, \ldots, Z)$

16a. $f(X, \bar{X}, Y, \ldots, Z) = X \cdot f(1, 0, Y, \ldots, Z) + \bar{X} \cdot f(0, 1, Y, \ldots, Z)$

16b. $f(X, \bar{X}, Y, \ldots, Z) = [X + f(0, 1, Y, \ldots, Z)][\bar{X} + f(1, 0, Y, \ldots, Z)]$

1b. $1 + X = 1$

2b. $0 + X = X$

3b. $X + X = X$

4b. $X + \bar{X} = 1$

5b. $X + Y = Y + X$

6b. $X + Y + Z = (X + Y) + Z$
$\qquad\qquad = X + (Y + Z)$

7b. $\overline{X + Y + \cdots + Z} = \bar{X}\bar{Y} \cdots \bar{Z}$

9b. $(X + Y)(X + Z) = X + YZ$

10b. $(X + Y)(X + \bar{Y}) = X$

11b. $X(X + Y) = X$

12b. $X(\bar{X} + Y) = XY$

appendix D

Gate Types and Characteristics

Characteristics	Gate Types				
	RTL	DTL	TTL	ECL	FET
Basic logic function	+NOR	+NAND	+NAND	+OR, NOR	−NOR
Interconnection type	Current sourcing	Current sinking	Current sinking	Current mode	Voltage
Logic levels	0.2 V = 0 1.6 V = 1	0.2 V = 0 3.0 V = 1	0.2 V = 0 3.3 V = 1	−1.6 V = 0 −0.75 V = 1	−2 V = 0 −10 V = 1
Fan-out	5	8	10	25	5
Propagation delay	25 nsec	25 nsec	13 nsec	3 nsec	~0.5 μsec
Noise immunity, mV	~200	750	1000	~200	
Supply voltage, V	+3.6	+5	+5	−5.2	−20
Primary advantages	Low cost	Moderate cost Package variety	High speed Noise immunity High fan-out Versatile Package variety	Highest speed Package variety	Low powers Large-scale integration

appendix E

Logic Functions and NAND/NOR Implementation

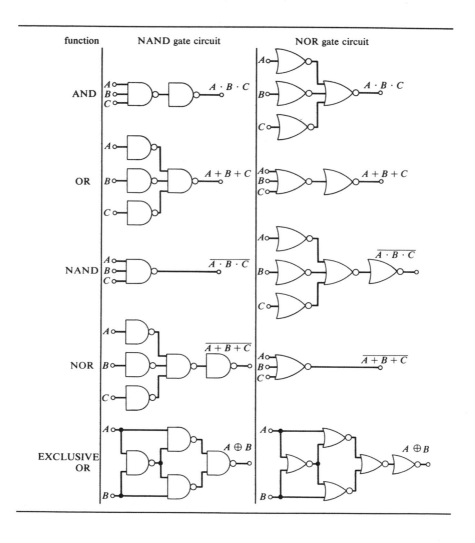

514

Inputs		Outputs					
A	B	AND $A \cdot B$	NAND $\overline{A \cdot B}$	OR $A + B$	NOR $\overline{A + B}$	EXCL.-OR $A \oplus B$	Eq. comp. $A \oplus B$
0	0	0	1	0	1	0	1
0	1	0	1	1	0	1	0
1	0	0	1	1	0	1	0
1	1	1	0	1	0	0	1

Resistor Color Code

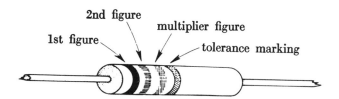

Black	0	Green	5
Brown	1	Blue	6
Red	2	Violet	7
Orange	3	Gray	8
Yellow	4	White	9

appendix G

Transistor and Diode Specifications and Base Diagrams

Diodes

IN2071
RECTIFIER DIODE

CATHODE

Peak inverse voltage	600 V
Forward current, max	1 A

IN4149
SILICON SIGNAL DIODE

CATHODE

Peak inverse voltage	75 V
Forward current, max	10 mA
Power dissipation, max	500 mW

IN710
6.8 V ZENER DIODE

BROWN VIOLET
CATHODE

Zener current	25 mA
Power dissipation, max	250 mW

VR-9.1
9.1 V ZENER DIODE

CATHODE

Zener current	25 mA
Power dissipation, max	1 W

Junction Transistors

2N2369
NPN HIGH SPEED SWITCH

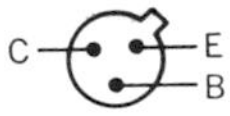

C E B

V_{CBO}, max	40 V
I_C, max	500 mA
Power dissipation	1 W
β	40–120
Frequency	300 MHz

2N2509
NPN HI VOLTAGE

V_{CBO}, max	90 V
I_C, max	150 mA
Power dissipation, max	200 mW
β	60–400
Frequency	150 MHz

2N3393
NPN AMPLIFIER

V_{CBO}, max	25 V
I_C, max	100 mA
Power dissipation, max	200 mW
β	100–200
Frequency	140 MHz

2N3702
PNP AMPLIFIER

V_{CBO}, max	−40 V
I_C, max	200 mA
Power dissipation, max	300 mW
β	60–300
Frequency	100 MHz

FET

MPF 105
N-CHANNEL FIELD EFFECT

Drain-to-source voltage, max	25 V
Gate current, max	10 mA
Power dissipation, max	200 mW

Unijunction

2N4894
SILICON UNIJUNCTION

Emitter reverse voltage, max	30 V
Emitter current, max	1 A
Power dissipation, max	300 mW

SCR

TIC-47   **SILICON**
    CONTROLLED RECTIFIER

CATHODE ━━━⟨• •⟩━ GATE
            ━ ANODE

Peak reverse voltage	200 V
Peak gate current	1 A

Photoconductor

Voltage, max	80 V peak
Dissipation, max	100 mW
Peak response	7200 A
R for 0.1 foot-candle	2.5 MΩ typ.
R for 100 foot-candle	2.6 kΩ typ.

appendix H

Circuit Card Top Layout

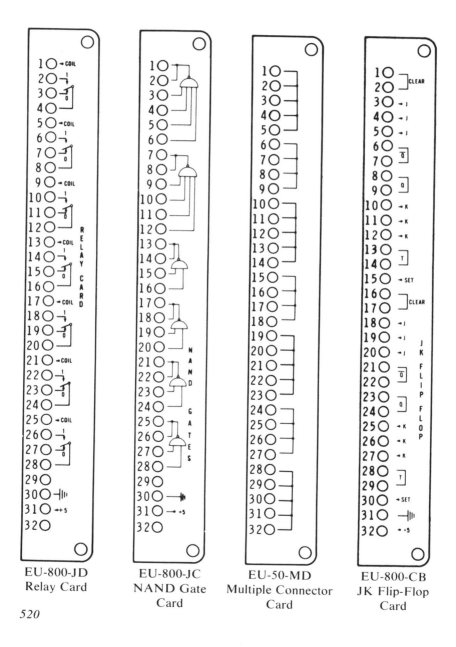

EU-800-JD Relay Card	EU-800-JC NAND Gate Card	EU-50-MD Multiple Connector Card	EU-800-CB JK Flip-Flop Card

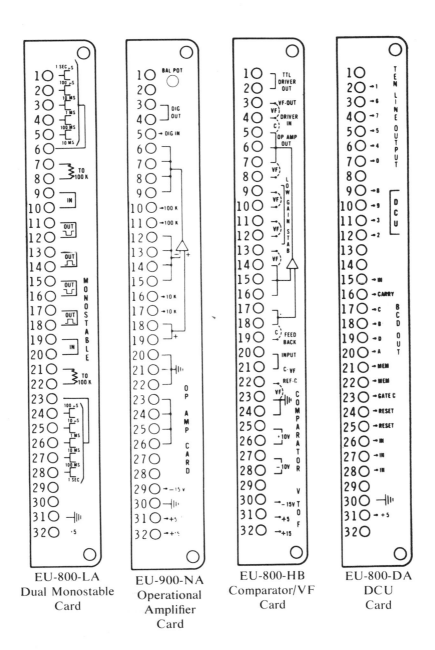

EU-800-LA
Dual Monostable
Card

EU-900-NA
Operational
Amplifier
Card

EU-800-HB
Comparator/VF
Card

EU-800-DA
DCU
Card

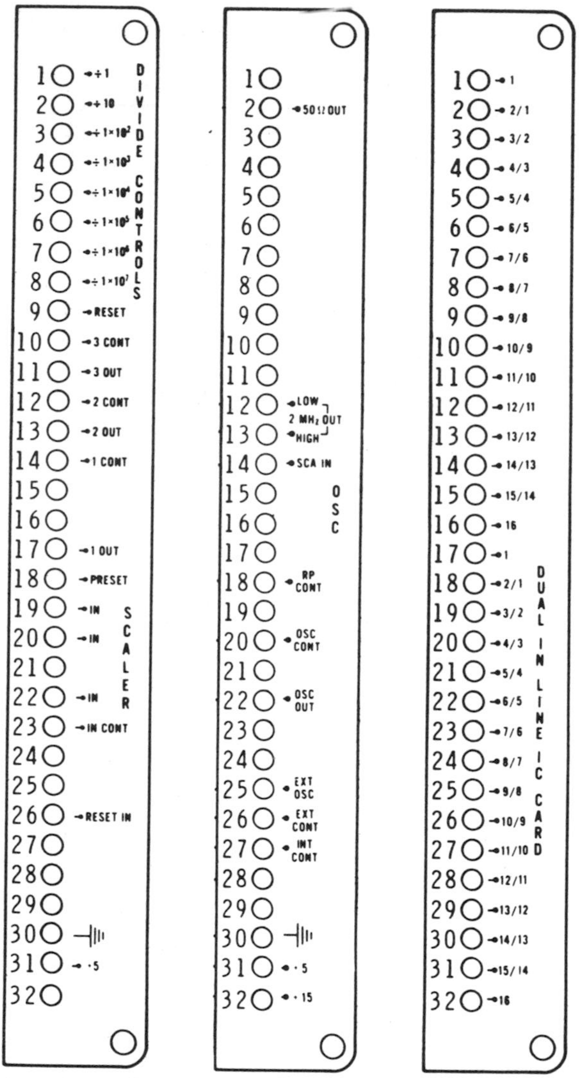

EU-800-CA
Decade
Scaler
Card

EU-800-KA
Crystal
Oscillator
Card

EU-50-MC
Dual-in-line
IC Card

Modules

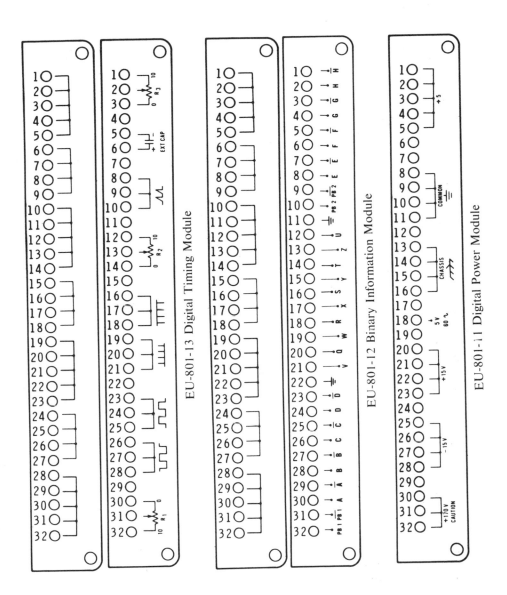

EU-801-13 Digital Timing Module

EU-801-12 Binary Information Module

EU-801-11 Digital Power Module

appendix I

Glossary of Terms[1]

Access time* (1) The time interval required to communicate with the memory or storage unit of a digital computer. (2) The time interval between the instant at which the arithmetic unit calls for information from the memory and the instant at which the information is delivered. Also, the time interval between the instant at which the arithmetic unit starts to transmit information to the memory and the instant at which the storage of information in the memory is completed.

Active elements Those components in a circuit that have gain or that direct current flow: diodes, transistors, silicon-controlled rectifiers, etc.

Adder Switching units that combine binary bits to generate the sum and carry of these bits. Takes the bits from the two binary numbers to be added (*addend* and *augend*) plus the carry from the preceding less significant bit and generates the sum and the carry.

Address Noun: a location, either name or number, where information is stored in a computer. Verb: to select or pick out the location of a stored information set for access.

AND A Boolean logic expression used to identify the logic operation wherein given two or more variables, all must be logical **1** for the result to be logical **1**. The AND function is graphically represented by the dot ($\cdot$) symbol.

Anticipated carry adder A parallel *adder* in which each stage is capable of looking back at all *addend* and *augend* bits of less significant stages and deciding whether the less significant bits provide a **0** or a **1** *carry in*. Having determined the *carry in* it combines it with its own *addend* and *augend* to give the *sum* for that bit or stage. Also called *fast adder* or look ahead *carry adder*.

Asynchronous inputs Those terminals in a flip-flop that can affect the output state of the flip-flop independent of the clock. Called set, preset, reset or dc set and reset, or clear.

Binary-coded decimal (BCD) A binary numbering system for coding decimal numbers in groups of 4 bits. The binary value of these 4-bit groups ranges from 0000 to 1001 and codes the decimal digits 0 through 9. To count to 9 takes 4 bits; to count to 99 takes two groups of 4 bits; to count to 999 takes three groups of 4 bits.

Binary logic Digital logic elements which operate with two distinct states. The two states are variously called TRUE and FALSE, HIGH and LOW, ON and OFF, or **1** and **0**. In computers they are represented by two different voltage levels. The level which is more positive (or less negative) than the other is called the

---

[1] Definitions for all terms except those marked with an asterisk were obtained from the *Pocket Dictionary of Integrated Circuit Terminology* (Sylvania Electric Products, Inc., Semiconductor Division, Woburn, Mass.).

524

high level, the other the low level. If the true **(1)** level is the most positive voltage, such logic is referred to as positive true or positive logic.

Bistable element   Another name for flip-flop. A circuit in which the output has two stable states (output levels **0** or **1**) and can be caused to go to either of these states by input signals, but remains in that state permanently after the input signals are removed. This differentiates the bistable element from a gate also having two output states but which requires the retention of the input signals to stay in a given state. The characteristic of two stable states also differentiates it from a monostable element which keeps returning to a specific state and from an astable element which keeps changing from one state to the other.

Bit   A synonym for binary numeral. Also refers to a single binary numeral in a binary word.

Boolean algebra   The mathematics of logic which uses alphabetic symbols to represent logical variables and **1** and **0** to represent states. There are three basic logic operations in this algebra: AND, OR, and NOT. (See also NAND, NOR, INVERT which are combinations of the three basic operations.)

Buffer   A circuit element which is used to isolate between stages to handle a large fanout or to convert input and output circuits for signal level compatibility.

Buried layer   A heavily doped (N+) region directly under the N-doped epitaxial collector region of transistors in a monolithic integrated circuit used to lower the series collector resistance.

Chip (die)   A single piece of silicon which has been cut from a slice by scribing and breaking. It can contain one or more circuits but is packaged as a unit.

Clear   An asynchronous input. Also called "reset." It is used to restore a memory element or flip-flop to a "standard" state, forcing the Q terminal to logic **0**.

Clock   A pulse generator that controls the timing of computer switching circuits and memory states and regulates the speed at which the computer central processor operates. It serves to synchronize all operations in a digital system.

Clock input   That terminal on a flip-flop whose condition or change of condition controls the admission of data into a flip-flop through the synchronous inputs and thereby controls the output states of the flip-flop. The clock signal performs two functions: (1) it permits data signals to enter the flip-flop; (2) after entry, it directs the flip-flop to change state accordingly.

CML (current mode logic)   Logic in which transistors operate in the unsaturated mode as distinguished from most other logic types which operate in the saturation region. This logic has very fast switching speeds and low logic swings. Also called ECL or MECL.

Command*   A pulse, signal, or set of signals that occur in a computer as a result of an instruction and which initiate one step in the process of executing the instruction.

Complement*   Verb: to reverse the state of a storage device or of a control level (e.g., changing a flip-flop from the **1** to the **0** state or vice versa). Noun: the complement of a number (base or base minus 1).

Counter A device capable of changing states in a specified sequence upon receiving appropriate input signals. The output of the counter indicates the number of pulses which have been applied. (See also Divider.) A counter is made from flip-flops and some gates. The output of all flip-flops is accessible to indicate the exact count at all times.

Counter, binary An interconnection of flip-flops having a single input so arranged as to enable binary counting. Each time a pulse appears at the input, the counter changes state and tabulates the number of input pulses for readout in binary form. It has 2^n possible counts, where n is the number of flip-flops.

Counter, ring A special form of counter sometimes called a Johnson or shift counter which has very simple wiring and is fast. It forms a loop or circuits of interconnected flip-flops so arranged that only one is **0** and that as input signals are received, the position of the **0** state moves in sequence from one flip-flop to another around the loop until they are all **0,** then the first one goes to **1** and this moves in sequence from one flip-flop to another until all are **1.** It has $2 \times n$ possible counts, where n is the number of flip-flops.

Data Term used to denote facts, numbers, letters, symbols, binary bits presented as voltage levels in a computer. In a binary system data can only be **0** or **1.**

DCTL (direct-coupled transistor logic) Logic employing only transistors as active circuit elements.

Decimal A system of numerical representation which uses ten numbers 0, 1, 2, 3, . . . , 9. Each numeral is called a digit. A numbering system to the radix 10.

Delay The slowing up of the propagation of a pulse either intentionally, such as to prevent inputs from changing while clock pulses are present or unintentionally, as caused by transistor rise and fall time pulse response effects.

Die See Chip.

Diffusion A process used in the production of semiconductors which introduces minute amounts of impurities into a substrate material such as silicon or germanium and permits the impurity to spread into the substrate. The process is very dependent on temperature and time.

Digital circuit A circuit that operates in the manner of a switch, that is, it is either ON or OFF. More correctly should be called a binary circuit.

Discrete circuits Electronic circuits built of separate, individually manufactured, tested and assembled diodes, resistors, transistors, capacitors, and other specific electronic components.

Divider, frequency A counter that has a gating structure added which provides an output pulse after receiving a specified number of input pulses. The outputs of all flip-flops are not accessible.

Driver An element that is coupled to the output stage of a circuit in order to increase its power or current handling capability or fanout; for example, a clock driver is used to supply the current necessary for a clock line.

DTL (diode-transistor logic) Logic employing diodes with transistors used only as inverting amplifiers.

Enable   To permit an action or the acceptance or recognition of data by applying appropriate signals (generally a logic 1 in a positive logic) to the appropriate input. (See Inhibit.)

Epitaxial growth   A chemical reaction in which silicon is precipitated from a gaseous solution and grows in a very precise manner, that is monocrystalline, upon the surface of a silicon wafer placed in the solution.

Exclusive OR   A logical function whose output is 1 if either of the two variables is 1 but whose output is 0 if both inputs are 1 or both are 0.

Fall time   A measure of the time required for the output voltage of a circuit to change from a high-voltage level to a low-voltage level once a level change has started.   Current could also be used as the reference, that is, from a high current to a low current level.

Fan-in   The number of inputs available to a specific logic stage or function.

Fan-out   The number of input stages that can be driven by a circuit output.

Fast adder   See Anticipated Carry Adder and Adder.

FEB (functional electronic block)   Another name for a monolithic integrated circuit or thick film circuit.

Flip-flops (storage elements)   A circuit having two stable states and the capability of changing from one state to another with the application of a control signal and remaining in that state after removal of signals.   (See Bistable Element.)

Flip-flop, D   D stands for delay.   A flip-flop whose output is a function of the input that appeared one pulse earlier; for example, if a 1 appeared at the input, the output after the next clock pulse will be a 1.

Flip-flop, JK   A flip-flop having two inputs designated J and K.   At the application of a clock pulse, a 1 on the J input and a 0 on the K input will set the flip-flop to the 1 state; a 1 on the K input and a 0 on the J input will reset it to the 0 state; and 1's simultaneously on both inputs will cause it to change state regardless of the previous state.   J = 0 and K = 0 will prevent change.

Flip-flop, RS   A flip-flop consisting of two cross-coupled NAND gates having two inputs designated R and S.   A 1 on the S input and 0 on the R input will reset (clear) the flip-flop to the 0 state, and 1 on the R input and 0 on the S input will set it to the 1.   It is assumed that 0's will never appear simultaneously at both inputs.   If both inputs have 1's it will stay as it was; 1 is considered nonactivating.   A similar circuit can be formed with NOR gates.

Flip-flop, RST   A flip-flop having three inputs, R, S, and T.   This unit works as the RS-flip-flop except that the T input is used to cause the flip-flop to change states.

Flip-Flop, T   A flip-flop having only one input.   A pulse appearing on the input will cause the flip-flop to change states.   Used in ripple counters.

Full-adder   See Adder.

Gate, AND   All inputs must have 1 level signals at the input to produce a 1 level output.[2]

Gate, NAND   All inputs must have 1 level signals at the input to produce a 0 level output.

[2] The gate definitions given here assume positive logic.

Gate, NOR Any one input or more than one input having a 1 level signal will produce a 0 level output.

Gate, OR Any one input or more than one input having a 1 level signal will produce a 1 level output.

Gates (decision elements) A circuit having two or more inputs and one output. The output depends upon the combination of logic signals at the input.

Half-adder A switching circuit that combines binary bits to generate the sum and the carry. It can only take in the two binary bits to be added and generate the sum and carry. (See also Adder.)

Half-shift register Another name for certain types of flip-flops when used in a shift register. It takes two of these to make one stage in a shift register.

HIGH See Binary Logic.

Hybrid A method of manufacturing integrated circuits by using a combination of monolithic and thick film techniques.

Inhibit To prevent an action or acceptance of data by applying an appropriate signal to the appropriate input (generally a logic 0 in positive logic). (See Enable.)

Integrated circuit (EIA definition) "The physical realization of a number of electrical elements inseparably associated on or within a continuous body of semiconductor material to perform the functions of a circuit." (See Slice and Chip.)

Inverter A circuit whose output is always in the opposite state from the input. This is also called a NOT circuit. (A teeter-totter is a mechanical inverter.)

Linear circuit A circuit whose output is an amplified version of its input or a predetermined variation of its input.

Logic diagram A picture representation for the logical functions of AND, OR, NAND, NOR, NOT.

Logic swing The voltage difference between the two logic levels 1 and 0.

LOW See Binary Logic.

Monolithic Refers to the single silicon substrate in which an integrated circuit is constructed. (See Integrated Circuit.)

NAND A Boolean logic operation that yields a logic 0 output when all logic input signals are logic 1.

Negative logic Logic in which the more negative voltage represents the 1 state; the less negative voltage represents the 0 state. (See Binary Logic.)

Noise immunity A measure of the insensitivity of a logic circuit to triggering or reaction to spurious or undesirable electrical signals or noise, largely determined by the signal swing of the logic. Noise can be in either of two directions—positive or negative.

NOR A Boolean logic operation that yields a logic 0 output with one or more true 1 input signals.

NOT A Boolean logic operation indicating negation, not 1. Actually an inverter. If input is 1, output is NOT 1 but 0. If the input is 0, output is NOT 0 but 1. Graphically represented by a bar over a Boolean symbol such as A. A means "when A is not 1. . . ."

Offset   The change in input voltage required to produce a zero output voltage in a linear amplifier circuit.   In digital circuits it is the dc voltage on which a signal is impressed.

One (1)   See Binary Logic.

OR   A Boolean logic operation used to identify the logic operation wherein two or more TRUE 1 inputs only add to one TRUE 1 output.   Only one input needs to be TRUE to produce a TRUE output.   The graphical symbol for OR is a plus sign (+).

Overflow*   In a counter or register, the production of a number that is beyond the storage capacity of the counter or register.   The extra number may be held in an "overflow element."

Parallel   This refers to the technique for handling a binary data word that has more than one bit.   All bits are acted upon simultaneously.   It is like the line of a football team.   Upon a signal all line men act.   (See also Serial.)

Parallel adder   A conventional technique for adding where the two multibit numbers are presented and added simultaneously (parallel).   A ripple adder is still a parallel adder; the carry is rippled from the least significant to the most significant bit.   Another type of parallel adder is the "look ahead," or "anticipated carry" adder.   (See Ripple Adder, Anticipated Carry Adder, and Adder.)

Parallel operation   The organization of data manipulation within computer circuitry where all the digits of a word are transmitted simultaneously on separate lines in order to speed up operation, as opposed to serial operation.

Parallel storage*   Storage in which all bits, characters, or words are equally accessible in time; contrasts with Serial Storage.

Parity check*   Use of a digit, called the "parity digit," carried along as a check which is 1 if the total number of ones in the machine word is odd, and 0 if the total number of ones in the machine word is even (odd parity).   Even parity uses the reverse conditions.

Passive elements   Resistors, inductors, or capacitors; elements without gain.

Positive logic   Logic in which the more positive voltage represents the 1 state; e.g., 1 = +3.45 V, logic 0 = +0.45 V.   (See Binary Logic.)

Preset   An input like the "set" input and which works in parallel with the set.

Propagation delay   A measure of the time required for a change in logic level to be transmitted through an element or a chain of elements.

Propagation time   The time necessary for a unit of binary information (high voltage or low) to be transmitted or passed from one physical point in a system or subsystem to another.   For example, from input of a device to output.

Punch card*   A card of uniform size and shape, suitable for punching a pattern of holes that has meaning and can be sensed mechanically by metal fingers, electrically by wire brushes, photoelectrically, and in other ways.

Punched tape*   Paper tape punched in a pattern of holes that convey information.

Q output   The reference output of a flip-flop.   When this output is 1 the flip-flop is said to be in the 1 state; when it is 0 the output is said to be in the 0 state.   (See also State and Set.)

$\overline{Q}$ **output**   The second output of a flip-flop. It is always opposite in logic level to the Q output.

RCTL (resistor-capacitor-transistor logic)   Same as RTL except that capacitors are used to enhance switching speed.

Read*   To acquire information, usually from some form of storage.

Real time*   In solving a problem, a speed sufficient to give an answer in the actual time during which the problem must be solved.

Real-time operations*   Processing data in the time scale of a physical process so that the results are useful in guiding the physical process. Also, solving problems in real time.

Register   An interconnection of computer circuitry, made up of a number of storage devices (usually flip-flops) to store a certain number of digits, usually one computer word. For example, a 4-bit register requires four flip-flops.

Reset   Also called "clear." Similar to set except it is the input through which the Q output can be made to go to **0**.

Ripple   The transmission of data serially. It is a serial reaction analogous to a bucket brigade or a row of falling dominoes.

Ripple adder   A binary adding system similar to the system most people use to add decimal numbers, that is, add the units column, get the carry, add it to the 10's column, get the carry, add it to the 100's column, and so on. Again it is necessary to wait for the signal to propagate through all columns even though all columns are present at once (parallel). Note that the carry is rippled.

Ripple counter   A binary counting system in which flip-flops are connected in series. When the first flip-flop changes it effects the second which effects the third and so on. If there are ten in a row, the signal must go sequentially from the first flip-flop to the tenth.

Rise time   A measure of the time required for the output voltage of a state from a low-voltage level (**0**) to a high-voltage level (**1**) once a level change has been started.

RTL(resistor-transistor logic)   Logic is performed by resistors; transistors are used to produce an inverted output.

Serial   This refers to the technique for handling a binary data word which has more than one bit. The bits are acted upon one at a time. It is like a parade going by a review point.

Serial operation   The organization of data manipulation within computer circuitry where the digits of a word are transmitted one at a time along a single line. The serial mode of operation is slower than parallel operation, but utilizes less complex circuitry.

Serial storage*   Storage in which words, characters, or bits appear one after another in time sequence and in which the access time, therefore, includes a variable waiting (latency) time from zero to many word (character, bit) times. For example, magnetic drums are serial by word, but may be serial or parallel by bit, or serial by character and parallel by bit.

Set   An input on a flip-flop not controlled by the clock (see Asynchronous Inputs) and used to effect the Q output. It is this input through which signals can be entered to get the Q output to go to **1**. Note it cannot get Q to go to **0**.

Shift The process of moving data from one place to another. Generally many bits are moved at once. Shifting is done synchronously and by command of the clock. An 8-bit word can be shifted sequentially (serially), that is the first bit goes out, second bit takes first bit's place, third bit takes second bit's place, and so on, in the manner of a bucket brigade. Generally referred to as shifting left or right. It takes eight clock pulses to shift an eight-bit word or all bits of a word can be shifted simultaneously. This is called parallel load or parallel shift.

Shift register An arrangement of circuits, specifically flip-flops which is used to shift serially or in parallel. Binary words are generally parallel loaded and then held temporarily or serially shifted out.

Skewing Refers to time delay or offset between any two signals in relation to each other.

Slewing rate Rate at which the output can be driven from limit to limit over the dynamic range.

Slice A single wafer cut from a silicon ingot forming a thin substrate on which all active and passive elements for multiple integrated circuits have been fabricated, utilizing semiconductor epitaxial growth, diffusion, passivation, masking, photo resist, and metallization technologies. A completed slice generally contains hundreds of individual circuits. (See Chip.)

State This refers to the condition of an input or output of a circuit as to whether it is a logic **1** or a logic **0**. The state of a circuit (gate or flip-flop) refers to its output. The flip-flop is said to be in the **1** state when its Q output is **1**. A gate is in the **1** state when its output is **1**.

Synchronous Operation of a switching network by a clock pulse generator. All circuits in the network switch simultaneously. All actions take place synchronously with the clock.

Synchronous inputs Those terminals on a flip-flop through which data can be entered but only upon command of the clock. These inputs do not have direct control of the output such as those of a gate but only when the clock permits and commands. Called JK inputs or ac set and reset inputs.

Thick film A method of manufacturing integrated circuits by depositing thin layers of materials on an insulated substrate (often ceramic) to perform electrical functions; usually only passive elements are made this way.

Toggle To switch between two states as in a flip-flop.

Transfer* (1) To transfer information from one register to another without modifying it; i.e., to copy, exchange, read, record, store, transmit, transport, or write data. (2) To transfer control of a computer from one instruction to another.

Trigger A timing pulse used to initiate the transmission of logic signals through the appropriate circuit signal paths.

Truth table A chart that tabulates and summarizes all the combinations of possible states of the inputs and outputs of a circuit. It tabulates what will happen at the output for a given input combination.

TTL, T²L(transistor-transistor logic) A logic system which evolved from diode-transistor logic wherein the multiple diode cluster is replaced by a multiple-emitter transistor but is commonly applied to a circuit which has a multiple-emitter input and an active pullup network.

Turn-on time   The time required for an output to turn on (sink current, to ground output, to go to 0 V). It is the propagation time of an appropriate input signal to cause the output to go to 0 V.

Turn-off time   Same as Turn-on Time except the output stops sinking current, goes off and/or goes to a high-voltage level (logic 1).

Wired OR   Externally connected separate circuits or functions arranged so that the combination of their outputs results in an AND function. However, common usage is that the point at which the separate circuits are wired together will be 0 if any one of the separate outputs is a 0. The same as a dot AND.

Word*   An ordered set of characters which has at least one meaning and is stored, transferred, or operated upon by the computer circuits as a unit. Also called "machine word" or "information word." A word is treated as an instruction by the control unit, and as a numerical quantity by the arithmetic unit.

Zero(0)   See Binary Logic.

Index

Index